Physiological Ecology:

AN EVOLUTIONARY APPROACH TO RESOURCE USE

Physiological Ecology:

AN EVOLUTIONARY APPROACH TO RESOURCE USE

EDITED BY
COLIN R. TOWNSEND
School of Biological Sciences,
University of East Anglia, UK

AND

PETER CALOW
Department of Zoology,
University of Glasgow, UK

SINAUER ASSOCIATES, INC. · PUBLISHERS
SUNDERLAND, MASSACHUSETTS

Editorial Offices:
Osney Mead, Oxford, OX2 OEL
8 John Street, London, WC1N 2ES
9 Forrest Road, Edinburgh, EH1 2QH
52 Beacon Street, Boston
Massachusetts 02108, USA
214 Berkeley Street, Carlton
Victoria 3053, Australia

First published 1981

Distributed in the USA by
Sinauer Associates, Inc., Publishers
Sunderland, Massachusetts

Library of Congress
Cataloging in Publication Data

Main entry under title:
Physiological Ecology
Includes bibliographical
references and index.
1. Ecology 2. Physiology
3. Evolution 4. Bioenergetics
5. Resource partitioning (Ecology)
I. Townsend Colin R II. Calow Peter
QH541.P47 574.1 81–13559

ISBN 0–87893–827–3 AACR2
ISBN 0–87893–828–1 (pbk.)

Printed in Great Britain

Contents

List of Contributors

PETER CALOW Department of Zoology, University of Glasgow, Glasgow G12 8QQ, UK

L. MORRIS GOSLING Ministry of Agriculture, Fisheries and Food, Coypu Research Laboratory, Jupiter Road, Norwich NR6 6SP, UK

ROGER N. HUGHES Department of Zoology, University College of North Wales, Bangor, Gwynedd LL57 2UW, UK

DANIEL H. JANZEN Department of Biology, University of Pennsylvania, Philadelphia 19104, USA

THOMAS B. L. KIRKWOOD National Institute for Medical Research, The Ridgeway, Mill Hill, London NW7 IAA, UK
Formerly at National Institute for Biological Standards and Control, Holly Hill, Hampstead, London NW3 6RB, UK

JOHN PHILLIPSON Animal Ecology Research Group, Department of Zoology, Oxford University, Oxford OX1 3PS, UK

MARION PETRIE School of Biological Sciences, University of East Anglia, Norwich NR4 7TJ, UK

ERIC R. PIANKA Department of Zoology, The University of Texas at Austin, Austin, Texas 78712, USA

CAROLINE M. POND The Open University, Walton Hall, Milton Keynes MK7 6AA, UK

RICHARD M. SIBLY Department of Zoology, University of Reading, Reading RG6 2AH, UK

OTTO T. SOLBRIG The Gray Herbarium, Harvard University, 22 Divinity Avenue, Cambridge, Massachusetts 02138, USA

COLIN R. TOWNSEND School of Biological Sciences, University of East Anglia, Norwich NR4 7TJ, UK

HAROLD H. WOOLHOUSE John Innes Institute, Colney Lane, Norwich NR4 7UH, UK

Preface

Organisms can be considered as resource transformers that partition a finite input of food or photosynthate between the metabolic compartments of respiratory metabolism, defence, repair, storage, growth and reproduction. The way in which resources are partitioned is critical to the fitness of the individual. For example, more resource to active metabolism may increase the organism's chances of finding food or escaping predators; more resource allocated to storage may increase its ability to survive periods without food; more to growth may give a bigger body size which in turn may reduce the risk of predation, increase competitive ability and ultimately lead to greater reproductive returns; more resource to reproduction is likely to give greater rates of reproduction in the shorter term; and so on. Acquired resource that is allocated to any one metabolic compartment is not available to the others and the optimum pattern of partitioning is the one that most effectively transforms the resource input into viable reproductive products and thus transmits more genes to future generations. This will vary according to physical and biological constraints, but more particularly according to the environment in which they have evolved. The aim of this book is to consider, using optimality principles, which options are favoured under particular sets of environmental conditions.

Despite the fundamentally different ways in which plants, animals and micro-organisms obtain their 'food', the options relating to the partitioning of this energy along the various possible pathways are the same for all. Some general principles will be elucidated about these options and their consequences, with a range of examples from bacteria to elephants and Protozoa to trees. One important feature of this approach is that traditional divisions within biology are broken down. The book represents a synthesis of ecology, physiology and evolution.

This integrated collection of chapters is divided into four parts. The first part (Chapters 1 and 2) discusses the relationship between optimization theory, strategies of resource use and fitness, as well as outlining phylogenetic trends in bioenergetic processes. The second part (Chapters 3–5) examines the dynamics of resource gathering in plants and animals and discusses the alternative options and their adaptive significance with respect to the ecological circumstances in which they occur. The third part (Chapters 6–10) considers the way in which acquired resources are partitioned amongst the five metabolic compartments (defence, repair, storage, growth and reproduction) and a recurring theme is that natural selection has acted to optimize the form of the partitioning compromises in order to maximize individual fitness. The final part (Chapters 11–13) draws together the principles enunciated in the earlier essays by considering in-depth studies of plant and animal physiological ecology and animal social organization.

C. R. T.
P. C.

Acknowledgements

The generous help of the following is gratefully acknowledged: John Krebs, John Ollason, Richard Sibly (Chapter 4), Myrfyn Owen, David Houston, Fadhil Al-Joborae, Robert Moss (Chapter 5), W. Hallwachs and the National Science Foundation (NSF DEB 77–04889) (Chapter 6), Malcolm Maden, Robin Holliday (Chapter 7), P. Randolph (McClure), E. Marsh, R. N. Sinha, L. Fitzpatrick, N. Stenseth (Chapter 12). Mr. D. Baird kindly helped to prepare the subject index. We would especially like to thank Robert Campbell of Blackwell Scientific Publications Ltd for his enthusiastic guidance throughout the preparation of this book, and Erica Ison for editorial assistance.

Part 1

Chapter 1

Energetics, Ecology and Evolution

PETER CALOW AND COLIN R. TOWNSEND

1.1 INTRODUCTION

All of the chapters that are brought together in this book would be equally appropriate for texts on physiology, ecology or evolutionary biology. We are concerned with the physiological question of *how* organisms work and with the ecological and evolutionary questions of *why* they work in the way they do. The thread drawing the chapters together is the process on which they all focus: that of resource acquisition and its allocation. What factors determine the effectiveness with which resources are captured by organisms in a given environment? How are the materials and energy acquired in photosynthesis and feeding partitioned between the demands of maintenance, active metabolism, defence, repair, storage, growth and reproduction? Why are certain metabolic demands given priority over others and how do these priorities change with age and size? Given that certain constraints are inherent to biological systems, what are the best possible (optimal) allocation strategies in particular environments? Similar questions might be asked about the physiology of circulatory, endocrine and excretory systems and the fine details of intermediary metabolism (Hochachka & Somero, 1973), but since all aspects of the physiology of organisms are either powered by acquired resources or concerned with controlling resource allocation ours is, in a sense, the more general consideration. Futhermore, it is as 'resource-transformers' that organisms plug into ecosystems and so it should be relatively straightforward to appreciate this aspect of their physiology in ecological and evolutionary terms. Such an approach has its roots in ecological energetics (Phillipson, 1966) and the MacArthurian School of evolutionary ecology (MacArthur, 1962, 1972), both of which flourished in the mid-1960s.

A further crucial consideration underlies all that follows: the assumption that there is a more or less direct relationship between strategies of resource allocation and fitness. In this first chapter our aim is to sketch out what fitness means in a physiological context and to show how this physiological usage relates to the more traditional, genetical one. We will also attempt to justify two explicit biases which appear in many of the chapters that follow: the tendency to equate 'survival of the fittest' with 'optimization' and the tendency to replace 'resource' with 'energy'. But first we will address a fundamental criticism of evolutionary theory itself.

1.2 IS THE NEO-DARWINIAN THEORY OF EVOLUTION SCIENTIFIC?

Evolution by natural selection is the 'differential perpetuation of genotypes' (Mayr, 1963). Given two conditions: a population of organisms differentially susceptible to an environmental factor which influences fecundity or mortality (a selection pressure) and the fact that at least some of this susceptibility is inheritable; evolution by natural selection is inevitable. This has led some critics to propose that the theory of evolution is irrefutable and therefore unscientific. A related criticism has been the suggestion that general statements of evolution are not predictive since evolutionary theory is sufficiently general to accommodate any observation (Peters, 1976). For example, heavy predation pressure on a population of burrowing animals might be predicted to lead to a population of better burrowers in the future. However, the burrowers might become more noxious, resort to an arboreal life-style, etc. Peters argues that 'if the alternatives to a theoretical prediction also support the theory, the theory is not predictive'.

An answer to these criticisms is to recall that a hypothesis can be tested either by comparing its predictions with observation or by a direct test of the validity of the assumptions it incorporates (Mayard Smith, 1969, 1978a). The two major assumptions associated with neo-Darwinism (the genetical interpretation of Darwinian Evolution) are concerned with heredity and variability. These supporting assumptions can lead to non-obvious predictions which may be tested and therefore refuted in the real world. On *heredity* it is assumed that what is transmitted from generation to generation is a series of discrete, hereditary particles—the genes—which behave according to rules first enunciated by Mendel and first grafted onto Darwin's theory by Haldane (1924), Fisher (1930) and Wright (1931). On *variation* it is assumed that genes can be altered (they mutate); that this might be caused by environmental agents like irradiation, but that such alterations are never directed by the environment to specific ends. How might these assumptions be tested? It is possible, for example, that the bean-bag, genetical mechanisms of Mendel might be refuted in favour of the blending inheritance that Darwin favoured, or that mutations are directed in some way by the whims of organisms. In both cases survival of the fittest (as understood by neo-Darwinists) would collapse as a plausible mechanism for evolution. We are now reasonably confident, of course, from a wide variety of evidence derived from molecular biology, cytology and breeding experiments that genes are particulate and that they can only be modified through random mutation. Contemporary population genetics is therefore concerned more with predicting the details of changes and constancies in gene frequencies than with attempting to establish that evolution occurs by natural selection or that it is compatible with Mendelism (cf. the major concern of Haldane (1924) and Fisher (1930)).

1.3 TWO VIEWS OF FITNESS

The *fitness* of a genotype refers to the relative contribution which carriers of that genotype make to the gene pool of future generations. The link between the transmissibility of a gene and the gene itself is dependent on how its carrier interacts with the environment. This interaction does not occur directly, but through a gene-controlled intermediary, the phenotype.

The model summarizing the neo-Darwinian organism is as depicted in Fig. 1.1. There is a genotype (G) which controls the development of the phenotype (P) and the way that it interacts with the environment. The latter includes resource transactions (I, O) which influence the extent to which the genes in the genotype are copied and transmitted to the next generation. Note that the transfer of information is from G to P only and that the phenotype to genotype interaction (broken line) is of a completely different kind. This unidirectionality of information flow, if correct, precludes Lamarckian evolution. The hypothesis of unidirectionality is refutable, but hardly a shred of evidence has yet been brought forward which casts doubt on its validity. Note also that whole genotypes are transmitted intact only in asexual systems which reproduce by mitosis. The shuffling mechanisms associated with the production of gametes and subsequent fertilisation in sexual forms causes the genotype to be broken apart during the act of gametogenesis and mixed with other genes during the process of syngamy. In principle, because of sex, all genes are separable and independent so the units that are selected and transmitted are individual selfish genes (Dawkins, 1976) not whole genotypes. The collection of genes brought together in any one individual at one time is ephemeral and when discussing evolutionary change it becomes necessary to consider the gene pool of the population as a whole. In practice, however,

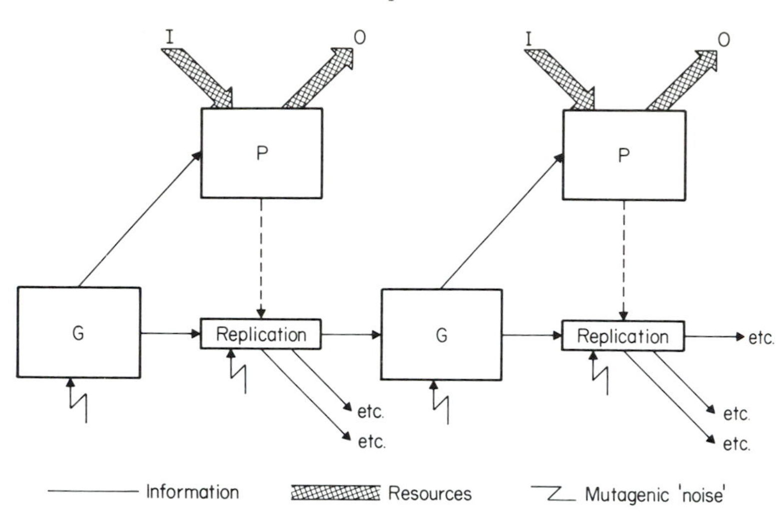

Fig. 1.1 Model of the neo-Darwinian organism. Information from the genotype (G) specifies the building and mode of operation of the phenotype (P). The latter involves the inputs and outputs of material and energy. The success of the phenotype determines the extent to which the genotype is replicated and 'passed on' to new, viable organisms.

strong linkage of a non-linear kind may occur between particular gene groupings so that concepts such as 'selfish gene' and 'gene pool' may be somewhat abstract. In this event, there is an as yet unresolved difficulty in objectively deciding what the units of selection are (Brandon, 1978).

Once the phenotype and the environment are brought into focus and once it is realised that the success of gene transmission depends upon the interaction between the phenotype and its environment then it becomes important to consider why certain phenotypes are better at promoting gene transmission (i. e. are fitter) than others under given environmental circumstances. This is usually the way in which 'whole-organism biologists' consider fitness and we refer to it as the adaptationist approach. There are, therefore, at least two ways in which the word fitness can be used. *Neo-Darwinian fitness* measures the *effects* of a selective advantage on the abundance of gene(s) coding for it, relative to other genes that do not. It can be measured precisely and rigorously as a gene frequency or more technically as a coefficient of selection (Haldane, 1924), Malthusian parameter (Fisher, 1930) or fitness space (Wright, 1968). Alternatively, the *adaptationist's fitness* is concerned with the *causal* basis of a selective advantage—why one character should promote gene transmission better than others—in terms of the interaction between the phenotype and its environment. This is more difficult than neo-Darwinian fitness to measure and to express precisely.

A further problem for the adaptationist is the fact that he can rarely, if ever, consider the demographic effects of traits directly in his investigations. Various assumptions need to be made about the way traits expressed at any point during the life cycle of an organism influence its chances of survival and future reproductive performance. The phenotype is essentially the gene's way of making more genes. It is a unit which must acquire resources to transform into reproductive products while maintaining itself in the face of predation and other environmental risks (Fig. 1.2). We believe it can safely be assumed

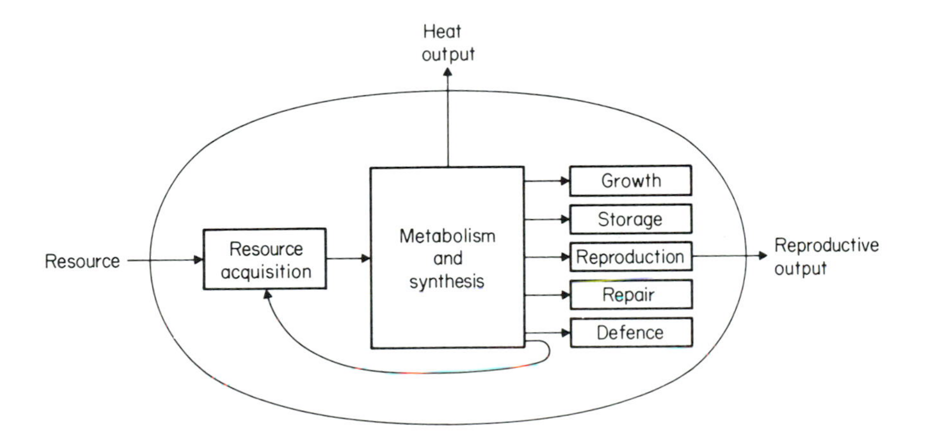

Fig.1.2 The phenotype as a resource transformer: schematic representation of resource acquisition and allocation.

that natural selection will tend to produce organisms which are maximally effective at propagating genes, and therefore at doing other activities which subserve the same function in the end, such as foraging, defending themselves, storing fat, growing, etc. (in so far as these are compatible). These are subcomponents of fitness and can be referred to as *phenotypic measures of fitness*.

1.4 METHODOLOGY

Adaptationists ask two similar but distinct kinds of question about the phenotype–environment interaction and its relationship to gene transmission.

1 Given that certain phenotypic characters are correlated with particular ecological circumstances, what is it that renders them more appropriate for gene transmission?
2 What kinds of characters might be expected to be more appropriate as aids for gene transmission in prescribed ecological circumstances?

Question 1 is the more traditional one. It is an *a posteriori* or natural historical approach which often proceeds in a comparative fashion by analysing characters of different species, or populations of the same species living in different or similar ecological circumstances. Question 2 is a more recent one. It is *a priori* or predictive in that it attempts to anticipate, using simple conceptual models, computer simulations or models based on complex algebra, what characters ought to promote fitness and hence what characters ought to be selected in carefully defined ecological conditions.

1.4.1 THE *A POSTERIORI* APPROACH

A pre-requisite for this kind of approach is a method for distinguishing between ecological circumstances with differing selection pressures and hence different effects on the evolution of populations within them. Environmental physiologists have traditionally classified habitats on the basis of temperature, humidity, wave action, pH, oxygen concentration, availability of nutrient salts and a host of other physical and chemical factors (see textbooks on 'Environmental Physiology', e.g Bligh, Cloudsley-Thompson & Macdonald, 1976). With variation of this kind it has been possible to discover correlated inter- and intraspecific variation in physiological traits which can be explained as products of natural selection. Sometimes quite wide-ranging 'laws' have been formulated in this way; for example, Bergmann's law which recognises a recurrent positive correlation in the animal kingdom between body size and latitude (Mayr, 1956).

More recently, Grime (1979) has attempted to formulate an even more general system of environmental classification which at the same time relates more obviously and directly to evolutionary theory. Grime bases his comprehensive classification of vegetation on the assumption that the external factors limiting plant biomass in any habitat fall into one of two basic categories; conditions which restrict production and which are referred to as stressful (e.g. shortage of light, water, minerals and sub-optimal temperatures) and conditions which effect the total or partial destruction of plant biomass and which are referred to as disturbances (e.g. herbivores, pathogens, trampling by man). There are four permutations of high and low stress and high and low disturbance (Table 1.1), but only three are habitable by plants (because in highly disturbed habitats severe stress prevents recovery and re-establishment of vegetation). Grime suggests that each of the three remaining categories has promoted a distinct kind of selection pressure and hence vegetative strategy; low stress with low disturbance—competitive plants; high stress with low disturbance—stress-tolerant plants; low stress with high disturbance—ruderal plants. These three primary strategies are of course extremes and the majority of plants probably compromise between the conflicting pressures of competition, stress and disturbance. Nevertheless, using this framework of classification as a foundation, Grime has been able to perceive interesting trends with an apparent evolutionary basis in what otherwise might have been an unco-ordinated collection of natural history observations. Some of these correlations between vegetation strategies and habitat types are listed in Table 1.2 (Grime, 1978).

Table 1.1 Suggested basis for the evolution of three strategies in vascular plants.

Intensity of disturbance	Intensity of stress	
	Low	High
Low	Competitive strategy	Stress-tolerant strategy
High	Ruderal strategy	No viable strategy

Table 1.2 Some characteristics of competitive, stress-tolerant and ruderal plants.

Characteristic	Competitive	Stress-tolerant	Ruderal
Morphology of shoot	High dense canopy of leaves, extensive lateral spread above and below ground	Extremely wide range of growth forms	Relatively small stature, limited lateral spread
Leaf form	Robust, often mesomorphic	Often small or leathery, or needle-like	Various, often mesomorphic
Litter	Copious, often persistent	Sparce, sometimes persistent	Sparce, not usually persistent
Maximum potential relative growth rate	Rapid	Slow	Rapid

Table 1.2 (*contd.*)

Characteristic	Competitive	Stress-tolerant	Ruderal
Life-forms	Perennial herbs, shrubs and trees	Lichens, perennial herbs, shrubs and trees (often very long-lived)	Annual herbs
Longevity of leaves	Relatively short	Long	Short
Phenology of leaf production	Well-defined peaks of leaf production coinciding with period(s) of maximum potential productivity.	Evergreens, with weakly defined patterns of leaf production	Short period of leaf production in period of high potential productivity
Phenology of flowering	Flowers produced after (or, more rarely, before) periods of maximum potential productivity	No general relationship between time of flowering and season	Flowers produced at the end of temporarily favourable period
Proportion of annual production devoted to seeds	Small	Small	Large
Response to stress	Rapid, morphogenetic responses (root–shoot ratio, leaf area, root surface area) maximizing vegetative growth	Morphogenetic responses slow and small in magnitude	Rapid curtailment of vegetative growth, diversion of resources into flowering
Photosynthesis	Strongly seasonal, coinciding with long continuous period of vegetative growth	Opportunistic, often uncoupled from vegetative growth	Opportunistic, coinciding with vegetative growth
Storage of photosynthate, mineral nutrients	Most photosynthate and mineral nutrients are rapidly incorporated into vegetative structure but a proportion is stored and forms the capital for expansion of growth in the following growing season	Storage systems in leaves, stems and/or roots	Confined to seeds
Palatability	Various	Low	Various

From Grime, J. P. (1978).

There are a number of problems with the *a posteriori* approach, however, which are tied up with the inductive philosophy upon which it is based (Popper, 1959). In the first place, it is tempting to assume (erroneously we believe) that all the characteristics observed in plants and animals are adaptive. Maynard Smith (1978a) has pointed out that neutral or maladaptive traits may occur because evolutionary change is not instantaneous and species may lag behind a changing environment. In addition, it is quite plausible that some characters may not be directly adaptive but may arise as side-effects of others which are. For example, Weisman thought that ageing was selected because it brought the direct advantage of removing the old and decrepit individuals from populations, thereby making way for the young and more virile (Weisman, 1891). However, this argument is circular in that it assumes what it sets out to prove, i.e. that organisms become more decrepit with time or, in other words, age! It seems much more likely that ageing has arisen as a side-effect of selection for other characters which bring more definite and direct effects on the fitness of the bearer (Calow, 1978a). Given a sufficiently inventive mind it is always possible to imagine what the selective advantage of any trait might be. The tendency to imagine adaptations where they do not exist is often referred to as the Panglossian falacy (Gould & Lewontin, 1979), after the philosopher in Voltaire's Candide who argued that 'since everything is made for an end, everything is necessarily for the best end . . . noses were made to wear spectacles . . . legs are visibly instituted to be breached . . . stones were formed to be quarried and build castles.'

Another potential problem with inductive science is that it often confuses correlations with cause–effect relationships and results in the formulation of law-like statements when less strong inferences are more appropriate. Bergmann's rule (defined above), Cope's Law which claims an orthogenetic trend in size with evolutionary time and Dollo's Law which argues for the irreversibility of evolutionary trends all suffer from this kind of problem.

The inductive, *a posteriori* method is, therefore, an important first step in the adaptationist's approach (Clutton-Brock and Harvey, 1979), giving order to observation and teaching us how organisms actually meet the challenge posed by different kinds of ecological circumstance. It is important to tread cautiously, however; to be suspicious of Panglossian optimism, and to be vigilant against the danger of confusing causal and correlational statements.

1.4.2 THE *A PRIORI* APPROACH

This orientation is of a more predictive kind. On the basis of one of several different kinds of deductive scheme it is possible to formulate hypotheses on phenotypic 'design' which can be tested in true Popperian fashion against real-world phenomena (Popper, 1959). Our concern here will not be with the precise form of these hypotheses since this will be considered in greater detail later and is, indeed, the main theme of the book. Instead we will be more concerned with the general way that such hypotheses might be formulated and with some of the problems that might arise in their testing.

A priori evolutionary arguments are usually based on the assumption that natural selection is an optimizing process; that there is some sense in which it can be said to result in the evolution of the best possible traits. Engineers and economists who are also concerned with selecting best solutions to particular technological or economic problems, have developed mathematical methods, collectively termed *optimality theory*, for dealing with this kind of problem and these are now being used in evolutionary biology. The main requirements for the application of optimality theory are first that all possible solutions to the problem are known and second that to each solution it is feasible to assign a number, or more complex function, which denotes either its value (v) or its cost c relative to some predetermined requirement. What the mathematics of optimality theory does is to search among the alternatives for the one which provides the largest v or smallest c. In the simplest case, each alternative may be expressed as a single number and the solution is trivial; for any distinct pair of real numbers x and y, we must have either $x > y$ or $y > x$. For more complex functional relationships, however, more complex techniques of ordination are needed. The reader will find these in any textbook on optimality theory but in particular should consult Rosen (1967) who still provides one of the most straightforward introductions to optimality principles and their use in biology.

The crucial requirement for the application of optimality theory is the measurement of profits and costs and this depends, in turn, on a clear, unambiguous definition of what is required from the system. Engineers and economists invariably have predetermined specifications against which the performance of any design can be compared and so also, in a sense, do biologists. According to neo-Darwinian theory, the phenotypic specifications of biological systems should tend to those that maximize the spread of descendents, genes and perhaps even gene complexes. Unfortunately, as we have pointed out, for all except a few traits it is usually impossible to compute this fitness value in a rigorous, quantitative fashion. However, as will become apparent from what follows, less rigorous, qualitative approaches which approximate to the programme we have outlined can, nevertheless, give a considerable insight into evolutionary processes.

The logic of the *a priori* approach can be summarised as shown in Table 1.3. Note that the translation of neo-Darwinian fitness into a phenotypic

Table 1.3 A summary of the ideal *a priori* approach.

1 Assume that evolution maximizes neo-Darwinian fitness.
2 Translate this into a phenotypic measure of fitness.
3 Using appropriate theoretical techniques find the trait which maximizes the phenotypic measure of fitness.
4 Compare this prediction with what is observed in nature or in contrived, experimental circumstances.

measure of fitness (related to foraging, defence, growth, etc.) involves an assumption about what is being maximized, or in other words, what selective forces have been responsible for the trait in question. The process of comparing the predicted optimal solution with the real world is essentially a test of this assumption. It is an attempt to sharpen our understanding of what fitness means in terms of the phenotype and the way it interacts with its environment. The role of optimality theory in biology is decidedly not to test the theory of evolution, for which genetic experiment is appropriate (see section 1.2).

There are a number of constraints that must be considered in the optimization approach. Indeed some authors (e.g. Gould & Lewontin, 1979) have proposed that more emphasis should be placed on the constraints themselves, rather than on the selection process, maintaining that the former are sometimes of greater interest in delimiting pathways of evolutionary change. The following factors are likely to be important.

1 Optimization models in biology cannot afford to consider individual factors in isolation. There are many components that should be simultaneously maximized if the overall neo-Darwinian fitness is to be maximized. One character may be constrained by others such that it is not possible for any to achieve the best conceivable state. For example, in principle fitness is maximized when reproductive output is maximized, but big-bang reproduction can have adverse consequences for the metabolism and hence survival of the parent. Therefore, where parental survivorship is an important component of fitness, as it is when juvenile survivorship is poor, reproductive output will operate at a sub-maximum level (Calow, 1979). Simultaneous adjustment of related phenotypic states is best dealt with using vector optimization techniques (e.g. Sibly & MacFarland, 1976) and the major challenge arising out of this kind of approach in biology is usually to establish a common currency in which each parameter can be measured and meaningfully related to fitness.

2 Developing this theme at the population and inter-population level, it is clear that conflicts between individuals may also impose constraints. The optimal strategy of one organism must take into account that of others. In a sense, organisms, or more abstractly, strategies, play existential games with each other (Slobodkin, 1968). A theoretical solution to such conflicts can be derived from game theory, a branch of optimality theory concerned with finding the best possible moves in any kind of game. It has been possible to define evolutionary stable strategies (ESS) which are those moves in the game that cannot be beaten by alternatives, that is that cannot be replaced by 'invading' mutants (Maynard Smith & Price, 1973). Between-individual conflicts are generally of most interest to behavioural ecologists (Krebs & Davies, 1978), and will not be a prominent feature of the chapters that follow (but see Chapter 13).

3 Stearns (1977), following Lewontin (1974), suggests that a full account of evolutionary change must be expressed in terms of the transformations shown in Fig. 1.3. viz.: T_1—the epigenetic laws which specify how a given genotype is translated into a particular phenotype under specified environmental circumstances; T_2—the ecological laws which specify the survivorship and fecundity properties of a known array of phenotypes; T_3—the reverse epigenetic laws which enable a given array of phenotypes to be translated into an array of genotypes; T_4—the genetic laws which specify how after recombination etc, genotypes should be constituted in the next generation. Mendel's laws and their population equivalents compose T_4, but at this stage we have little or no information on T_1 and T_3, and only the outlines of T_2. The *a priori* optimization approach explicitly short-circuits between P_1 and P'_1 with the aim of obviating the need to have complete knowledge of all four transformations (but see chapter 3). The simple assumption made by adaptationists is merely that like begets like. However, as well as drawing our attention to the complexity of the evolutionary process, the 'Lewontin model' reminds adaptationists that what is achieved phenotypically does depend ultimately on genotypic potentialities or constraints. Does sufficient genetic flexibility always exist to realise optimal phenotypic strategies? Are the laws of genetics and epigenetics always compatible with the optimal solution? It is frustratingly difficult to give answers to these questions even for the simple one-locus, two-allele case (Oster & Wilson, 1978), and at this stage it is virtually impossible to be explicit about the consequences of genetic constraints in polygenic systems, which are probably responsible for the control of complex ecological strategies (Law *et al.*, 1977).

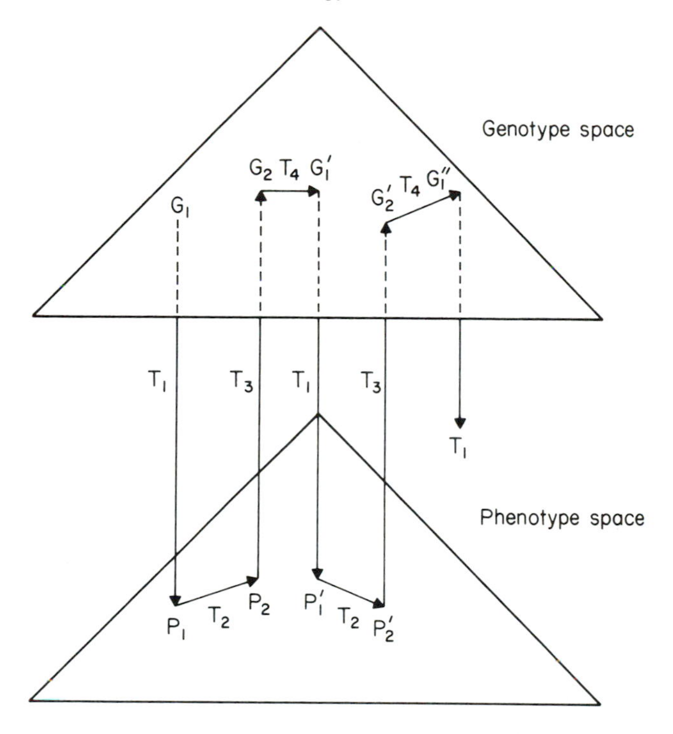

Fig. 1.3 A population geneticist's view of evolution. The cycle starts with an initial distribution of genotypes (G_1) proceeds through a series of epigenetic and selective transformations, and finishes with a new distribution of genotypes (G_1').

[We should note that much of genetics also suffers from a one-sided treatment of the 'Lewontin model' this time ignoring T_1, T_2 and T_3 and short-circuiting between G_1 and G_1' (Waddington, 1968)].

4 Finally, perhaps the most fundamental reason why the *a priori* approach cannot be used without reservation concerns the first requirement for the application of optimality theory; the specification of all possible solutions to the problem. Since selection can only 'choose' from that which is generated by chance it might only be possible to define options on the basis of comparative information on characters that have evolved in the past and, if this is the case, the *a priori* method depends on information that can only be derived *a posteriori*. Thus, stage 3 in the logic of the *a priori* approach (Table 1.3) becomes use optimality theory, to choose the best strategy from those *known to be feasible on the basis of biological experience*. This is certainly less rigorous than the originally specified

procedure because models built on the basis of biological experience are tested against that experience, raising the spectre of circular argument. Such a feature is a real weakness for which the only excuse is that all hypothesis formulation is based on some kind of experience (the *tabula rasa* does not exist; Popper, 1972), and for which the only antidote is the continued realization that the weakness exists.

1.5 FITNESS AND STRATEGIES OF RESOURCE USE

The most unequivocal measure of the *effect* of a particular selective advantage is in terms of the spread of genes. The *cause* of a particular selective advantage, on the other hand, can only be understood in terms of the way the phenotype interacts with its environment. A major part of this interaction involves the uptake and utilization of resources, and depends to a large extent on the physiological, and to some extent behavioural functioning of the organism. Our concern is to judge physiological and related behavioural processes involving resource transactions in terms of their effect on gene transmission.

The assumption that the way organisms make use of resources is critical to their success as gene transmitters is underlain by a further assumption: that these resources are limited and must be apportioned 'wisely' to obtain the best result. In many situations, resource availability in the environment is known to be limiting and the assumption is clearly valid. In other situations, resources appear to be superabundant and the assumption can be questioned. We maintain, however, that even in these circumstances the assumption holds, because an organism's 'machinery' for uptake will always limit its capacity for uptake, and thus the amount of resource it can use in a day or a life-time is limited, even if the supply in the environment is potentially unlimited.

'Resources' comprise a variety of nutrients (e.g. carbon, nitrogen, protein, essential amino acids, vitamins, mineral salts, etc.) as well as potential energy, and the material resources may sometimes be the limiting factor rather than energy. For example, nitrogen is often limiting for herbivores and detritivores (White, 1978), and essential salts for autotrophs (Harper, 1977). Nevertheless, much of what follows will be expressed in terms of potential energy and the reasons for this are both pragmatic and philosophic. First, potential energy is usually more easily measured in biological tissues than the other resources and so more data are available on budgets of energy than of matter. Second, it can be argued that energy is a more general measure of resources than anything else; for example, all nitrogen compounds have a potential energy but not all biochemical molecules contain nitrogen. It is also the case, as for resources in general, that the input of energy to organisms must be limited and those organisms which optimize their use of energy should also optimize their fitness. There are a number of complications which can, in fact, usually be ignored either because they are unimportant or because they assume equal importance in the systems being compared. Nevertheless, these underlying assumptions should be made explicit. In particular, there should be no incompatability between optimal energy allocation and the optimal organization or form of the organism. For example, maximizing the allocation of potential energy to the gonads may be selectively advantageous but only as long as the gonads are not too large to be supported by the body of the parent. Furthermore, it is assumed that the qualitative form of the resource input puts no constraints on the way that it is allocated between growth, storage, reproduction, etc., and yet this will clearly depend on the composition of the input in terms of amino acids, fatty acids, carbohydrate, and so on.

The question now is: what general criteria of energy transactions ought to be important in maximizing neo-Darwinian fitness? Clearly, the ideal would be to calibrate them in terms of their effects on survival, the speed at which organisms mature and reproductive success. However, sometimes it is not possible to determine the influence of minute by minute energy transactions on demographic events that might take place at some time in the future, and then it is necessary to define more immediate characteristics that should be strongly correlated with the rate and success of gene transmission. Obvious possibilities are the efficiency and/or rate with which energy is made available for the production of new protoplasm by an acquisition or allocation strategy. These will be seen to figure prominently throughout the book.

1.6 HISTORICAL PERSPECTIVE AND RATIONALE OF THIS BOOK

In 1930 Fisher wrote, 'It would be instructive to know not only by what physiological mechanism a just apportionment is made between the nutriment devoted to the gonads and that devoted to the rest of the parental organism, but also what circumstances in the life-history and environment would render profitable the diversion of a greater or lesser share of the available resources towards reproduction'.

From this beginning there has grown an appreciation for the organism as an adapted complex, not simply at the level of biochemistry, organ function, anatomy etc., but also at the level of whole life-histories (Lack, 1948; Svardson, 1949; Cole, 1954; Williams, 1966a) and related complex strategies (e.g. defence strategies, Feeny, 1975; foraging strategies, Schoener, 1971; optimal behaviour sequences, Sibly & McFarland, 1976).

A crucial piece of work in the development of this holistic approach was that of MacArthur & Wilson (1967). They recognized that environments could be classified in a continuum, one extreme representing a perfect ecological vacuum with no density effect and no competition, the other a completely saturated ecosystem where density is high and competition intense. In the former (generally understood to be unpredictable or seasonal environments), it is supposed that such life-history characteristics as high maximum rate of increase (r), early reproduction and large litter or clutch size are likely to be optimal since species occupying such environments probably often experience periods of population expansion. These environments are referred to as r-selecting and the associated traits r-selected. At the latter extreme, in stable environments, the premium is supposed to be on competitive ability. Such environments are called K-selecting, and the associated traits K-selected after the term describing the upper limit to population growth in the logistic equation.

Some studies on r and K selection have followed the *a priori* approach by comparing the allocation of resources to reproductive and competitive functions in contrasting environments. For example, McNaughton (1975) was able to show for various species of *Typha* that, as predicted, plants occupying environments with a short growing season (r-selecting) had a faster development and a higher fecundity and showed less evidence of selection for traits increasing competitive ability, such as increased height and rhizome production. Perhaps even better examples, because they study intraspecific variation, are the investigations of Gadgil & Solbrig (1972) on dandelions and of Law (1979b) on the grass *Poa annua* (and see Chapter 11). Much other work, however, has tended towards a more correlational, *a posteriori* approach (Pianka, 1970), in the same way as Grime's classification of vegetation (section 1.4.1), which can be considered as an extension of the r/K approach (Grime, 1979).

The r/K concept has been enormously influential in the development of ideas but it is clearly inadequate as a complete explanation of life-history phenomena. It identifies resource availability as a single selection pressure capable of explaining many life-history differences within and between species. However, it has been suggested (Wilbur, Tinkle & Collins, 1974) that attempts to explain life-histories as outcomes of a single selection pressure may sometimes have obscured rather than elucidated their evolution, and Stearns (1976), for example, notes that the traits associated with stable and fluctuating environments would be expected to be reversed if juvenile mortality fluctuates more than adult mortality. In fact, MacArthur (1972) himself pointed out that the r/K division is by no means the only possible one: 'We could speak of youth vs old age selection; or predator escape selection vs selection for feeding ability, and so on.'

The rationale of this book is based partly on MacArthur's contention that looking at other possible selection pressures should prove rewarding. In addition, we are seeking to bring together all the different strategies which rely on the optimal use of limited resources; resource gathering, defence, storage, repair, growth and reproduction.

Our objective in Parts 2 and 3 is to explore each kind of resource acquisition and allocation strategy in turn, to describe the theoretical basis for predictions about optimal resource use and to determine how far predictions are supported by observation in nature or the results of experiments. In some chapters the hypotheses will be derived from relatively simple conceptual models, in others from more rigorous mathematical models. Sometimes the hypotheses will be tested using the comparative approach; given a functional hypothesis there are usually testable qualitative predictions about the form of a trait in different species or in different environments—McNaughton's work on *Typha* was of this kind. In other chapters, quite precise quantitative predictions will have been tested by experiment.

Our final section is based upon the recognition that the adaptational features of one physiological system depend upon others—the organism is a tightly integrated and precisely controlled system which has evolved as a whole. Hence, there is a need for consideration of intensive, in-depth studies on all aspects of the bioenergetics of individual species. The three chapters in this section will deal with the range of allocation options as they operate in selected species. These chapters will draw together the principles enunciated in Parts 2 and 3.

In essence, the aim of this book is to attempt to explain the diversity of strategies of resource acquisition and allocation in terms of a minimum number of selection pressures.

Chapter 2

Bioenergetic Options and Phylogeny

J. PHILLIPSON

2.1 INTRODUCTION

Over geological time large numbers of different kinds of organisms have appeared on earth and it is not surprising that man, with his tendency toward orderliness, has attempted to describe and classify them. One result of these endeavours has been the construction of phylogenetic trees to represent their evolutionary history. Recognition of an evolutionary history (= phylogeny) implies acceptance of the idea that living organisms are related to each other and have arisen from a simple ancestry by a long sequence of structural and functional differentiation. That evolution has occurred can hardly be doubted although exactly how it happened is still subject to debate.

Not subject to debate is the fact that all living organisms expend energy to survive and reproduce. Any characteristics which reduce the energy requirement per unit mass of an organism should be considered evolutionarily advantageous in that a greater proportion of the available energy can be directed towards producing offspring. Nevertheless, one would do well to remember that an overall increase in energy expenditure can lead to a greater energy intake per unit mass which may in itself be advantageous. In evolutionary terms the question arises as to whether identifiable relationships exist between phylogenetic status and different levels of energy utilization; it is this aspect which forms the subject matter of this chapter.

2.2 PHYLOGENY, PHYLOGENETIC STATUS, ENERGY SOURCES AND BODY SIZE

The majority of phylogenetic trees, an example of which can be seen in Fig. 2.1, imply that:

1 a common stock of single-celled organisms (Protista) gave rise to both multicellular plants (Metaphyta) and multicellular animals (Metazoa);

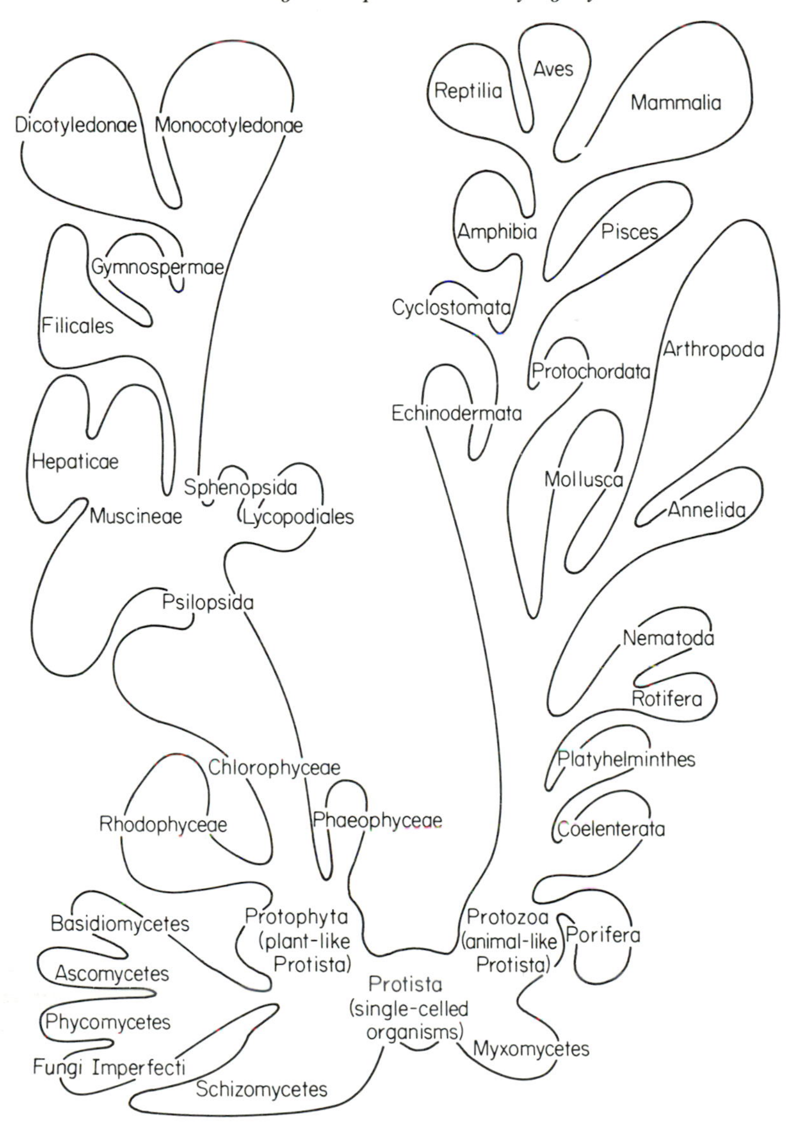

Fig. 2.1 One version of a phylogenetic tree.

2 there is a clear divergence of the two major stems which respectively represent Metaphyta and Metazoa.

Current theories of the origin of life suggest that it occurred under the conditions of a reducing atmosphere—an atmosphere containing high proportions of hydrocarbons and no oxygen. The energy-obtaining processes of early unicellular organisms were undoubtedly anaerobic and probably similar to those found in present day anaerobes which gain their energy by substrate-linked phosphorylation (= fermentation). However, not all of the available substrates can be used for fermentation and Broda & Peschek (1979) have argued that an ability to exploit additional resources arose when a redox reaction was developed by a cell that had lived previously by fermentation. They suggest that in one and the same cell the reversible proton-translocating ATPase of the fermenter combined with a proton-translocating redox system which evolved from a precursor process they term 'Prerespiration'. This postulated evolutionary step gave rise to a situation in which the ATPase of the fermenter was turned into ATP synthase, the overall result being the production of energy conserving ATP. One should not imagine that the suggested change in metabolic pathways was selected for in all early fermenters, indeed fermentation is characteristic of present day slime fungi (Myxomycetes), many bacteria (Schizomycetes) and fungi (Fungi Imperfecti, Phycomycetes, Ascomycetes and Basidiomycetes).

Among the early ATP-producing unicellular organisms there occurred further evolution of metabolic pathways. It is generally believed that the ancestors of the photosynthetic organisms developed a chlorophyll-like photocatalyst which combined with the electron transport phosphorylation system that had arisen with the evolutionary step from prerespiration. The new phototrophic forms, of which green and purple photosynthetic bacteria are representative, derived their energy from light and were able to metabolize most of the end products of anaerobic fermentation. In turn, the tissues they synthesized were themselves future substrates for the fermenters. Clearly, the two types of organism could, and still can, coexist in an oxygen-free environment.

The predominant energy generating process of the green and purple photosynthetic bacteria is cyclic photophosphorylation; it is assumed that photosynthesis with chlorophylls as photocatalysts began with this efficient ATP-producing system. Later developments, which precluded the necessity of using the energy of ATP for NADH production, gave rise to a non-cyclic photophosphorylation. Such a system is the main photosynthetic process in blue-green algae and all higher plants.

Following the appearance of photosynthesis and the consequential enrichment of the biosphere with nutrients and oxidants (including oxygen) of biological origin the adoption of aerobic respiration (oxidative phosphorylation) and heterotrophy became a distinct possibility. Because of the close similarity of the mechanisms in photophosphorylation and oxidative phosphorylation Broda & Peschek (1979) have concluded that a common phylogenetic origin of photosynthesis and respiration must be postulated; convergence from separate origins is unlikely. It is thought that many photosynthetic prokaryotes abandoned photosynthesis by converting their photosynthetic machinery to that for respiration and its associated ATP production, at the same time they maintained the capacity for fermentation. Such changes most probably occurred among those organisms with better access to organic reductants than to light.

The above conversion hypothesis is in keeping with the situation as seen in the phylogenetic tree of Fig. 2.1. The common stock of single-celled organisms with its admixture of fermenters, photosynthesizers and respirers reflects the close relationships of the early evolutionary forms while the dichotomy with respect to later autotrophs (Protophyta and Metaphyta) and heterotrophs (Protozoa and Metazoa) mirrors their different and further evolved energy obtaining processes of non-cyclic photophosphorylation and oxidative phosphorylation.

Not obvious in a phylogenetic tree diagram, but of considerable importance, is a trend towards general increase in body size during phylogeny. Recognition of this tendency has been with us since the time of Lamarck and is frequently referred to under the general epithet of Cope's Rule or Law (Boucot, 1978). Acceptance of the existence of this trend carried with it the implication that large body size *per se* confers an evolutionary advantage.

Among others Calow (1977a) states that the major advantages of being big are that

1 larger females tend to produce larger numbers of offspring;
2 bigger organisms tend to have an advantage in competitive situations involving direct aggression;
3 size may confer greater immunity against predation.

It is suggested that all these factors increase the chances of an organism reaching maturity, thereby increasing its reproductive output. There has been a trend towards increasing size during the course of evolution although the findings of Stanley (1973) show that in a number of taxa the size frequency distribution of members has a skewed form with the mode towards the lower end of the size spectrum. It would appear that the optimum size is not the maximum size and the paradox arising from these contrasting conclusions requires an explanation. A probable solution can be found in a consideration of the energy requirements and utilization of different-sized organisms which, by inference from Cope's Rule, are of different phylogenetic status.

2.3 ENERGY UTILIZATION

Of the potential energy assimilated by organisms a proportion is directed into new tissues or secretory products as a consequence of work being performed. Additionally the acquisition of raw materials and energy is itself an energy-consuming process and a further proportion of the acquired energy must be utilized for such activity. Energy expenditure on the acquisition and re-ordering of molecules leads to the transformation of the energy of chemical bonds (chemical energy) into heat energy. In large measure the heat energy is not put to any useful purpose by the majority of organisms producing it and is lost to the external environment. We can thus envisage a partitioning of energy by organisms where a proportion is directed into what might be considered a useful end product (e.g. new tissues including gametes and young) and the remainder to a much less useful product (heat) (see Fig. 2.2).

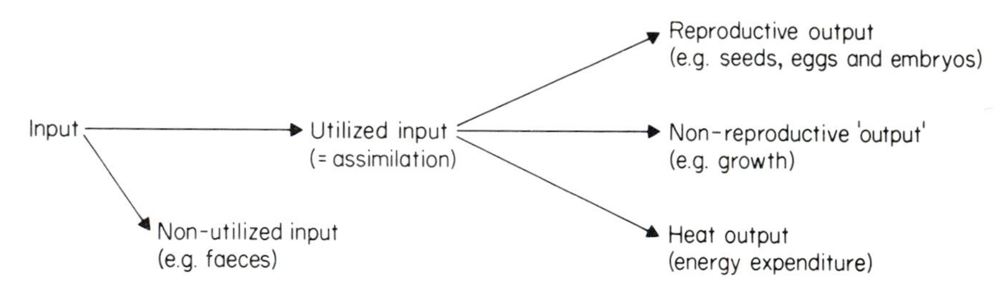

Fig. 2.2 Partitioning of food energy by organisms.

It is well known that of the energy assimilated by organisms, by far the largest proportion is dissipated as heat. Heat output is notoriously difficult to measure with accuracy and metabolic activity, usually determined as oxygen consumption, is frequently adopted as an indirect measure. On average, the complete combustion of 1 mg of carbohydrate releases approximately 23.64 J and requires an oxygen uptake of 1.119 ml, hence the energy equivalent of 1 ml of oxygen when used in the combustion of carbohydrate is 21.13 J. Because heat is proportionally the largest output from assimilated energy, and because oxygen uptake is an indirect measure of it, it is not difficult to see why oxygen consumption has been used as a vehicle for exploring energy expenditure in relation to body size and phylogenetic status.

2.4 ENERGY EXPENDITURE, BODY SIZE AND PHYLOGENETIC STATUS

Zeuthen (1947, 1953, 1970) and Hemmingsen (1950, 1960) were among the first to explore the idea that energy expenditure patterns evolved as a function of phyletic increase in body size. Both authors, although not in agreement regarding detail, are in accord with the view that among living organisms there exist three major groupings which can be distinguished easily by their different levels of maintenance energy expenditure. The first category contains only the Protista, the second includes all non-protistan invertebrates plus the 'cold-blooded' vertebrates and Metaphyta, while the third accommodates the 'warm-blooded' vertebrates (Aves and Mammalia). A major distinction between the Protista and the other two groups is unicellularity while the Aves and Mammalia differ from most other taxa in being endothermic as opposed to ectothermic*. Knowledge of these differences allows us to describe the three groupings in terms other than names of taxa, e.g.: (i) unicellular ectotherms; (ii) multicellular ectotherms; and (iii) multicellular endotherms, such descriptions reflecting both the structural and functional organization of descendent lines.

The levels of maintenance energy expenditure exhibited by the three groupings are shown graphically in Fig. 2.3 where log oxygen consumption per individual per hour has been plotted against log body mass. This figure is, in effect, a check on the earlier noted general conclusions of Zeuthen (1970) and Hemmingsen (1960) as the information has been presented in a similar diagrammatic form but data have been used which, in the main, appeared after the publication of their accounts. The number of cases involved was 554 and following Hemmingsen (1960) all unicellular and other ectotherm data were corrected to a single temperature; in this instance 10° C which approximates the mean annual temperature of the environments experienced by most of the organisms.

For endotherms the temperature corrections were related to 39° C as this reflects the internal environment of the animals concerned. It was not practical to represent every measurement in Fig. 2.3 and the small numerals indicate the

* An endotherm derives bodily heat from internal sources (generally metabolic or muscular heat) and an ectotherm from external sources (generally solar radiation, directly or indirectly). The two terms should not be confused with homeothermy where the organism maintains a relatively constant body temperature and poikilothermy where the organism has little or no control over body temperature.

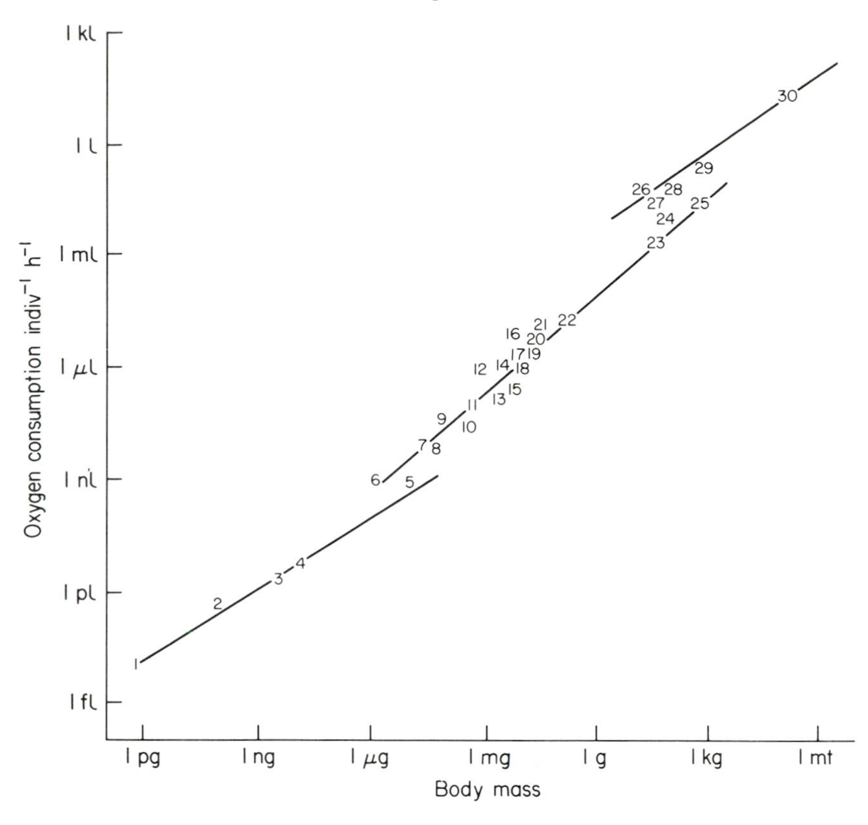

Fig. 2.3 Double logarithmic plots of oxygen consumption against body mass for unicellular and multicellular organisms at 10°C and multicellular endotherms at 39°C. The numerals 1–30 represent the pivotal points of the regression lines of each of the taxa listed in Table 2.1.

pivotal points of the regression lines drawn for each of the coded taxa listed in Table 2.1. The regression lines constructed for each of the three major groupings do, however, take into account all measurements and, as can be seen from Table 2.2, the slopes of these lines are significantly different from zero and their regression coefficients are significantly different from each other.

The major points of interest to be noted in Fig. 2.3 are:

1 the upper size limit of unicellular ectotherms overlaps the lower size limit of multicellular ectotherms;
2 multicellular ectotherms of a given body mass have a maintenance energy

Table 2.1 List of taxa used in compiling Fig. 2.3 and the slopes (*b*) of their least squares regression lines.

Code number	Group	*n*	*b*	Code number	Group	*n*	*b*
Unicellular ectotherms				*Multicellular ectotherms cont.*			
1	Bacteria	5	0.68	17	Diplopoda	77	0.79
2	Fungi	2		18	Aranea	6	0.81
3	Flagellata	4	1.33	19	Isopoda	40	0.69
4	Ciliata	5	0.28	20	Mollusca	6	0.76
5	Rhizopoda	5	0.93	21	Coleoptera (adults)	14	0.81
Multicellular ectotherms				22	Lumbricidae (adults)	18	0.76
6	Nematoda	24	0.82	23	Macrocrustacea	3	0.81
7	Microcrustacea	12	0.91	24	Pisces	6	0.69
8	Acari	71	0.61	25	Reptilia	5	0.66
9	Collembola	29	0.74	*Multicellular endotherms*			
10	Isoptera (larvae)	4	0.75	26	Small mammals	9	0.58
11	Enchytraeidae	61	0.87	27	Chiroptera	8	0.70
12	Coleoptera (larvae)	17	0.67	28	Aves	19	0.45
13	Isoptera (adults)	21	0.94	29	Primitive mammals	13	0.73
14	Formicidae (workers)	23	1.14	30	Large mammals	14	0.77
15	Lumbricidae (cocoons)	3	1.00				
16	Phalangiida	30	0.69				

Table 2.2 Regression equations from Fig. 2.3 relating respiration individual^{-1} h^{-1} (y) to size (x), also given are the levels of significance of the slopes from zero and from each other, the latter in parentheses.

Group	*n*	$y = bx + a$	95% limits of *b*	*P*
Unicellular ectotherms	21	$\log_{10} y = 0.66 \log_{10} x + 1.41$	0.09209	< 0.001
				(< 0.001)
Multicellular ectotherms	470	$\log_{10} y = 0.88 \log_{10} x + 0.49$	0.00002	< 0.001
				(< 0.001)
Multicellular endotherms	71	$\log_{10} y = 0.69 \log_{10} x + 4.83$	0.00165	< 0.001

expenditure which is 6–10 times higher than that of unicellular ectotherms of the same size;

3 the upper size limit of multicellular ectotherms overlaps the lower size limit of multicellular endotherms;

4 multicellular endotherms of a given body mass have a maintenance

energy expenditure which is 10–30 times higher than that of multicellular ectotherms of the same size;

5 the slopes of the regression lines are all less than one, indicating that within each category maintenance energy expenditure per unit mass decreases with increasing body size.

Earlier in this account we noted two attributes considered to be evolutionarily advantageous to organisms, namely reduction in maintenance energy requirement per unit mass (section 2.1) and increase in body size (section 2.2). It is not difficult to envisage these trends within each of the three major functional groupings but it is more difficult to recognize them *between* groups. Why do the size ranges of members of the different categories overlap? Why should similar-sized organisms from the different groups have a 10–30-fold difference in energy expenditure per unit mass? Exploration of these questions provides possible answers to the paradox of why the optimum size for a set of closely related organisms is not necessarily the maximum size.

2.5 UNICELLULAR ECTOTHERMS AND ENERGY EXPENDITURE

The distinctly different regression lines for unicellular ectotherms, multicellular ectotherms and multicellular endotherms illustrate quite clearly that identifiable relationships between phylogenetic status and different levels of energy expenditure do exist. However, as Table 2.3 shows, the regression lines of Hemmingsen (1960), Zeuthen (1970) and those of the present study differ in their slopes. The findings of the latter tend not to support the general contention of Hemmingsen that all three lines have a similar basic slope approximating 0.75 but are more in keeping with Zeuthen's conclusions that the slope changed during the course of evolution.

Over short ranges of body size (e.g. within taxa such as those listed in Table 2.1) it is a well-established fact that the slopes of regression lines of

Table 2.3 The values of the slopes of the regression lines relating energy expenditure individual^{-1} h^{-1} to size as calculated by different authors.

	Hemmingsen (1960)	Zeuthen (1970)	Phillipson (present study)
Unicellular ectotherms	0.756 ± 0.021	0.70	0.66 ± 0.09209
Multicellular ectotherms	0.738 ± 0.0095	$0.95 \rightarrow 0.80$	0.88 ± 0.00002
Multicellular endotherms	0.739 ± 0.010	0.75	0.69 ± 0.00165

log oxygen consumption ($\equiv$ log energy expenditure) against log body size vary quite markedly, usually between 0.5 and 1.0. Even so, over greater ranges of body size (> 3–5 logarithmic decades), there has been much speculation as to why a given slope should be characteristic of a particular set of organisms (see Hemmingsen, 1960; Kleiber, 1961; Zeuthen, 1970).

The size spectrum of the unicellular ectotherms in this study ranges from 700 fg ($= 0.7 \times 10^{-12}$g) in bacteria to 886 μg ($= 8.86 \times 10^{-4}$g) in rhizopods and covers some 8 log decades. Over this size range oxygen consumption spans 5 log decades and the slope of 0.66 implies that metabolism is proportional to an almost 2/3 power of the body mass; put another way, for every 100 % increase in body mass energy expenditure increases by only 66 %. If all the unicellular ectotherms were spherical in shape and had a non-crenulated plasma membrane then it is clear that a slope of 0.66 will also represent a metabolism which increases in direct proportion to surface area. The metabolism/surface area hypothesis for unicellular ectotherms is very appealing if one accepts that the plasma membrane influences everything that enters or leaves the cell, a situation that can only be achieved if the membrane is relatively impermeable to small molecules but at the same time not so impermeable that the passage of respiratory gases is restricted.

It can be argued that higher values of b than 0.66–0.67 do not negate the general conclusion that metabolism is related to external surface area, for instance not all cells are spherical nor are their surfaces smooth. The gross configuration of the plasma membrane ranges from the relatively regular outline of certain cells (e.g. mammalian red blood cells) to a multiple infolded surface as exemplified by the cells of kidney tubules. Moreover, in some organisms (e.g. the cnidarian *Cordylophora*) certain cells have the ability to change shape and thereby alter the proportion of the surface exposed to the environment (Kinne, 1964).

In addition to the plasma-membrane features, all cells have numerous metabolically active surfaces not related directly to the external surface, e.g. nuclear envelopes, endoplasmic reticulum, Golgi apparatus, lysosomes and mitochondria. Weiner, Loud, Kimberg & Spiro (1968) have shown that a single rat liver cell of 5100 μm^3 has a total membrane surface area of 94 470 μm^2, of which the inner mitochondrial membranes contribute 39 600 μm^2. Both of these values greatly exceed the 3400 μm^2 of external surface membrane necessary to satisfy the 2/3 power of body mass relationships and so there still remains the enigma of why values of b generally vary between 0.5 and 1.0.

The energy expenditure of any organism depends upon the work it has to perform, this in turn is reflected in the surface area of its metabolically active membranous organelles. The amount of oxygen entering an organism is governed by the surface area of its plasma membrane exposed to the external medium. It does not seem unreasonable to suggest that variation in values of b is a result of the balance between oxygen demands of the inner membranes and its supply and transport as governed by the outer plasma membrane.

In the case of unicellular organisms, where there is protoplasmic movement but no complex circulatory system, the transport of oxygen is mainly by simple diffusion. Krogh (1941) calculated that for a smooth surfaced spherical organism of $< 500\,\mu g$ simple diffusion alone would be sufficient to provide an adequate supply of oxygen for maintenance purposes. At high activity levels and/or above $500\,\mu g$ it is clear that oxygen demands could only be met by an improved circulatory system and/or an increase in the surface area of the plasma membrane relative to body mass. In keeping with these calculations the body mass of the majority of unicellular organisms is below 1 mg and in the present review the largest individual for which results were available was $886\,\mu g$. Of those forms exceeding $500\,\mu g$ (some Ciliata and Rhizopoda) none is of spherical shape and, in association with infolded plasma membranes, they tend towards elongation and flattening. Moreover, the largest (e.g. *Chaos chaos*) possess additional organelles by virtue of being multinucleate. The indisputable importance of plasma-membrane surfaces, and the length of diffusion pathways, in supplying the energy demands of different organelles of unicellular organisms offers an acceptable explanation as to why the slope of b can vary but most closely approaches 0.67.

As the size of an organism increases so does its total energy requirement, it follows that for survival the advantages accruing from size must outweigh the cost of acquiring additional energy. Clearly, there is a metabolic limit to the size a unicellular organism can achieve without loss of mobility and presumably fitness in other respects. It would appear that the optimum size for most unicellular organisms is likely to be below the size at which simple diffusion is only just able to meet the maintenance requirements. This becomes even more obvious when one takes into account the fact that active, as opposed to maintenance, metabolism can be 10 or more times demanding of oxygen. If an organism is to satisfy its metabolic needs during periods of high activity it would seem that a likely optimum size for the more active unicellular organisms would be nearer to 50 than $500\,\mu g$. In a situation such as this one might expect the size frequency distribution of unicellular organisms to be skewed towards the lower end of their size range, such a feature would be in accord with the more general findings of Stanley (1973) and place in question any universal application of the dictum 'bigger is better'. Whilst it is difficult to apply the 'bigger is better' concept to the largest of the unicellular ectotherms it can be considered in a broader evolutionary context. One means of doing this is to look for alternatives to size increase by cell dimensions alone, the most obvious of these is multicellularity.

2.6 MULTICELLULAR ECTOTHERMS AND ENERGY EXPENDITURE

Major genetic changes must have been associated with the origin of multicellularity and the consequent appearance of Metaphyta and Metazoa. However, as we can see from Fig. 2.3, multicellular ectotherms with body sizes smaller than those of the largest Protista do occur. In such circumstances interplay between genetic and selective factors must be assumed such that for some organisms multicellularity was selected for while for others it was increased cell dimensions. Which selection pressures operated to produce the different outcomes remain unknown but energy and oxygen requirement must have had a significant effect.

Among multicellular ectotherms, just as in unicellular ectotherms, values of b over short size ranges usually vary between 0.5 and 1.0. Possible reasons for this variation can again be sought in the quality and quantity of the membrane surfaces of different cells, and in the efficiency of the respiratory gas transport systems of the organisms. Of greater interest in the context of energy expenditure and phylogeny are the values of b over more extensive size ranges, also the overall higher metabolic rate per individual multicellular ectotherm when compared with a unicellular ectotherm of the same size.

Adopting the reasoning employed earlier it can be argued that the value of 0.88 found for b in the present study of metazoan ectotherms indicates a plasma-membrane surface per unit mass greater than could be reached by the non-folded membrane of a spherical unicellular organism. On the basis of this argument one would expect that on average the plasma membranes of the cells of multicellular ectotherms will exhibit more extensive folding than those of unicellular ectotherms. Whether this is so is not known.

The effect of multicellularity *per se* on metabolic potential can be illustrated by reference to a hypothetical example. Given two organisms of the same size, say a cube of $1\,cm^3$, then a unicellular organism would have a 'plasma-membrane' surface area of $600\,mm^2$ whereas a multicellular organism with cells each of $1\,mm^3$ would consist of 1000 cells with a total 'plasma-membrane' surface of $6000\,mm^2$. It is not difficult to comprehend that providing oxygen can reach all surfaces then the multicellular organism is capable of receiving ten times as much oxygen per unit time as the unicellular one. The degree of difference demonstrated in this hypothetical example approximates the 6–10-fold difference in energy expenditure between unicellular and multicellular ectotherms of the same size and within the overlap range of 1–900 μg shown in Fig. 2.3.

Below a certain critical size, 500 μg according to Krogh (1941) and 1 mg according to Goddard (1947), simple diffusion supplies the oxygen requirements of organisms without a complex circulatory system. This condition holds not only for unicellular ectotherms but also for the smallest of the Metazoa, e.g. free-living Nematoda and Rotifera. In these cases the minimum diffusion path remains short because of either elongation or flattening of the body. Nevertheless, with a 6–10-fold increase in oxygen supply over similar-sized unicellular organisms the early metazoans would have had a greater potential for evolutionary advancement. Additional oxygen would have provided the basis for an active as opposed to a sedentary mode of life, more work could be performed and directed towards increased food capture and its processing. Clearly, such food would provide additional energy and raw materials for reproduction and growth. There can be little doubt that the major genetic shift as represented by multicellularity conferred considerable potential for survival in the face of the forces of natural selection.

Above the critical body size of 500 μg to 1 mg, realization of the full potential afforded by multicellularity and the associated increase in oxygen supply required further changes in body design. The changes favoured by natural selection were probably those most closely associated with food capture, its processing and transport. As organisms increase in size there is a need for skeletal support and also for an outer covering to protect the body as well as prevent excessive water loss. The latter need requires impermeable outer coverings which in turn require the development of special respiratory surfaces with a means of transporting the respiratory gases and nutrients. Such morphological and anatomical changes are the ones we most readily observe, e.g. primitive tracheae in well-armoured mites, lungs and vascular systems in amphibians, no legs and eight legs as exemplified by nematode worms and arachnids, respectively. It is features of this type which are employed in grouping similar organisms into the phylogenetically arranged taxa of evolutionary trees, their similarities and differences are part of the evidence that evolution has occurred. Without multicellularity, enhanced oxygen supply and the associated opportunities for increased activity and size, none of these features could have arisen.

Multicellular ectotherms thus appeared in great variety and ranged in size from the < 1 μg of some nematodes to the estimated 50 mt of the extinct *Brachiosaurus*. The inclusion of this last genus among multicellular ectotherms follows Benton (1979) who is of the opinion that the large dinosaurs maintained a relatively constant body temperature via inertial homeothermy (see footnote on p. 25) and not via endothermy. If this is the correct interpretation then it is reasonable to suppose that dinosaurs would have fitted the extrapolated multicellular ectotherm regression line of Fig. 2.3. The slope of this line has a value of 0.88 and indicates that for every 100 % increase in mass there is an increase in energy expenditure of 88 %. Considering only the size range of multicellular ectotherms shown in Fig. 2.3 (1 μg to 10 kg) it can be demonstrated that the energy expenditure per unit mass of the largest organisms will be 3.5 % of that shown by the smallest. If *Brachiosaurus* is included then its energy expenditure would have been a mere 0.7 % of the equivalent measure for the smallest metazoans.

The question arises as to whether, as in unicellular organisms, there are metabolic limitations to the size a multicellular ectotherm can reach. Lack of an adequate oxygen/energy supply to the inner membrane surfaces of the cells is clearly one possibility, particularly when one realises that the energy expenditure per unit mass of the largest unicellular and multicellular ectotherms is less than 1 % of that of their smallest counterparts. Less obvious, but nevertheless important, are the physical constraints on structural design, e.g. problems of skeletal support where, with increasing size, limb bones must become relatively as well as absolutely more massive (Newell, 1949). Another problem which, despite decreased energy expenditure per unit mass, required a solution if multicellular ectotherms were to increase in size during the course of evolution was the one of thermoregulation.

Bligh (1976) states that the rate of heat production by extant ectotherms is too low to have very much influence on tissue temperatures and that these vary more or less passively with the environment. However, studies of large living reptiles (see Benton, 1979) have shown that rates of internal temperature change are very slow during normal subtropical diurnal temperature fluctuations. By extrapolation, temperatures of large ectotherms (e.g. dinosaurs) living under subtropical conditions would remain constant to within 1 or 2° C inertially without internal heat production (McNab, 1978). With internal heat production due to the metabolism of activity, large ectotherms would suffer overheating unless methods for dissipating the excess heat developed. Effective heat dissipation is only achieved when the value of the regression slope of log metabolism versus log size (Fig. 2.3) is < 1.0.

Hemmingsen (1950) calculated that for a python with a body mass of 31.8 kg, a surface area of 1.4 m^2 and a metabolism proportional to surface area a difference of less than 0.1° C between body surface and external environment was all that was necessary to prevent overheating. However when the metabolism was taken to be proportional to body mass (i.e. $b = 1.0$) loss by radiation alone would require a surface temperature rise above ambient of between 25 and 63° C to achieve the same end.

Heat loss by radiation can obviously be enhanced by increasing the surface area relative to body mass. This procedure appears to have been adopted by some of the largest multicellular ectotherms, e.g. the well-vascularized dorsal plates of *Stegosaurus* (Farlow, Thompson & Rosner, 1976). It should not be forgotten, however, that such structures would act as heat absorbers in the cooler morning hours, thereby aiding overall homeostasis of body temperature. Of course, radiation is not the sole means of heat dissipation; other factors, especially evaporation (= insensible perspiration), are very important. Although in the case of the python mentioned earlier the heat balance could only be achieved by an evaporative loss of 2.4 kg of water per hour, i.e. an impractical 7.5 % of body mass each hour. The importance of body mass/surface relationships in governing what is evolutionarily possible, and the corollary that there are metabolic limits to the size a multicellular ectotherm can reach, cannot be denied.

The optimal size and structure of a multicellular ectotherm will clearly depend on the operative temperature of its body tissues and the environmental temperatures to which it is subjected. The largest extant forms occur in the relatively stable climates of the oceans but of all the ectotherms known to have existed the largest were the terrestrial dinosaurs. Benton (1979) has argued that dinosaur ectothermy, in association with a low energy requirement per unit mass, uricotelism and tolerance of a wide range of temperature conditions, was particularly advantageous in the arid subtropical environments of the Late Triassic. In the somewhat less arid Jurassic and Cretaceous, development of large body size, and thus inertial homeothermy, assured their continuing success.

Towards the end of the Cretaceous a probable temperature decrease, increasing seasonality and the appearance of a sparser flora militated against large size and undoubtedly contributed to the eventual extinction of the dinosaurs. In a cooling and variable climate the body temperature of large ectotherms would, despite inertial homeothermy, gradually fall as night-time heat loss came to exceed daytime heat absorption. Eventually body temperatures would reach a dangerously low level and the animals would become torpid and die since they lacked dermal insulation, temperature homeostasis or the ability to hibernate, being generally too large to find hibernation sites (Cys, 1967). Once again it would seem that in broad evolutionary context 'bigger' is not necessarily 'better'. For any given set of environmental conditions one might reasonably expect, as indicated by Stanley (1973), that the size frequency distribution of multicellular ectotherms would be skewed towards the lower end of their size range.

The advantages of optimum size being lower than the maximum attainable clearly relate in large degree to homeostasis with respect to body temperature. Among the smaller multicellular ectotherms tissue functions and activities, as well as body temperature, rise and fall in keeping with variations in ambient thermal conditions. Bligh (1976) has argued that 'thermal conformers' of different sizes exhibit a phase differential in their daily periods of high activity and that this has provided a selective pressure for the evolution of behavioural characteristics by which ectotherms can optimize the periods of high body temperature and high activity. If the body temperature can be maintained somewhere in the upper half of the thermal range compatible with life then there would clearly be a competitive advantage in the search for food and escape from predators. Being 'conformers' ectotherms cannot achieve this 'goal' at all times, especially during the cold of winters. The advantage in a competitive situation of an animal with a well-regulated body temperature (thermal regulator) over a thermal conformer is obvious and hence the acquisition of this ability constituted an evolutionary advance.

2.7 MULTICELLULAR ENDOTHERMS AND ENERGY EXPENDITURE

There is no clear dividing line between ectotherms and endotherms since endothermy is not an all or nothing characteristic. Many living mammals (e.g. monotremes and didelphid marsupials) regulate their body temperatures at relatively low levels; numerous reptiles and fishes (e.g. lizards, leatherback turtles, tuna and Porbeagle sharks) generate heat by muscular activity, as do large flying insects (e.g. butterflies, bees and beetles). Different forms of endothermy have risen independently several times but the 'temperature regulators' *par excellence* are the birds and higher mammals because the majority of them are able to sustain homeothermy. In those cases where the birds and mammals experience difficulty in coping with temperature fluctuations in their environment there is seasonal reversion to poikilothermy (heterothermy) but, as Bligh (1973, 1976) has pointed out, all the available evidence indicates that the various kinds of seasonal and nycthemeral heterothermy (e.g. torpidity and hibernation) in mammals are environmental adaptations of erstwhile fully established homeotherms.

Heinrich (1977) has suggested that the appearance of endothermy was particularly advantageous, allowing the maintenance of high levels of activity over extended periods of time. Indeed, fixed high body temperatures (40° C or more) may have evolved as the result of an inability to dissipate rapidly all the heat produced from high activity rates. Constant high body temperatures probably invoked changes in biochemistry which led to a more rapid substrate turnover by enzymes.

Clearly, the difference between an endotherm and an ectotherm need not be too great, the only basic change necessary in the evolution of endothermy was an increase in the concentration of mitochondria (Bennett & Dawson, 1976). Other changes associated with the ability to regulate body temperature are new neural circuitry, insulating layers of fur, feather or subcutaneous fat, the development of involuntary responses like shivering, panting or sweating and the regulation of peripheral circulation.

Turning once more to Fig. 2.3 the energy costs of active homeothermy are seen to be 10–30 times that of multicellular ectotherms of the same size. This difference is not due solely to the effect of the higher temperatures at which the oxygen consumption of the endotherms was measured. Even with a temperature correction to 10° C the metabolism remains higher than that of similar-sized ectotherms. As with the metazoan ectotherms the increased potential afforded by enhanced metabolism can be directed into activity, increase in size and reproduction.

There is a size overlap between the poikilotherms and homeotherms in the size range from 3 g to 3 mt and as in the case of unicellular versus multicellular ectotherms it is reasonable to assume that interplay between genetic and selective factors resulted in the selection of endothermal homeothermy in some forms and larger body size with inertial homeothermy in others. It is quite likely that climate constituted a major selection pressure and strongly influenced which option was selected for. The success of the thecodonts over the possibly endothermic theraspids during the Middle and Late Triassic has been attributed to the better adaptations of these ectotherms to higher temperatures, increasing aridity and lower food and water availability (Benton, 1979). Any advantages that endothermy may have had for the Late Permian and Early Triassic theraspids in the wetter, more equable conditions of those times would not have applied in the hot, seasonally arid Late Triassic. It is possible that endothermy did precede the evolution of mammals by 25 million years but Crompton, Taylor & Jagger (1978) believe that it arose somewhat later in small Mesozoic mammals. Whatever its origin, endothermy would have had considerable survival value for small insulated animals with a large surface area relative to body mass and which were either nocturnal (e.g. early mammals) or required large amounts of energy for flight (e.g. birds, pterosaurs).

The maintenance of set, high body temperatures in small animals is hindered by problems of heat conservation and the lower size limit of an endothermic homeotherm is probably fixed by the amount of effective insulation it can sustain while still being able to remain active. Conversely, large endothermic homeotherms face the same problems of heat dissipation as do large multicellular ectotherms.

With the simplifying assumptions that all heat is lost via radiation and that the emissivity is that of an absolute black surface, the temperature of a radiating surface can be calculated according to the Stefan–Boltzmann equation:

$$\text{Radiation in cal h}^{-1}\,\text{m}^{-2}\text{ body surface} = 4.96 \times 10^{-8}\,(T_1^{\,4} - T_2^{\,4})$$

where T_1 = absolute temperature of the radiating surface and T_2 = absolute environmental temperature.

Using these figures for a 100 g rodent one can show that at an environmental temperature of 28° C the 'Critical temperature' (i.e. the temperature at which heat production is at a minimum) of the rodent's surface will be 33.6° C. With body size increase and no drop in metabolism per unit mass, heat production per unit of surface area will increase. Effective heat dissipation can only be achieved by raising the temperature of the surface, i.e. by reducing insulation etc., but surface temperature cannot exceed that of the body (37° C) and hence a limit is set to the maximum size the rodent can reach. Without reduced metabolism per unit mass this is 338 g for the rodent under consideration. Clearly, over a limited range of body size metabolism per unit mass can remain the same, any increase in size above the upper limit requires a reduction in mass specific metabolism. That such a reduction does occur in the endothermal homeotherms can be inferred from the regression slope of 0.69, which indicates that for every 100 % increase in mass there is an increase in energy expenditure of only 69 %. From these values it can be calculated that the energy expenditure per unit mass of a 3 mt elephant is < 0.1 % of that of a small mammal of 3 g and may be near the limits at which lack of an adequate oxygen/energy supply to the membrane surfaces of the cells would preclude effective tissue metabolism. Certain whales, e.g. the blue whale at 120-mt body mass, reach a much greater size than do elephants but the two are exposed to different environmental conditions so that whales possess thick insulating layers of blubber. In keeping with whales, the small mammals of cooler climes require effective heat conservation while the large terrestrial forms need to effect heat dissipation via cooling structures such as the well-vascularized ears of African elephants. Clearly, there are metabolic limitations on both the minimum and maximum sizes which can be attained by endothermal homeotherms and judging by the large number of species of rodents and smaller birds relative to elephants and ostriches it would seem that the optimal size for a multicellular endotherm is not the maximum size.

2.8 PHYLOGENY AND ENERGY UTILIZATION

During the course of evolution the change to multicellularity facilitated enhanced activity and size increase while the change to endothermal homeothermy permitted further increase in activity and allowed a greater proportion of the total time to be devoted to essential activities. In an evolutionary context it is to be expected that organisms capable of directing the greatest proportion of their acquired energy into useful end products (growth and reproduction) were the most successful (see section 1.5). On this basis it might be anticipated that the additional metabolic costs associated with multicellularity and endothermic homeothermy were met by increased energy assimilation without detriment to growth and reproductive output.

The relationships between energy assimilation, utilization and expenditure are neatly summarised by the energy budget equation:

$$A = P + R = C - FU$$

where A = assimilated energy, P = energy of production due to growth (P_g) and reproduction (P_r), R = respiratory energy, C = consumed energy and FU = egested (faecal) and urinary energy. Here, we focus our attention on $A = P + R$.

A number of factors, especially age and physiological state, are known to affect the relative values of P and R in the energy budget of a single species as shown in Fig. 2.4. It will be obvious that to investigate the possible existence of relationships between A, P and R and phylogenetic status the influence of within species variation should be reduced to a minimum. This is best achieved by considering either individuals of similar developmental status or whole species populations which, by virtue of their age–class structure, integrate the within-species variation. Unfortunately there are, as yet, too few data on

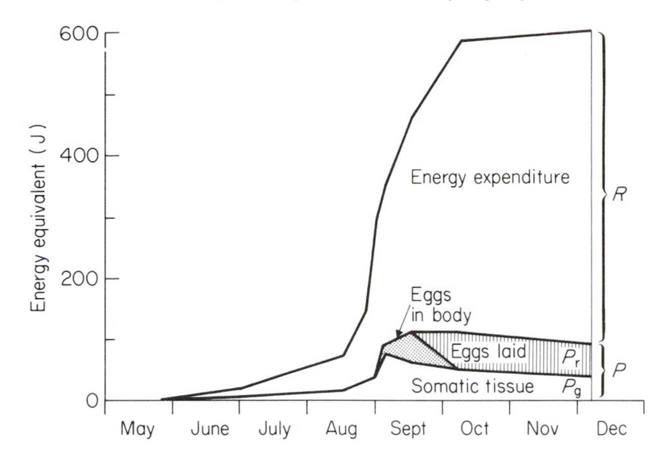

Fig. 2.4 The cumulative energy budget of a female *Oligolophus tridens* (Phalangiida, Arachnida) throughout its life-history. Note the differential allocation of energy to P_g, P_r and R with time. (Based on data supplied by J. Phillipson, after Klekowski & Duncan, 1975).

populations of unicellular organisms but comparative studies within the multicellular forms provide some insight into these relationships.

A number of studies of this kind have been attempted (Engelmann, 1966; McNeill & Lawton, 1970) but the most comprehensive is that of Humphreys (1979). Humphreys (1979) explored the relationships between annual production (P) and respiration (R) by analysing 235 population energy budgets taken from the literature. The data sets were grouped initially into 14 taxonomic categories and least squares regressions were calculated for each category treating both P and R (log cal m^{-2} yr^{-1}) as the dependent variable. Regressions were compared with each other and pooled if not significantly different in slope and intercept until the minimum number of separate groups was found. Irrespective of whether P or R was the dependent variable the same seven groups separated out; three of these (non-social insects, invertebrates other than insects, social insects plus fish) belong to the multicellular ectotherms while the remaining four (mammals other than insectivores, 'small mammal communities', birds, insectivores) are all endothermic homeotherms. The regression lines for the seven derived groups are shown in Fig. 2.5 and their regression equations in Table 2.4. The 'small mammal communities' line is possibly anomalous in that it is based on an admixture of 8 non-species specific energy budgets, two of which included insectivores.

From Fig. 2.5 it can be seen that the regression lines of the multicellular

Table 2.4 Geometric mean (GM) regression equations relating respiration (log R cal m^{-2} yr^{-1}) to production (log P cal m^{-2} yr^{-1}) in seven groups of animal populations (after Humphreys, 1979).

Group	n	GM regression equations
Multicellular ectotherms		
1 Non-social insects	61	$P = 1.000R - 0.144$
2 Invertebrates other than insects	73	$P = 1.068R - 0.820$
3 Social insects and fish	22	$P = 1.042R - 1.234$
Multicellular endotherms		
4 Insectivores	6	$P = 0.636R - 0.952$
5 Birds	9	$P = 0.790R - 1.055$
6 'Small mammal communities'	8	$P = 1.356R - 3.236$
7 Mammals other than insectivores	56	$P = 0.938R - 1.259$

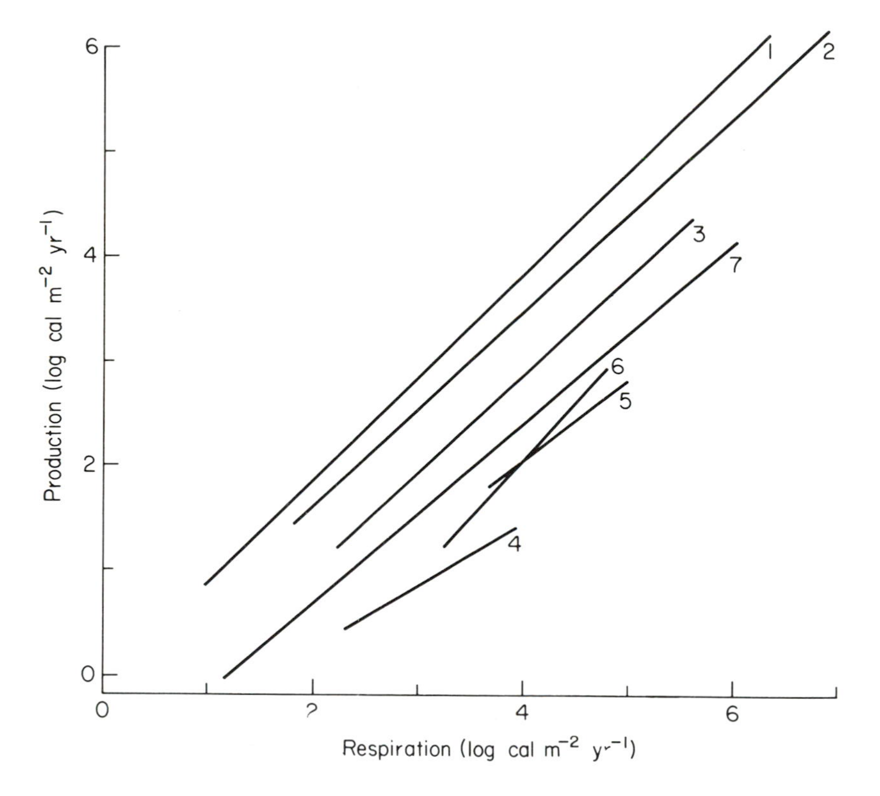

Fig. 2.5 Double logarithmic plots of annual production against annual respiration in seven groups of animal populations. The numerals 1–7 accord with the numbered groups listed in Tables 2.4 and 2.5.

ectotherms lie above those of the endothermic homeotherms. In a separate analysis Humphreys (1979) substantiated the findings of earlier authors that multicellular ectotherms (≡ poikilotherms) and endothermic homeotherms (≡ homoiotherms) could be represented by statistically different regression lines. The separation of these major phylogenetic categories according to their production and respiration characteristics is reminiscent of their separation on the basis of size and respiration (Fig. 2.3). Production and size are not the same parameter however and it is unlikely that the observed separations are the result of the same or related factors. In fact, production to biomass (P/B) ratios of 3.0 for invertebrates, 2.0 for small mammals, 0.2 for larger mammals and 0.05 for elephants (see Phillipson, 1973) indicate a general inverse relationship between size and production. Moreover, Humphreys (1979) was unable to find any significant relationship between maximum live weight and P and R, as reflected by production efficiency ($P/A = P/P+R$) in non-social insects, invertebrates other than insects, fish, mammals other than insectivores and insectivores.

The nature of the distribution of the data for the seven phylogenetic categories (Fig. 2.5 and Table 2.5) led Humphreys (1979) to conclude that there is no quantum jump in the P/A ratios of multicellular ectotherms and endothermic homeotherms. There does indeed appear to be a relatively smooth transition between the two. It would be of interest to know where the appropriate regression line for reptiles would lie in relation to the other lines but in the absence of such information it should be noted that the line

Table 2.5 Mean values for respiration (R), production (P) and assimilation (A) in kJ m^{-2} yr^{-1} for seven groups of animal populations. Also the R/P, R/A and P/A ratios (based on Humphreys, 1979).

Group	n	Mean R	Mean P	Mean A	R/P	R/A	P/A
Multicellular ectotherms							
1 Non-social insects	61	11.945	8.574	20.519	1.39	0.58	0.42
2 Invertebrates other than insects	73	135.574	41.608	177.182	3.26	0.77	0.23
3 Social insects and fish	22	23.077	1.850	24.927	12.47	0.93	0.07
Multicellular endotherms							
4 Insectivores	6	5.918	0.047	5.565	125.62	0.99	0.01
5 Birds	9	41.800	0.532	42.332	78.52	0.99	0.01
6 'Small mammal communities'	8	30.212	0.415	30.627	72.78	0.99	0.01
7 Mammals other than insectivores	56	11.700	0.387	12.087	30.20	0.97	0.03

representing the earliest true mammals (insectivores) is situated furthest away from those representing the multicellular ectotherms. Could it be that during the period when endothermal homeothermy evolved there was a quantum jump which created a gap since occupied by the later mammals and birds? By extending this reasoning to the multicellular ectotherms one might well ask whether the position of the fish and social-insect line reflects the effects of evolution in size and social organization? If so we might then postulate that evolution in both multicellular ectotherms and endothermic homeotherms has created a situation where the relationships between relative energy utilization and expenditure are very similar. This is illustrated in Fig. 2.6 where the seven derived groups of Humphrey (1979) have been ranked in an approximate evolutionary sequence to show the changing relationships between *R*, *P* and *A*. Using this figure it can be inferred that:

1 within the multicellular ectotherms production efficiency (P/A) fell during the course of evolutionary advancement while the maintenance costs, as a proportion of total assimilation (R/P), increased;
2 the advent of endothermic homeothermy was accompained by a fall in

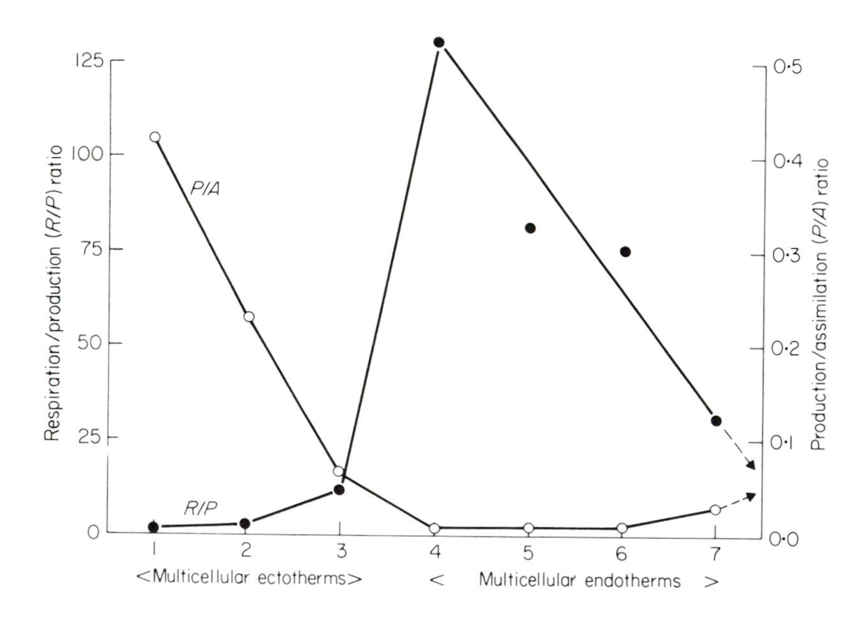

Fig. 2.6 The relationships of P/A and R/P in seven groups of animal populations. The numerals accord with the numbered groups listed in Tables 2.4 and 2.5.

production efficiency and an order of magnitude increase in relative maintenance costs;

3 evolution among the endothermic homeotherms was associated with a slight improvement in maintenance costs as a proportion of assimilation;
4 in both multicellular ectotherms and endothermic homeotherms the P/A and R/P ratios converged on the point where respiration accounts for 93–97 % of assimilation and production accounts for 3–7 %.

The changing relationships of P/A and R/P within the multicellular ectotherms are undoubtedly related to increasing body size and the associated reduction of maintenance costs per unit body mass (section 2.6). Among the early endothermic homeotherms the order of magnitude increase in relative maintenance costs (R/P) is clearly accompanied by an even greater increase in energy expenditure per unit mass over multicellular ectotherms of the same size (section 2.7). In later endothermic homeotherms increase in body size, improved insulation and similar adaptive features (section 2.7) no doubt led to the marked reduction in relative maintenance costs (R/P) and the marginally improved production efficiency (P/A). The convergence by both ectotherms and endotherms towards similar R/P and P/A ratios could signify that, irrespective of the type of organism, there is an evolutionary endpoint in the relationships of P and R. The majority of organisms, however, do not exhibit the fixed relationship and it is reasonable to argue that their specific R/P and P/A ratios are optimal for the environmental conditions they experience.

On a size for size basis Table 2.6 shows that such 'optimization' in an average insectivore requires the assimilation of 7 times more energy than an average fish if it is to achieve the same level of production, while a non-insectivorous mammal requires to assimilate only 2.3 times as much. However, allowing for the actual 10–30-fold difference in respiration per unit mass between fish and mammals production of the latter exceeds that of the former by a factor of 4.

Resolution of the various 'optimization' problems is not to be found in either differential assimilation efficiencies (assimilation/consumption = A/C) with respect to the same amount of food or differing energy contents of food. It is true that different trophic types (e.g. detritivores, herbivores and carnivores) exhibit different assimilation efficiencies but examples of each are found in almost all major taxa; the various assimilation efficiencies have little significance in relation to phylogenetic status.

The most obvious attribute underlying the continued existence of all living organisms is the ability to obtain and process sufficient food to meet the demands of maintenance (R), growth (P_g) and reproduction (P_r). In this

Table 2.6 Evaluation of assimilation (A), respiration (R) and production (P) for organisms of the same size when: (i) P is fixed at 10 arbitrary units; and (ii) the estimated value of R is adjusted to conform with the known 10–30-fold differences in respiration per unit mass.

			(i)		(ii)		
Group	Fixed P	Known P/A (%)	Estimated A	Estimated R	Adjusted R	Adjusted A	Adjusted P
Fish	10	7	142.86	132.86	132.86 (132.86×1)	142.86	10.00
Insectivore	10	1	1000.00	990.00	3985.80 (132.86×30)	4026.06	40.26
Non-insectivorous mammal	10	3	333.33	323.33	1328.60 (132.86×10)	1369.69	41.09

context it should be recalled that within each of the three metabolic categories identified (unicellular ectotherms, multicellular ectotherms and endothermal homeotherms) there are constraints on the size organisms can attain. The optimal balance between R which reduces per unit mass with size, P_g which increases chances of survival to maturity, and P_r which ensures continuity of the species, varies with the size, physiological type and environment of the organism. The largest size is not necessarily the optimum one, 'optimality' depending on the competitive ability and survival probabilities of the various ages and sizes. These characteristics are not related to phylogenetic status and their effectiveness is largely dependent on the differential allocation of energy of assimilation into R, P_g and P_r in relation to habitat conditions. Southwood's (1976) hypothetical 'block-fish' and his, and Western's (1979) discussion of the r–K selection continuum reinforce this conclusion. Optimization of food intake and the subsequent partitioning of this resource by individual organisms form the subject matter of the chapters which follow.

Part 2

Acquiring the Resource Input

Introduction

The benefits derived from resource gathering are clearly the energy and materials gained from the environment. These are then available for metabolism and synthesis. Not always so obvious are the associated costs, such as those of building and running the machinery of photosynthesis and digestion or the expenditure involved in foraging activity. The phenotype is essentially the genotype's way of transforming environmental resources into more genotypes. One way that this will be promoted is by maximizing *the rate at which a phenotype acquires resources.* Hence, this can be taken as our *phenotypic measure of fitness.*

The first chapter in this section (Chapter 3) is concerned with the factors determining the efficiency with which radiant energy is captured by plants. It provides the metabolic details of various alternative photosynthetic pathways and discusses their adaptive significance with respect to the ecological circumstances in which they occur.

Attention then switches to two aspects of resource acquisition by animals. In Chapter 4 the basic assumption is that the foraging behaviour which endows an animal with the greatest fitness (optimal foraging) is that which maximizes the net rate of energy gain. An *a priori*, predictive approach is adopted to discover whether real predators behave like their theoretical optimal counterparts in terms of how they exploit a patchy environment and how they choose their diet. Since the optimal solution often appears to require sophisticated computational powers, attention is focussed on the means by which the solution may actually be accomplished or approximated. The energy maximization premise is also applied to food selection by microphagous filter-feeders and sediment-feeders.

The *a priori* approach continues in the final chapter in this section (Chapter 5) which is devoted to strategies of digestion and defecation. The efficiency of absorption of a meal is inversely related to the speed with which it is passed through the gut, so the rate of absorption represents a compromise between speed of food passage and efficiency of digestion. The chapter examines optimal digestive compromises in organisms which exploit different kinds of food and which face different levels of food availability.

Chapter 3

Aspects of the Carbon and Energy Requirements of Photosynthesis Considered in Relation to Environmental Constraints

H. W. WOOLHOUSE

Grey is all theory
Green is life's growing tree

Goethe

3.1 INTRODUCTION: THE FRAME OF REFERENCE

In the opening paragraph of this volume the editors write 'we are concerned with the physiological question of how organisms work and with the ecological and evolutionary questions of why they work in the way they do'. In their subsequent statement of the problems underlying the understanding of adaptation to environment, the editors distinguish two methods of approach:

1 the *a posteriori*, in which intra- or interspecific variations are correlated with generally accepted categories of habitat and so 'explained' as products of natural selection, for an example of this approach see Grime (1979);
2 the *a priori*, a newer method which works from the premise that natural selection is an optimizing process (Rosen, 1967). Appropriate tools of mathematical modelling are now available for the application of optimizing procedures; see for example Maynard Smith (1974) and Chapter 10.

In the account which follows I imply no adherence to either of these approaches, rejecting the former as a philosophically unsound use of the scientific method (Popper, 1959) and the latter as leading inevitably to naïvety when applied to systems as complicated and imperfectly understood as the photosynthetic apparatus. There is another way to study adaptation, which is to proceed from measurements of physiological responses to the environment to an analysis of the underlying biochemistry and from these to the genetic control of the biochemical characteristics and from the biochemistry back to classical population genetics. This must inevitably be a painstaking and slow approach which will probably, in the short term, have to take second place to the grand designs for the rationalization of a mass of field data afforded by the *a posteriori* and *a priori* methods; but it is an approach with an honourable history (Turesson, 1922, 1925, 1931; Clausen, Keck & Hiesy, 1948; Jones & Wilkins, 1971) and offers a sure route to a proper answer to the 'why they work in the way they do' sort of question.

3.2 BASIC ELEMENTS OF PHOTOSYNTHESIS

Photosynthesis, the conversion of energy from sunlight to a chemical form, was an early event in the evolution of life. In all photosynthetic organisms studied up to the present time, the light-absorbing pigments and several other components necessary for the process are located in specialised membrane systems. There are probably several reasons why membranes should be an essential feature of the photosynthetic apparatus. In photosynthesis absorbed light quanta drive a flow of electrons from a donor molecule to an acceptor in a direction opposite to that which one would predict from the standard redox potentials of donor and acceptor, i.e. the electrons are driven in the direction of an acceptor having a more electronegative standard potential; a membrane system is probably essential to provide the precise requirements for spatial association between the light-absorbing pigment molecules, electron donors and acceptors. From present knowledge it also seems likely that the organization of the photosynthetic apparatus has to be based upon a membrane system if it is to yield chemical energy in the form of ATP. The transport of electrons from donor to acceptor in the membrane generates a counter movement of protons to the opposite side of the membrane so that the drop in free energy during the subsequent electron transport is conserved in the form of a gradient of hydrogen ions across the membrane. This proton gradient generated by electron transport provides the energy for the formation of ATP from ADP and inorganic orthophosphate (P_i) in a condensation reaction in which the elimination of water is catalysed by an enzyme system known as the coupling factor (CF_1) (Fig. 3.1, Equation 3.1; (Mitchell, 1961; Jagendorf & Uribe, 1967)).

$$H^+ + ADP + P_i \xrightarrow{CF_1} ATP + HOH \qquad (3.1)$$

In its earliest manifestations, this light-driven synthesis of ATP probably occurred by a cyclic process in which the phosphorylation was coupled to a flow of electrons from the excited donor molecule to a primary acceptor and back along a chain of electron-carrier molecules to the electron hole left in the de-excited or ground-state donor (Fig. 3.1). The next major development in

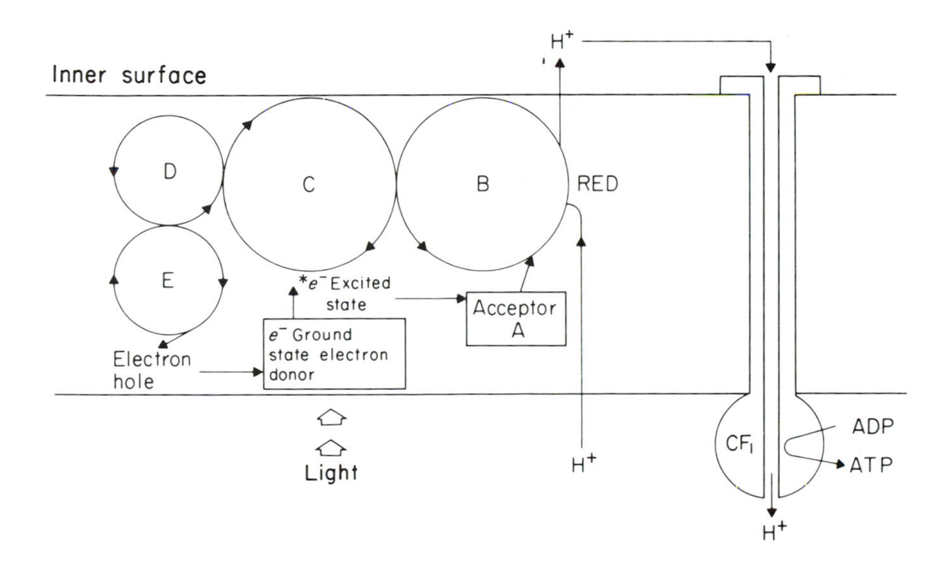

Fig. 3.1 A rudimentary photosynthetic reaction sequence accommodated within a membrane structure. High-energy electrons leaving the excited donor molecule, reduce an electron acceptor A. From A the electrons pass to a sequence of redox acceptors forming an electron-transport chain back to the positive hole created in the donor by the initial excitation. The drop in free energy in the course of the electron transport is conserved in the form of a gradient of hydrogen ions transported from the outside to the inside of the membrane via the shuttle B. The proton gradient formed in this way drives the formation of ATP catalysed by the enzyme complex denoted CF_1.

the early evolution of the photosynthetic process probably involved the opening of the chain of electron flow in the membrane so that the electron hole in the de-excited pigment is filled by an external donor (D) and the chain of electron flow leads to an external acceptor (A), Equation 3.2, (Fig. 3.2).

$$H_2D + A \xrightarrow{\text{light}} H_2A + D \tag{3.2}$$

Amongst the photosynthetic bacteria there are species able to use such substances as Fe^{2+}, H_2S and isopropanol as electron donors. The first stable electron acceptor is usually NADP and this is used by different species for a variety of reducing reactions including CO_2 to carbohydrate, nitrate to nitrite,

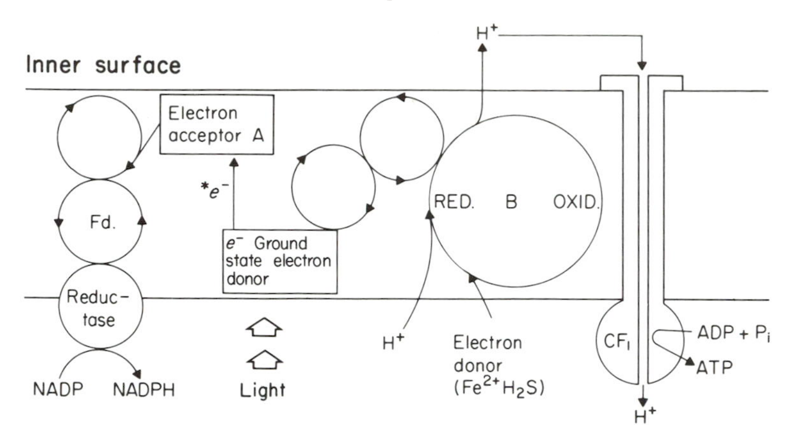

Fig. 3.2 A photosynthetic reaction sequence accommodated within a membrane system in which an external reductant replenishes the oxidized electron donor. The excited electrons pass via an electron acceptor A through a sequence of redox acceptors including ferrodoxin (Fd) and a reductase enzyme, to bring about the reduction of NADP. Electron flow from the external donor takes place via the shuttle B which releases protons to the inside of the membrane upon oxidation.

nitrite to ammonia, molecular nitrogen to ammonia and hydrogen ions to molecular hydrogen. These reactions, which are still to be found in micro-organisms from localised anaerobic environments today, are generally supposed to have been the prevalent form of photosynthesis in the reducing atmosphere of the primitive Earth. The final major elaboration of the photochemical side of photosynthesis was the adoption of water as the electron donor, a development which necessitated the aquisition of a second light-absorbing system to drive the electron flow to NADP and led to the release of oxygen into the atmosphere of the Earth for the first time. A reaction sequence which takes into account many of the attributes of this type of photosynthetic electron-transport chain is shown in Fig. 3.3, (Trebst, 1974).

I have laboured these elementary generalities concerning the basic events of photosynthesis for several reasons. First, we shall see that in respect of adaptive features of photosynthesis, and particularly with respect to temperature, the properties of the chloroplast membranes are of crucial importance. Second, the use of water as a source of electrons with concomitant release of oxygen has raised major adaptive problems concerned with photoxidative reactions which degrade the photosynthetic pigments, and third, the phenomenon known as the Warburg effect (Warburg & Negelein,

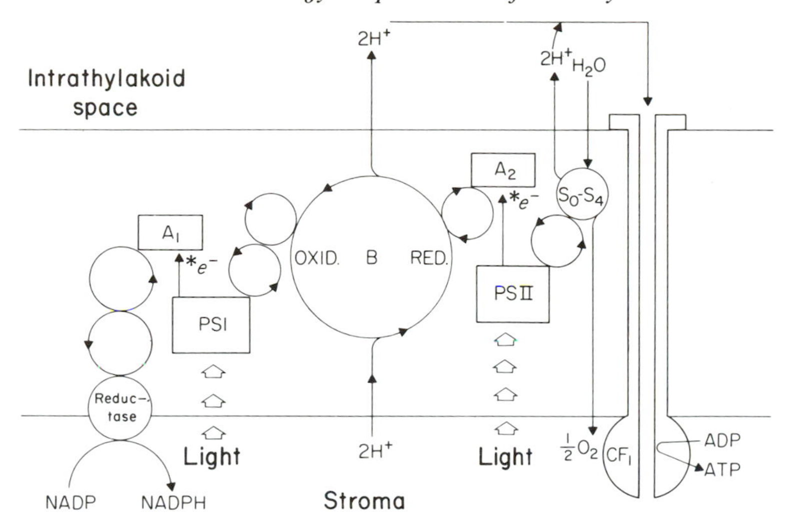

Fig. 3.3 A model depicting the photosynthetic electron transport chain accommodated within the chloroplast thylakoid membrane of a green plant. The system has become elaborated over that shown in Fig. 2 by the adoption of water as the primary electron donor to the system, the addition of a second light reaction to drive the oxidation of water and the addition of an enzyme system (S_0–S_4) to catalyse this oxidation with the release of oxygen. PSI and PSII are the two photosystems and A_1 and A_2 their respective electron acceptors. The shuttle B now links the two photosystems.

1920), with which we shall be much concerned, has its origins in the interference by oxygen in the carbon dioxide assimilation process.

Only a relatively small proportion of the total energy incident at the Earth's surface is available for the production of dry matter by plants because less than half lies within the photosynthetically active waveband and, of this, only about one-seventh is received at times and in places where other aspects of the climatic conditions permit it to be used for photosynthesis (Table 3.1). Actual values for rates of dry-matter production for some major categories of crops and natural communities are shown in Table 3.2. The higher values for the tropical crops shown in Table 3.2 do not necessarily imply any intrinsic superiority in the photosynthetic mechanism of these species. The values for the species from temperate regions are depressed because a proportion of the radiation falls at a time when the plants are dormant or when other conditions, most notably temperature, are limiting for growth.

We could explore the ways in which plants adapt their photosynthesis to environmental conditions by considering a vast compendium of structural

Table 3.1 A broad summary of the energy balance sheet of the earth showing proportions available for photosynthesis.

	Joules annum^{-1}	Percentage of incident energy
Energy incident at the Earth's surface	20×10^{23}	100
Energy available in the photosynthetic waveband	7×10^{23}	35
Maximum energy available for primary production	1×10^{23}	5
Estimate of actual energy stored in primary production	23×10^{21}	0.1

Table 3.2 Rates of dry matter production for ten types of plant community in various parts of the world.

	Rate of dry matter production (tonnes/annum)	Percentage of the photosynthetically active radiation converted
Tropical		
Napier grass	88	1.6
Sugar cane	66	1.2
Reed swamp	59	1.1
Temperate		
Perennial crops	29	0.5
Annual crops	22	0.4
Grassland	22	0.4
Evergreen forest	22	0.4
Deciduous forest	15	0.3
Savanna	11	0.2
Desert	1	0.02

and biochemical changes in relation to the major environmental variables. In an attempt to avoid the encyclopaedic flavour which such an approach would inevitably convey, I have chosen to limit discussion of the adaptive problem to the general standpoint of energy conservation.

3.3 QUANTUM REQUIREMENTS

The formulation of the electron-transport chain as shown in Fig. 3.3 requires four electrons donated by photosystem I (PSI) to form two NADPH molecules and four electrons to replace them from photosystem II (PSII) to yield one molecule of oxygen (Equation 3.4); thus, as one light quantum is required to displace each electron there is a requirement of eight quanta, four in PSI and four in PSII, to yield one molecule of oxygen (Equation 3.4), and if, as seems probable, there is one ATP molecule formed per electron flowing from PSII to PSI then we may formulate the process as in Equation 3.4.

$$8hV + 2NADP^+ + 4ADP + 4P_i + 2H_2O \longrightarrow 2NADPH + 4ATP + 2H^+ + O_2 \quad (3.4)$$

In all higher plants the assimilation of CO_2 into carbohydrate uses the ATP and NADPH generated photochemically according to the overall stoichiometry shown in Equation 3.5.

$$CO_2 + 3ATP + 2NADPH + 2H^+ \longrightarrow [CH_2O]_n \quad (3.5)$$

We may note in passing that according to the stoichiometry of Equations 3.4 and 3.5 there would be ATP available for processes other than CO_2 reduction from the standpoint of the relative proportions of ATP to NADPH formed and this takes no account of the possibility that further ATP may be formed by cyclic phosphorylation along the lines shown in Fig. 3.1. In all plant species so far investigated the assimilation of CO_2 ultimately proceeds through a cyclic process which commences with the carboxylation of the 5-carbon compound ribulose 1,5-bisphosphate (RuBP) which is cyclically regenerated as hexose sugar is produced via 3PGA, the initial product of the carboxylation (Fig. 3.4): this we shall refer to as the C_3 pathway, or the reductive pentosephosphate cycle of photosynthesis. Thus if we consider hexose sugar as the basic product of CO_2 assimilation, the overall reaction may be formulated as in Equation 3.6.

$$6RuBP + 6CO_2 + 18ATP + 12NADPH_2 + 12H^+ \longrightarrow Hexose + 6RuBP + 18P_i + 12NADP^+ \quad (3.6)$$

From the stoichiometries of Equations 3.4 and 3.6 we can see that 48 quanta are required per molecule of hexose ($6CO_2$) synthesised: thus if we take the energy of combustion of glucose as 2.82 MJ mol^{-1}, and the energy per mole of photons in the photosynthetically active waveband as 0.176 MJ then the efficiency of energy conversion $= 2.83 \times 100/48 \times 0.278 = 27\%$. Thus we see that even the highest-yielding plant communities listed in Table 3.2 have an efficiency of energy conversion which is less than one-twentieth of that predicted from the above analysis. Many factors contribute to the

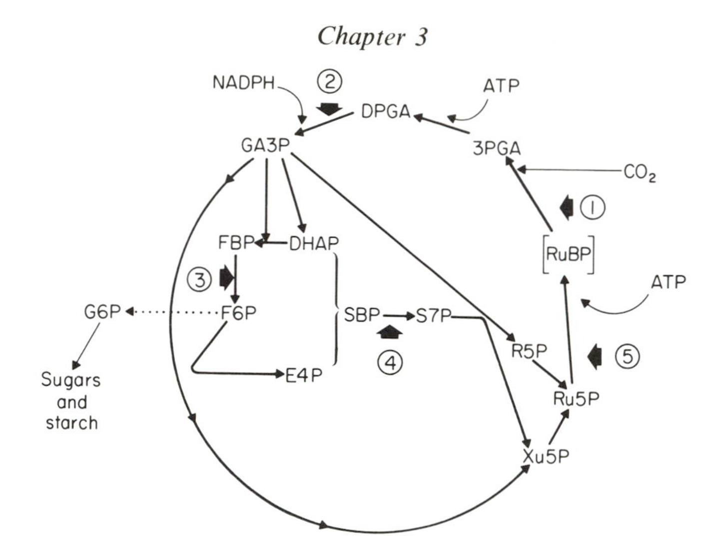

Fig. 3.4 The reductive pentose phosphate cycle (RPC) of photosynthesis. Note that a central feature of the scheme is the consumption of ribulose 1,5-bisphosphate (RuBP) in CO_2 fixation and its regeneration via a sequence of interconnected reactions. The heavy arrows → 1–5 indicate enzymes of the cycle which are activated by light; they are: 1 ribulose 1,5-bisphosphate carboxylase; 2 glyceraldehyde 3-phosphate dehydrogenase; 3 fructose 1,6-bisphosphate phosphatase; 4 sedoheptulose 1,7-bisphosphate phosphatase and 5 phosphoribulokinase. Abbreviations: RuBP, ribulose 1,5-bisphosphate; 3PGA, 3-phosphoglycerate; DPGA, 1,3-diphosphoglycerate; GA3P, glyceraldehyde 3-phosphate; FBP, fructose 1,6-bisphosphate; F6P, fructose 6-phosphate; DHAP, dihydroxyacetone phosphate; G6P, glucose 6-phosphate; E4P, erythrose 4-phosphate; SBP, sedoheptulose 1,7-bisphosphate; S7P, sedoheptulose 7-phosphate; Xu5P, xylulose 5-phosphate; R5P, ribose 5-phosphate; Ru5P, ribulose 5-phosphate.

shortfall in yields of plant dry matter from those which are theoretically possible; they include losses arising from reflection and imperfect interception of light by the canopy, light-saturation phenomena in the photosynthetic apparatus and factors which limit the supply of CO_2—the primary substrate for the reaction. There are also losses at the biochemical level arising from wasteful reactions and the physiological state of the plant which is influenced by such factors as the stage of development, nutrient status and water supply. Several environmental factors evoke adaptive responses in the plant and some of the more important of these will be referred to in subsequent sections. For present purposes it is convenient to consider particularly the losses at the biochemical level, since these strike directly at the very stoichiometric relations on which our overall estimate of potential efficiency was based; central to this problem is the phenomenon of photorespiration.

3.4 PHOTORESPIRATION

There have been many difficulties in arriving at the stoichiometry of photosynthesis shown in Equation 3.6, of which perhaps the greatest was the accurate measurement of the total CO_2 fixation. First, there was the question of re-fixation of endogenous CO_2 formed in respiration and worries as to whether respiration rate was the same in light and dark. The problem was further complicated by the discovery that CO_2 fixation was inhibited by oxygen (Warburg, 1920), a phenomenon which is commonly referred to as the Warburg effect.

Intense effort in the study of these problems led to the discovery that the rate of CO_2 evolution in the light is indeed greater than in darkness; using modern methods it can conveniently be measured by allowing a leaf to photosynthesize in a stream of $^{14}CO_2$ and then switching to a stream of unlabelled CO_2 of the same concentration, when it is found that $^{14}CO_2$ continues to be emitted from the photosynthesizing leaf at a high rate for several minutes before dropping to a lower rate which may continue for several hours. This light-dependent evolution of CO_2 was shown to be quite different to ordinary respiration on at least two counts: it did not saturate at low concentrations of oxygen but increased right up to 100% CO_2 (Fig. 3.5) and it showed a much greater increase in rate with rising temperature (Fig. 3.6); the process was therefore distinguished as photorespiration.

Studies of the biochemistry of photorespiration have revealed a situation of remarkable complexity; it was found that the substrate for photorespiration is none other than the CO_2 acceptor molecule of the reductive pentose phosphate pathway, RuBP, and the enzyme which catalyses the carboxylation of RuBP also has an oxygenase function in kinetic competition with the carboxylation reaction (Fig. 3.7) (Bowes, Ogren & Hageman, 1971; Badger & Andrews, 1974; Laing, Ogren & Hageman, 1974). The oxygenase reaction yields one molecule of PGA and one molecule of phosphoglycollate; the PGA is returned to the RPC cycle whilst the phosphoglycollate gives rise to the photorespiratory CO_2 via the cyclic sequence of reactions shown in Fig. 3.8 which we shall refer to as the photosynthetic carbon oxidation cycle (PCO) (Lorimer, Woo, Berry & Osmond, 1978). If this scheme is correct, (and it should be noted in passing that there are still some dissenting voices concerning such matters as the source of glycollate (Zelitch, 1975; Kelly, Latzko & Gibbs, 1976) and the oxidation of glycine to serine (Halliwell & Butt, 1974; Grodzinski & Butt, 1976)), then it follows from the competitive nature of the oxygenase and carboxylase reactions that the proportion of

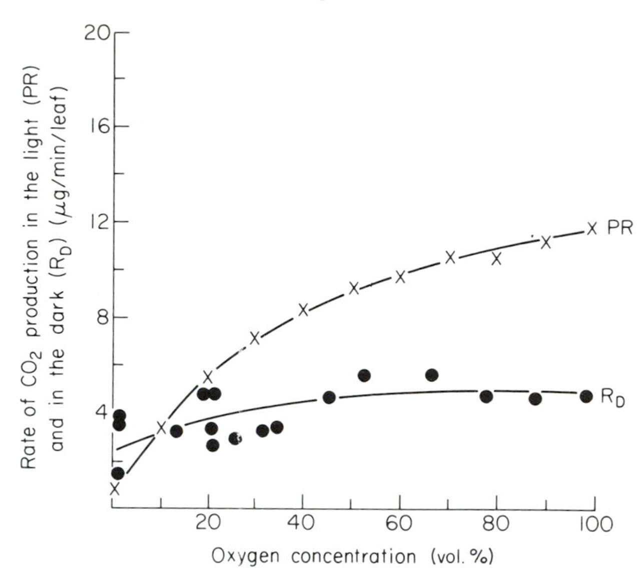

Fig. 3.5 The effect of oxygen concentration on the rate of photorespiration in detached leaves of soybean. Note that dark respiration is saturated at very low concentrations of oxygen but photorespiration does not saturate. After Forrester, Krotkov & Nelson (1966).

carbon flowing through the PCR or PCO cycles will depend upon the relative rates of the oxygenase and carboxylase reactions, which will in turn depend upon the relative concentrations of oxygen and CO_2 at the catalytic site of RuBPCase and the affinity of that enzyme for oxygen versus CO_2. This relationship of oxygenase to carboxylase activity (Φ) has been expressed in terms of a kinetic model (Farquhar, van Caemmerer & Berry, 1980) (Equation 3.7).

$$\Phi = \frac{V \text{ oxygenase}}{V \text{ carboxylase}} = \frac{V_{\text{Max}} \text{ oxygenase}}{V_{\text{Max}} \text{ carboxylase}} \times \frac{K_m CO_2}{K_m O_2} \times \frac{[O_2]}{[CO_2]} \quad (3.7)$$

Where V oxygenase and V carboxylase are the rates of the oxygenase and carboxylase reactions, respectively, V_{Max} oxygenase and V_{Max} carboxylase are the maximum velocities of these reactions, $K_m CO_2$ and $K_m O_2$ are the Michaelis constants of the enzyme for O_2 and CO_2, respectively, and $[O_2]$ and $[CO_2]$ are the concentrations of O_2 and CO_2 at the catalytic site of the enzyme.

Although there are still uncertainties regarding the accurate measurement of rates of photorespiration it may confidently be said that it often involves

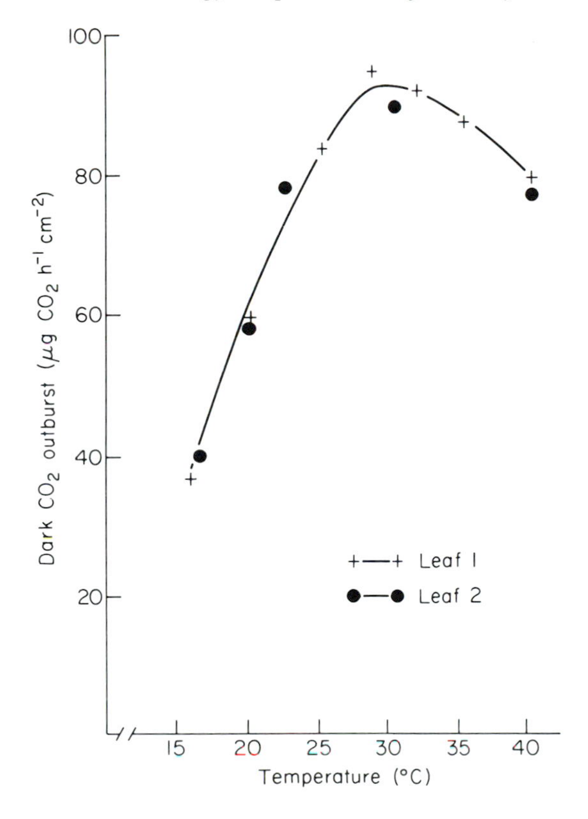

Fig. 3.6 Effect of temperature on photorespiration in attached leaves of sunflower. Photorespiration was measured in terms of the magnitude of the burst of CO_2 production in the dark period immediately following a period of illumination. After Hew, Krotkov & Canvin (1969).

losses of CO_2 of up to 50% of that fixed (Table 3.3.). These losses increase drastically with rise in temperature as judged from measurements of CO_2 evolution (Hew, Krotkov & Canvin, 1969), and O_2 inhibition of photosynthesis (Jolliffe & Tregunna, 1968; Ku & Edwards, 1977). If the rate of the oxygenase reaction is set at 40% of the rate of the carboxylase reaction and due account is taken of the energy requirements for re-assimilation of the ammonia released in the course of the PCO cycle (Fig. 3.9), then the stoichiometry of Equation 3.6 is profoundly altered to a utilisation of approximately 28 ATP and 23 $NADPH_2$ per molecule of hexose ($6CO_2$) produced. Under such conditions the quantum yield of photosynthesis is decreased (Ehleringer & Björkman, 1977).

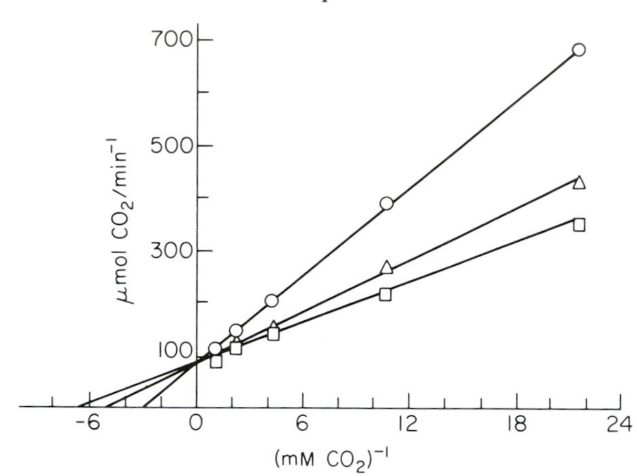

Fig. 3.7 A double reciprocal plot of the rate of incorporation of CO_2 (μmol min^{-1}) by ribulose 1,5-bisphosphate carboxylase as a function of CO_2 concentration in 0% (□), 21% (Δ) and 100% (o) oxygen. The position of intersection of the three lines is indicative of competitive inhibition. After Ogren & Bowes (1971).

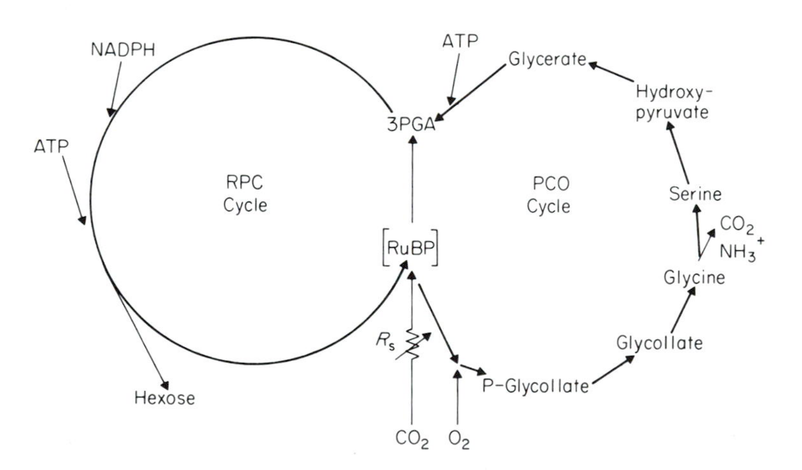

Fig. 3.8 The proposed photosynthetic carbon oxidation (PCO) cycle, coupled to the reductive pentose phosphate cycle through the competitive carboxylase/oxygenase reactions (Φ). The coupling of the two cycles is completed as phosphoglycollate is metabolized through the sequence of intermediates shown in the diagram leading to the regeneration of 3PGA, with the release of CO_2 and ammonia. The variable resistance R_s, represents the diffusive resistance to CO_2 uptake afforded by the stomata; P-glycollate, phosphoglycollic acid.

Table 3.3 Rates of photorespiration and concomitant rates of net photosynthesis for a range of crops.

Species	Method of assay	Temperature (°C)	Net photosynthesis in normal air (mg $CO_2/dm^2/h$)	Photorespiration percent of net photosynthesis in normal air[a]
Soybean[†]	CO_2 release, CO_2-free air	26	35.2	46
Soybean	Post-illumination CO_2 burst	25	11	75
Soybean	$^{14}CO_2$ release, CO_2-free air	24		
Soybean	CO_2 release, CO_2-free air	30	18	42
Sunflower	Short time uptake, $^{14}CO_2$ minus $^{12}CO_2$	25	25	60
Sunflower	$^{14}CO_2$ release, CO_2-free air	25	28	27
Sugar beet	CO_2 release, CO_2-free air	25	25.2	47
Sugar beet	CO_2 release, CO_2-free air	25	26	40
Tobacco	$^{14}CO_2$ release, CO_2-free air	30	17–25	
Tobacco	CO_2 release, CO_2-free air	25	11	55
Tobacco	Extrapolation of net photosynthesis to 'zero' CO_2	25	13.7	25
Tobacco	Post-illumination CO_2 burst	25.5	16.9	45
Tobacco	Post-illumination CO_2 burst	35.5	14.8	66
Maize	CO_2 release, air passed through leaf	30		0
Maize	Uptake of $^{18}O_2$	23–34	11.5	5.7
Maize	$^{14}CO_2$ release, CO_2-free air	30		
Maize	$^{14}CO_2$ release, CO_2-free air	24		
Maize	CO_2 release, CO_2-free air	35		0
Maize	CO_2 release, CO_2-free air	35	50	0

[a] These values are minimal and underestimates because photorespiration is assayed under conditions of high light intensity where the main flux of the gas (CO_2 or O_2) is in the opposite direction. Dark respiration contributes somewhat to the photorespiration measured.

[†] Results recalculated by the authors considering internal diffusive resistances; results are the mean values of 20 varieties.

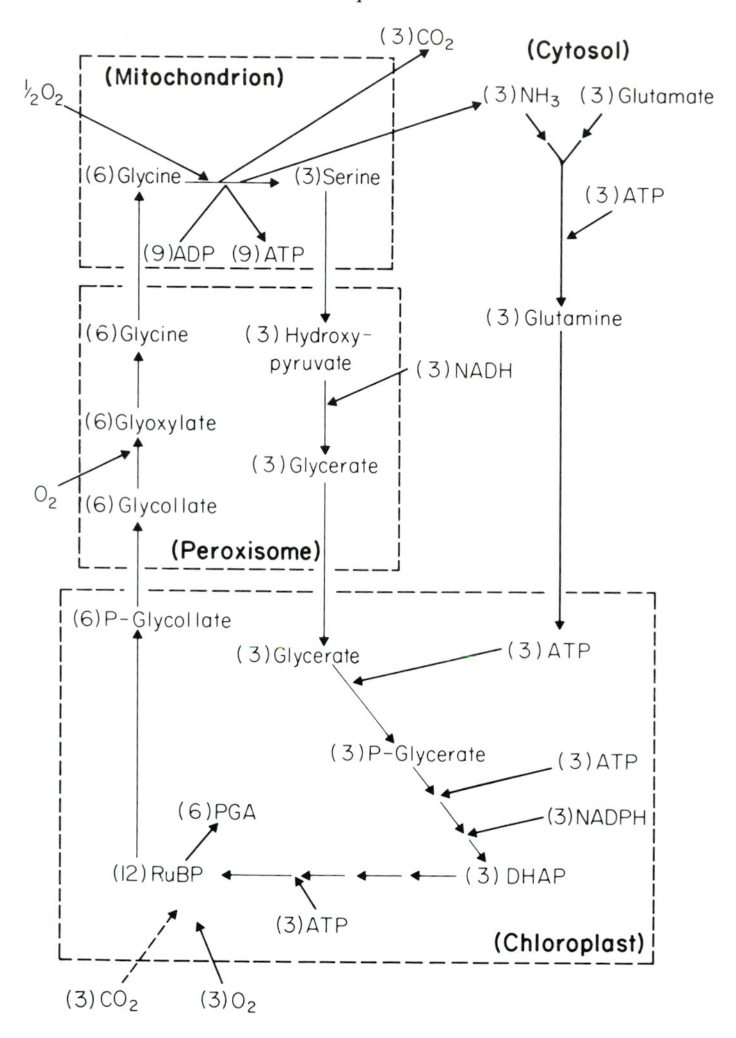

Fig. 3.9 A scheme involving expansion of the PCO cycle of Fig. 3.8, to show the stoichiometry of extra NADPH and ATP consumption in the cycle and in the re-assimilation of ammonia released in the course of the oxidation of glycine. The scheme also indicates the complex compartmentation of reactions between chloroplasts, peroxisomes, mitochondria and the cytosol.

In attempting to understand the adaptation of photosynthesis to environmental conditions we are therefore led to ask why the oxygenase reaction and the attendant carbon oxidation cycle should occur at all if they lead to such losses of recently assimilated CO_2 and a corresponding increase in load upon the energy-transducing system.

3.5 THE BIOLOGICAL SIGNIFICANCE OF PHOTORESPIRATION

There can be little doubt that photorespiration represents a potentially serious loss of energy for a plant, as may be inferred from the information in Fig. 3.9 (Krauss, Lorimer, Heber & Kirk, 1978) and Table 3.3. From Fig. 3.9 it can be seen that for every mole of CO_2 lost in photorespiration there is a consumption of six moles of ATP and three moles of NADPH, to which must be added a further ATP and two moles of NADPH for reassimilation of the NH_3 (Miflin & Lea, 1976). There are two schools of thought concerning why this waste of energy should occur. The first is that the oxygenase reaction is an inevitable consequence of the active site chemistry of RuBPCase (Andrews & Lorimer, 1978); with the corollary that the energy-consuming PCO cycle is the most effective way of salvaging 75 % of the carbon and putting it back into the profitable stream of the RPC cycle. Andrews & Lorimer note that RuBPCase possesses oxygenase activity in all the organisms in which it has been studied, including anaerobic bacteria which are devoid of oxygen-generating systems. Against this might be set the fact that there are a good many other carboxylating enzymes in living organisms and none of these have oxygenase activity. It has been pointed out (D. A. Walker, Pers. commun.), however, that RuBPCase is unique amongst carboxylases in its position in a cyclic process which is essential for the regeneration of its substrate (RuBP) and that this may impose constraints upon the design of the active centre of the enzyme which makes the oxygenase reaction inevitable. The greater activity of the oxygenase relative to the carboxylase reaction at higher temperatures may be in part accountable by a greater solubility of O_2 than CO_2 (Ku & Edwards, 1977) but there seems little doubt that an increased affinity of the enzyme for oxygen ($1/K_m O_2$) is also an important factor in diverting more carbon into the PCO cycle (Badger & Collatz, 1977; Ogren & Hunt, 1978).

The alternative view of photorespiration is that it is a mechanism for the dissipation of energy without which the photosynthetic apparatus would be fatally unstable. On land and in water it frequently happens that the supply of CO_2 for photosynthesis becomes limiting when other conditions, particularly the level of irradiance, are favourable. On land, for example in dry weather, water deficits may cause stomatal closure leading to decreased uptake of CO_2 and under water, particularly where the vegetation is dense, the solubility and rate of diffusion of CO_2 may become limiting for photosynthesis. Under conditions such as these one may frequently encounter circumstances in which the electron-transport system of the chloroplasts gets out of step with the carbon-assimilating enzymes, so that the pools of ATP and $NADPH_2$ are fully charged and the reducing traps (primary electron acceptors) in the electron-transport chain are filled. In these circumstances the chloroplasts may be at risk. For example, in algal mutants lacking carotenoids, in which the photosynthetic membranes are very labile, the cells may be cultured successfully in dim light or at higher irradiances under reduced P_{O_2}, but are lethally photobleached under normal conditions of illumination and at atmospheric P_{O_2}. It is suggested that when the electron traps for PSI are filled, other compounds, oxygen in particular, may accept electrons, forming superoxide radicals and hydrogen peroxide as shown in Fig. 3.10.

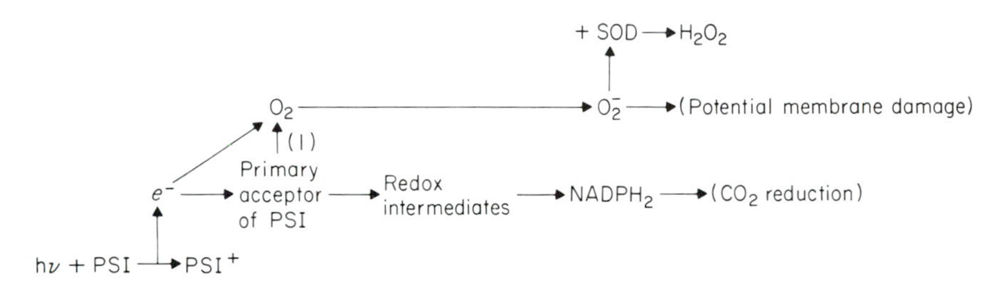

Fig. 3.10 Suggested reaction sequence for generation of free radicals causing damage to the thylakoid membranes. The reaction marked ($\xrightarrow{(1)}$) represents the suggested flow of electrons from a primary acceptor for PSI to O_2 when the acceptor is fully reduced.

Further circumstantial evidence of a propensity for the formation of superoxide radicals in chloroplasts is afforded by the presence of high levels of the enzyme superoxide dismutase (SOD). The presence of the PCO cycle arising from the oxygenase reaction provides a mechanism whereby, under conditions of limiting CO_2 supply, the tendency to saturate the electron traps and to elevate the superoxide load can be obviated by providing a route for the continued consumption of reductants. The consequence of this arrangement is, as we have seen, a wasteful 'burning off' of carbon, but this may be a low price to pay if the alternative is an oxidative destruction of the thylakoid membrane system.

Attempts have been made to test this hypothesis with intact leaves by illuminating them in the absence of CO_2 and suppressing photorespiration by imposing a low ($< 2\%$) P_{O_2} and examining the consequent photosynthetic activity of the leaf. Leaves of *Sinapis* treated for 60 min in this way showed subsequent inhibition of CO_2 fixation (Cornic, 1976), whilst similar treatment of leaves of *Vigna*, *Atriplex* and spinach led to an increased quantum requirement for photosynthesis, implying that damage to the thylakoid membranes had been incurred (Lorimer, Woo, Berry & Osmond, 1978). More experimental work is needed to test this hypothesis but for present purposes it is reasonable to suppose that whatever interpretation of the significance of photorespiration prevails, whether it is inevitable or essential for the plant, the process is wasteful of energy and carbon. The question therefore arises as to how natural selection may have operated to circumvent photorespiratory losses in plants?

3.6 THE BALANCING OF CO_2 UPTAKE AND LOSS

3.6.1 General considerations

The majority of plants do not avoid photorespiratory loss of CO_2 but live with it. This category includes almost all trees and the great majority of shrubs in all parts of the world, most of the herbaceous species in the temperate zones and quite a lot of those in the tropics as well. Plants which show pronounced modification to their carbon metabolism, in order to minimise the effects of photorespiration, are found particularly amongst the herbaceous floras of arid areas, savannahs, as epiphytes in many regions, in tropical swamps, salt marshes and in sub-aquatic habitats. At first sight this may seem an unlikely collection of bedfellows; what they have in common is that whilst all ultimately fix CO_2 via the RPC cycle (Fig. 3.4), they have opted in one form or another for the introduction of an additional carboxylation, located in the cytoplasm of the cell, usually involving PEP as the substrate and oxaloacetate as the first product of the carboxylation (Fig. 3.11). The oxaloacetate is usually further metabolised and then transported to a vacuole within the cells (Fig. 3.12) or to another group of cells (Fig. 3.14a) where the products are then decarboxylated and the CO_2 released is re-fixed via the RPC cycle. For convenience I shall refer to these various forms of photosynthesis collectively as acid-mediated.

It will be evident immediately, even before we consider the selective advantages of an acid-mediated step in photosynthesis, that it involves fixing the CO_2 twice, to this extent it requires extra energy, at least two moles of ATP per mole of CO_2 for the carboxylation (Hatch, 1970) and probably more when energy for transport requirements and maintenance of the extra enzymes involved are taken into account. The fact that species possessing acid-mediated

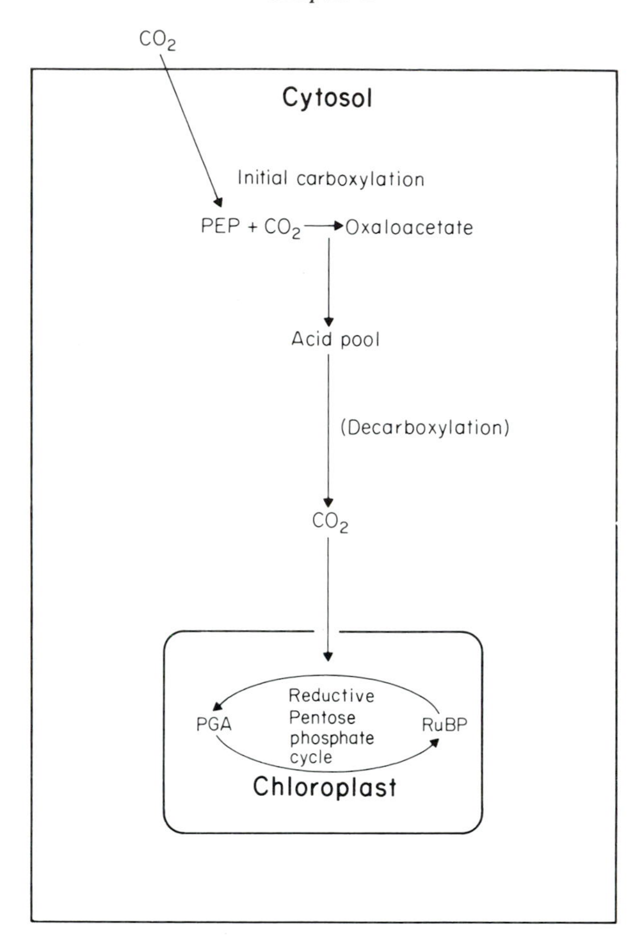

Fig. 3.11 A generalized scheme of acid-mediated photosynthesis in which an additional carboxylation reaction located in the cytoplasm is interposed between the atmospheric source of CO_2 and the ultimate carboxylation reaction involving the RPC cycle in the chloroplast. Specific variants of this scheme are exemplified in Figs 3.12 and 3.13.

photosynthesis rarely, if ever, thrive in conditions of shade is probably a consequence of the extra light requirements for this type of photosynthesis. What then are the various forms of acid-mediated photosynthetic systems and what do we know of the environmental conditions in which they have been selected? It is convenient to approach the second question first by considering how plants may obtain CO_2, the primary substrate for their photosynthesis.

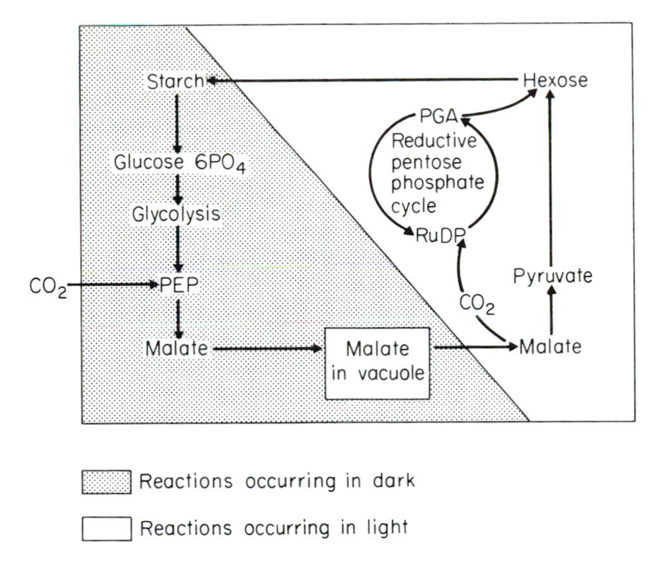

Fig. 3.12 Diagrammatic representation of the CAM variation of acid-mediated photosynthesis. The reaction sequence indicated in the shaded portion of the diagram indicates events which take place in the dark period; the reactions which occur in the light are shown in the unshaded portion.

Terrestrial habitats

All land plants have to live out a compromise between the amount of carbon dioxide which they are able to take up for photosynthesis and the amount of water which they must expend in the process. One could imagine that an ideal solution to this problem would have been for plants to evolve a membrane at the evaporating surface which could be modulated to optimise the amount of CO_2 passing inwards and the amount of water passing out. In the absence of such a structure the first lines of compromise are to be found in the size, shape and texture of the leaves (see Chapter 10) and at the stomatal pores through which most of the gas exchange of the leaf is conducted. The guard cells of stomata are amongst the most sensitive of living cells; as summarised in Fig. 3.13, they respond to most environmental stimuli through an elaborate system of feedback controls (Raschke, 1975). We may suppose that for the species carrying out normal photosynthesis (Fig. 3.4) these features of leaf design and stomatal control are sufficient to ensure a balance of CO_2 uptake versus water loss which is adequate for their overall fitness in a particular environment. What then are the circumstances in which plants exhibit more extreme

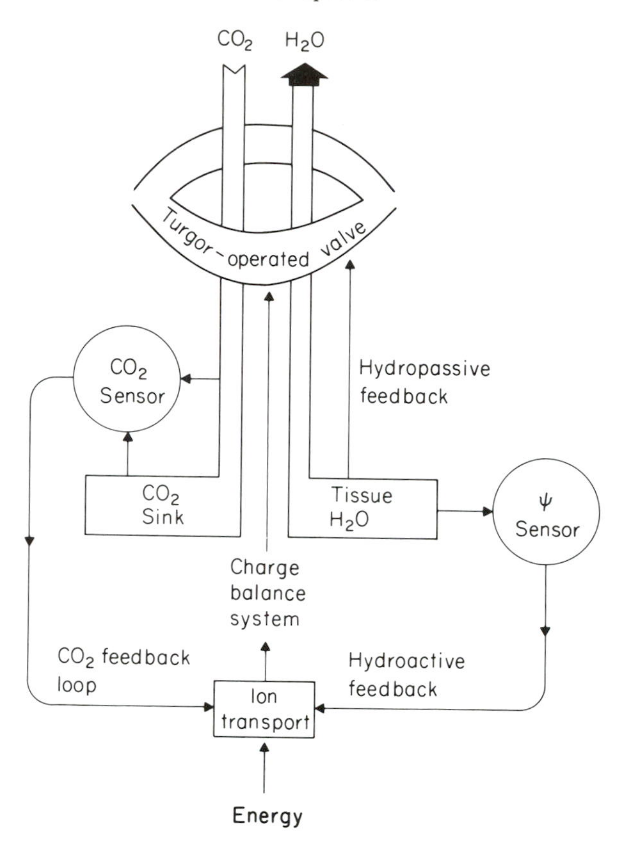

Fig. 3.13 A generalized scheme depicting the feedback controls which contribute to the regulation of stomatal aperture is based on the ideas of Raschke (1975).

adaptations in this aspect of their physiology? To understand the quantitative nature of this problem it is convenient to use an analogy with an electrical circuit, substituting fluxes, concentration gradients and resistances to diffusion for currents, potential differences and electrical resistances (Gradmann, 1926). Using this notation for CO_2 entering the leaf, the rate of uptake is given as:

$$\text{Rate} = \frac{\left[[CO_2]_{\text{air}} - [CO_2]_{\text{sub-stomatal cavity}}\right]}{r_a + r_s} . D_{CO_2} \tag{3.8}$$

and for water vapour leaving the leaf by:

$$\text{Rate} = \frac{\left[[H_2O]_{\text{sub-stomatal cavity}} - [H_2O]_{\text{air}}\right]}{r_a + r_s} \cdot D_{H_2O} \qquad (3.9)$$

Where r_a is the boundary layer resistance, r_s the stomatal resistance and D_{CO_2} and D_{H_2O} the diffusion coefficients for CO_2 and water vapour, respectively.

Thus the ratio of water loss to CO_2 uptake may be obtained by dividing Equation 3.9 by Equation 3.8, whence:

$$\frac{\text{Rate of } H_2O \text{ loss}}{\text{Rate of } CO_2 \text{ uptake}} = \frac{D_{H_2O}\left[[H_2O]_{\text{sub-stomatal cavity}} - [H_2O]_{\text{air}}\right]}{D_{CO_2}\left[[CO_2]_{\text{air}} - [CO_2]_{\text{sub-stomatal cavity}}\right]} \qquad (3.10)$$

Considering a leaf in an atmosphere containing 300 volumes per million CO_2 and a vapour pressure deficit of 1 kPa (10 mb), having a leaf temperature of 25°C at which the partial pressure of water vapour in the sub-stomatal air spaces of the leaf is assumed to be the saturated vapour pressure at 25°C and the CO_2 concentration is taken as 75 volumes per minute (an average sort of value for the CO_2 compensation point) and D_{H_2O}/D_{CO_2} as 1.56, then by converting these values to concentrations of substance and substituting in Equation 3.10 we obtain:

$$\frac{\text{g of water lost}}{\text{g of } CO_2 \text{ taken up}} = 1.56 \times \left[\frac{25.17 - 17.23}{0.54 - 0.14}\right]$$

$$= 30.4\,\text{g water vapour lost/g } CO_2 \text{ taken up.}$$

But in reality the position of the plant is much worse than this because the calculation has followed the CO_2 only as far as the sub-stomatal cavity, thereby ignoring the internal resistance to uptake of CO_2 (variously referred to as the mesophyll or residual resistance) and has taken no account of the additional driving force to water loss afforded by the heat load upon the leaf.

If we consider by way of an example events following sunrise on a sunny day in the context of the theory outlined above, and the diagram in Fig. 3.13, we see that as the air is warmed, the vapour pressure deficit will fall, transpiration will tend to increase, probably leading to partial closure of the stomata. As the air temperature rises, the energy load on the leaf rises and the stomatal closure reduces evaporative cooling, thus each of these factors conspires to raise the temperature of the leaf which in turn increases the rate of photorespiration as exemplified in Fig. 3.6. As a consequence of this increased efflux of CO_2 the PCO_2 in the intercellular air spaces of the leaf rises so that the concentration gradient of CO_2, the denominator in equation 3.10, becomes shallower, consequently uptake of CO_2 becomes slower and the expenditure of water per unit of CO_2 fixed is correspondingly increased. The effect of introducing an additional cytoplasmic carboxylation reaction is to mop up CO_2 as it is released from the mitochondria thereby maintaining the sub-stomatal concentration of CO_2 at a low value. In this way the concentration gradient of CO_2 is maintained so that the cost of CO_2 fixation, measured in terms of units of water lost, is minimized. The introduction of the decarboxylation steps affords the plant the option of so localising this reaction that the CO_2 is released close to the site of the RuBP carboxylase enzyme so that it may function as a CO_2-concentrating mechanism which pushes the value of Φ (Equation 3.7) in favour of the carboxylase reaction (see page 60). The available evidence suggests that natural selection has led to the development of a PEP-mediated acidification step in photosynthesis whenever the combination of elevated temperature, which favours water loss and photorespiration, and other environmental factors such as vapour pressure deficit and soil water potential, which affect the availability and rate of loss of water, have conspired to raise the cost of CO_2 fixation per unit of water lost.

Submerged aquatic habitats

The problems facing submerged aquatic plants with respect to their photosynthesis relate to the penetration of light, amounts and rates of diffusion of CO_2 and HCO_3 in water, and temperature. Generally speaking, temperature fluctuations are not so great for aquatic species except in the intertidal zones. Light levels may be adequate to saturate a normal photosynthetic system in the surface layers of clear waters but the attenuation of light with depth may be rapid depending on the amount of suspended matter in the water. Diffusion rates in water are slower than in air by a factor of approximately 10^4 and it is this which places a severe constraint on rates of photosynthesis of submerged aquatic plants. These species lack stomata and have a specialised leaf anatomy which includes a concentrated development of chloroplasts in the epidermal layer. It appears that the need to conserve carbon may have led to the development of a partial acid-mediated type of photosynthesis for CO_2 recycling in some species (Holaday & Bowes, 1980), notwithstanding the moderate temperatures and reduced levels of photosynthetically active radiation (PAR). It should be noted however that the photosynthetic pathways of aquatic macrophytes are a subject of controversy.

3.7 TYPES OF ACID-MEDIATED PHOTOSYNTHESIS

3.7.1 CRASSULACEAN ACID METABOLISM (CAM)

Fig. 3.12 shows, in a schematic form, the metabolic characteristics of a species possessing CAM. Typical CAM plants open their stomata at night and assimilate CO_2 into malate which is then stored in the cell vacuoles. Most CAM plants are usually of a succulent texture, probably in consequence of the vacuolar volume needed for this nocturnal storage of malate; there are, however, exceptions, such as Spanish moss (*Tillandsia usneoides*), which are not succulent (Kluge, Lange, von Eichmann & Schmid, 1973). By day the stomata close, malate moves back from vacuole to cytoplasm, is decarboxylated and the CO_2 re-fixed through the normal PCR cycle.

It is frequently found that CAM is a facultative form of metabolism which may be induced in species which would otherwise carry out the normal PCR cycle, when they are subjected to particular types of stress. This facultative behaviour may take many forms. *Agave americana* under drought is a typical CAM species but when well watered it retains some dark fixation of CO_2 but also carries out normal uptake and fixation of CO_2 during the day. In *A. desertii* on the other hand, watering induces a complete switch, the nocturnal opening of stomata and fixation of CO_2 is abolished and is replaced by normal photosynthesis (Nobel, 1976). In *Mesembryanthemum crystallinum* the induction of CAM has been found to depend on osmotic stress (Winter, 1974), although there is some suggestion that this may represent a specific salt requirement (Winter & Troughton, 1978).

Biochemical Regulation of CAM

It was generally held that starch formed the source of sugars from which PEP is generated in CAM plants during dark acidification. More recent work shows that in some species dextrins may account for up to half of the glucose required (Sutton, 1975a). The dextrins are metabolised through a glycolytic sequence which is compartmentally separated in the chloroplasts away from the main pool of soluble sugars (Sutton, 1975b). The PEP so formed in this system may be moved to the site of its utilization in the cytoplasm in an exchange reaction involving a P_i translocator (Walker, 1976).

Study of the regulatory properties of PEP carboxylase present formidable technical problems (Davies, 1979); suffice it to say for present purposes that feedback inhibition of the enzyme by malate is evident from *in-vitro* studies (Kluge & Osmond, 1972); changes in the 'capacity' of the enzyme in relation to CAM activity have been demonstrated (Queiroz, 1978) and a theoretical account of how such regulatory properties may operate is available in terms of a 'sensitivity coefficient' (Kacser & Burns, 1973). The question of how the PEP carboxylase remains active as malate accumulates is not entirely clear; transport to the vacuole offers one escape but this would have to be very efficient since the K_i(malate) for PEPCase is only five times higher than the K_m (PEP). The matter is further complicated by the finding that in *Mesembryanthemum crystallinum* the PEP carboxylase is only slightly inhibited, but on illumination of the leaves the enzyme is converted to a form which is sensitive to malate (Winter, 1980). This presumably affords a means of switching off the dark fixation mechanism by day. The inhibition of PEP carboxylase by malate is competitive so that it can be overcome by increased production of PEP, but PEP in turn is an inhibitor of phosphofructokinase (PFK) which occupies a central role in the regulation of the glycolytic pathway involved in the formation of PEP, thus if PEP were to build up, the glycolytic source of its formation could be turned off. The apparent conflict between these regulatory mechanisms may be avoided if the formation of PEP and the subsequent site of its carboxylation are separated, as the work of Sutton (1975a, b & c) would suggest. A further factor in the development of discriminating regulatory properties in this system is the observation that the PFK from CAM plants shows only 1% of the sensitivity to inhibition by PEP than that of other species (Sutton, 1975c).

The enzymes for the carboxylation of malate in CAM plants differ between species, some possess an NAD-malic enzyme others have a PEP carboxykinase and in each of these categories there is an array of isozymes each with different regulatory properties (Dittrich, Campbell & Black, 1973; Dittrich, 1976).

The impatient reader may be forgiven should he feel that the ecological path in this account is becoming obscured in a profusion of biochemical undergrowth. The digression has, however, been deliberate; in order to emphasise the range and complexity of biochemical, and by implication genetical, changes which must underlie the development of an adaptation such as CAM. Thus when we set out to discuss in general terms the 'energy costs' of these adaptative changes it is sobering to bear in mind not only the energy savings which the adaptations might achieve for the plant once they are in operation, but also the demands involved in their development in the first instance.

3.7.2 C_4 Photosynthesis

As with CAM, the C_4 modification of the pathway of CO_2 assimilation also centres on a cytoplasmically located carboxylation of PEP; but whereas in CAM the PEP and RuBP carboxylations are temporally separated night and day, in C_4 species they occur simultaneously but usually in spatially separated cells and tissues (Laetsch, 1974); the PEP carboxylation in the mesophyll cells and the RuBP carboxylation in chloroplasts located in the cells immediately surrounding the vascular bundles. This arrangement is referred to as the 'Kranz' anatomy (German—wreath) in reference to the green ring of cells as seen in transverse sections of the vascular bundles (Haberlandt, 1904).

The essence of the biochemical sequence in C_4 species is the formation of oxaloacetate as the primary carboxylation product in the mesophyll, which is then transaminated to aspartate or reduced to malate; the malate and aspartate are then transported to the bundle sheath. In the bundle sheath further metabolism of the aspartate and malate leads to decarboxylation with the CO_2 released being re-fixed via the RPC cycle. The details of the decarboxylation pathway also vary between species, the three main types are shown in Fig. 3.14(b, c, d) (Hatch, Kagawa & Craig, 1975). This spatial differentiation of the two carboxylation pathways in C_4 species is a consequence of a distinctive pattern of distribution of the enzymes by which they are catalysed, as between the mesophyll and the bundle sheath cells (Fig. 3.13; Ku & Edwards, 1975). In consequence of this arrangement it can be shown, for example, that addition of pyruvate to isolated mesophyll cells leads to the formation of malate or aspartate with a concomitant release of oxygen; addition of ^{14}C-labelled malate or aspartate to isolated bundle sheath cells in the light leads to the labelling of intermediates of the RPC cycle.

We have placed this brief description of C_4 photosynthesis in the context of a short-circuiting of the consequences of high rates of photorespiration under certain conditions—so that the CO_2 lost in this process is rapidly recycled, but the matter is not quite as simple as this. We must ask, for example, why it is that C_4 species do not show an oxygen-dependence in their quantum efficiency (Ehleringer & Björkman, 1977), or a Warburg effect, that is an enhanced rate of photosynthesis at low P_{O_2} (page 59) (Goldsworthy & Day, 1970), although the Warburg effect can be shown in bundle sheath cells

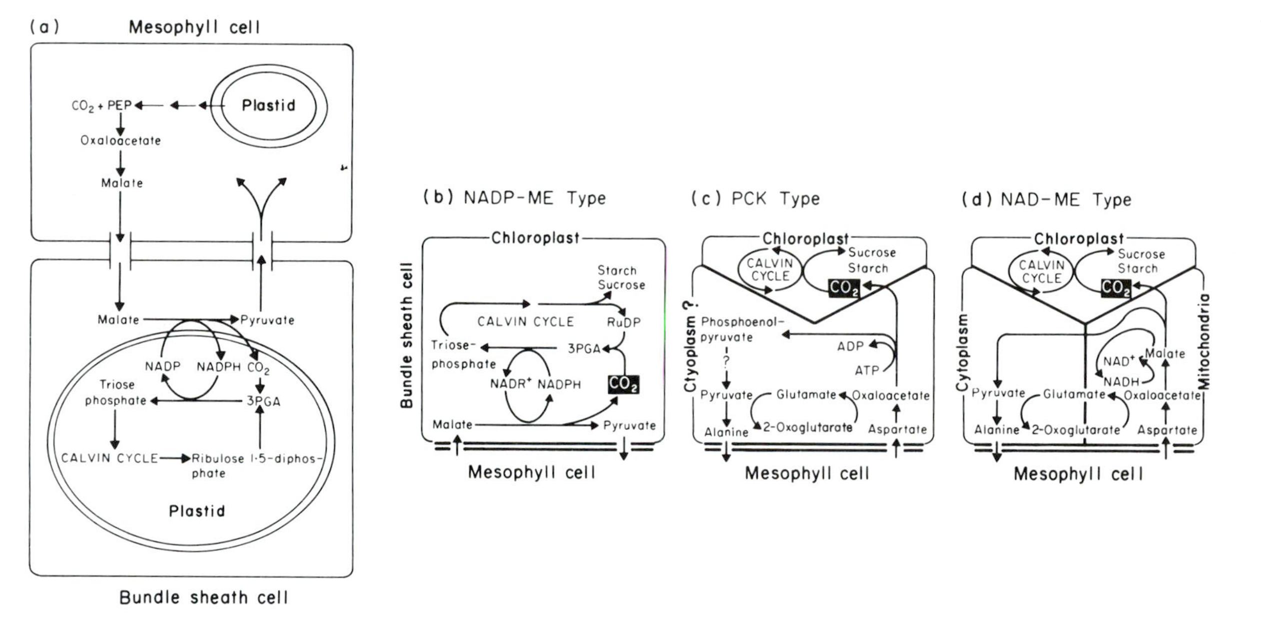

Fig. 3.14 (a) Diagrammatic representation of a generalized form of the C_4 variation of acid-mediated photosynthesis showing the separation of the two carboxylation reactions between cells of the mesophyll and the bundle sheath. (b, c, d) Specific variants of the C_4 pattern of acid-mediated photosynthesis; After Hatch, Kagawa & Craig (1975). The three variants are defined by the operation of: (b) the NADP malic enzyme (NADP–ME type); (c) PEP carboxykinase (PCK type) and (d) NAD malic enzyme (NAD–ME type) for C_4 acid decarboxylation.

isolated from such species (Chollet & Ogren, 1972). There are probably two factors which contribute to this disparity; first, it is now clear that although the enzymic apparatus for the photosynthetic carbon oxidation cycle is present in (and confined to) the bundle sheath cells of C_4 plants, the amounts of these enzymes are much less than in the mesophyll cells of C_4 species (Rehfeld, Randall & Tolbert, 1970). The second reason for this discrepancy is that the isolated bundle sheath is a physiological artefact which obviates a fundamental attribute of the C_4 syndrome because, in removing the mesophyll cells—the source of malate and aspartate—one is removing the fuel for what is essentially a pump which is functioning to elevate the concentration of CO_2 in the bundle sheath cells. This concept of the C_4 mechanism as a CO_2 pump was developed by Hatch, Kagawa & Craig (1975) who estimated that the CO_2 concentration in the mesophyll cells may be as high as 1–2 mM; it gains additional significance from the information which we now have of the dual oxygenase–carboxylase function of RuBP carboxylase, since these concentrations of CO_2 in the bundle sheath cells will be sufficient to saturate the carboxylase reaction and thereby minimize the value of Φ (Equation 3.7).

Finally it will be seen that the structure of the C_4 syndrome poses a further question concerning the role of photorespiration. If, as was postulated, the oxygenase reaction functions to sustain electron flow under conditions of CO_2 depletion and thereby minimise superoxide formation, this device would not be available in the mesophyll cells which lack RuBP carboxylase. One hypothesis which might provide a mechanism for such protection would be to link the electron flow in the chloroplasts of the bundle sheath to that of the chloroplasts of the mesophyll by means of 'reductant shuttle' as shown in Fig. 3.15. On this hypothesis, as stomatal closure leads to CO_2 depletion the oxygenase reaction in the bundle sheath would generate PGA as the carbon oxidation cycle continued to drive the flow of electrons in the thylakoid membranes of these cells. If some of the PGA generated via the PCO cycle was passed out to the mesophyll cells it could enter the mesophyll chloroplasts via the carboxylate transporting system, where it could undergo reduction to dihydroxyacetonephosphate before being returned to the bundle sheath, thereby sustaining electron flow in the mesophyll chloroplasts.

3.7.3 The variations in photosynthetic adaptations are not exclusive

We have referred to categories of photosynthetic behaviour, notably C_3, CAM and C_4 and a less well-defined system in some submerged aquatic plants; it is

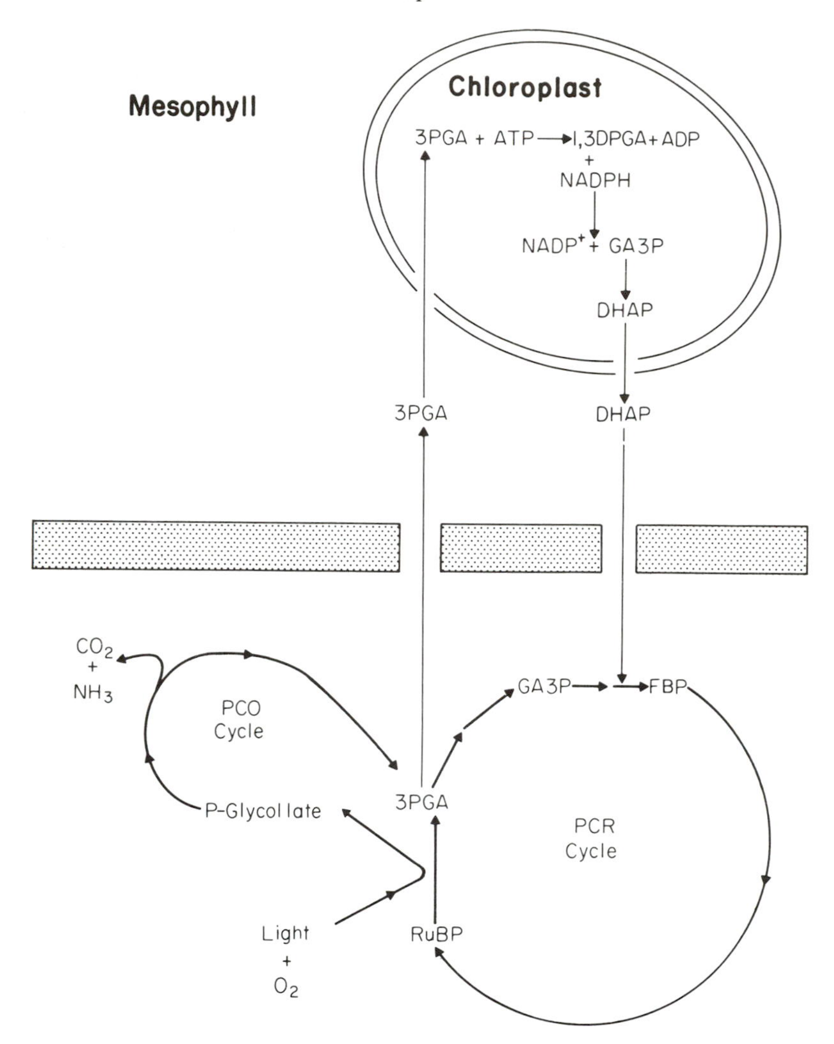

Fig. 3.15 A hypothetical scheme whereby a PGA–DHAP shuttle operates between bundle sheath and mesophyll cells in a C_4 plant. In the absence of CO_2 the RuBP-oxygenase reaction would generate PGA which passes to the mesophyll and enters the chloroplast where it is converted to DHAP with consumption of NADPH and ATP, thereby keeping the electron traps open in the thylakoids of the mesophyll cells. Maintenance of electron flow in the mesophyll chloroplasts would minimize the chances of formation of superoxide radicles (see text). Abbreviations: 3PGA, 3-phosphoglyceric acid; 1,3DPGA, 1,3-diphosphoglyceric acid; GA3P glyceraldehyde 3-phosphate; DHAP, dihydroxyacetone phosphate; FBP, fructose 1,6-bisphosphate; RuBP, ribulose 1,5-bisphosphate.

important to emphasize however that many plants cannot be fitted into one of these exclusive categories. Reference has already been made to species in which the CAM adaptation appears to be facultative, being variously evoked by salt (Ting & Hanscom, 1977), water stress (Neals, Patterson & Hartney, 1968; Szarek, Johnson & Ting, 1973), photoperiod (Quieroz, 1974), day and night temperature regimes (Nalborczyk, LaCroix & Hill, 1974) and flowering (Brulfert, Guerrier & Quieroz, 1975). Recent evidence suggests that in *Portulaca oleracea*, a C_4 species, short days and water stress may also induce a sort of CAM response (Koch & Kennedy, 1980); the compartmentative and enzymic controls underlying this double adaptation are unknown. *Frerea indica* is a stem-succulent CAM plant which produces leaves which carry out normal C_3 photosynthesis (Lange & Zuber, 1977) and circumstantial evidence suggests that there may be many other species of this kind.

The C_4 syndrome is best known in its fully developed form, but species possessing anatomical and physiological characteristics intermediate between those of C_3 and C_4 species have been described (Kennedy & Laetsch, 1974; Rathnam & Chollet, 1979). In *Mollugo verticillata* there is evidence of intraspecific variation in the development of a partial C_4 syndrome (Sayre, Kennedy & Pringnitz, 1979), which may well prove to be a more common phenomenon as more detailed work is carried out. It is clear that the C_4 mechanism is polyphyletic and occurs in many plant families; palaeontological evidence from the Late Tertiary indicates that the Kranz syndrome is of some antiquity in certain Graminae at least, but it is also increasingly evident that we are witnessing the current emergence of the syndrome in genera such as *Mollugo* (Nambudiri, Tidwell, Smith & Hebbert, 1978).

3.8 SOME ECOLOGICAL IMPLICATIONS IN THE ADAPTATION OF THE CARBON ASSIMILATING SYSTEM OF PHOTOSYNTHESIS

The immense complexity of the metabolic processes underlying the adaptive responses of photosynthesis described above will doubtless lead many to conclude that my plea for an eventual link to population genetics is an unrealistic aspiration. In reality the situation may not necessarily prove as difficult as might at first appear, for it may turn out that there are whole blocks of genetic attributes inimical to these adaptations which are closely linked. Nonetheless since there is as yet little progress in this direction we must look to other evidence of the ecological significance of these adaptations as may be available.

We have seen from our theoretical analysis that at the temperatures often prevailing in temperate regions (0–25°C) the quantum efficiency of net CO_2 fixation may be greater in C_3 than in C_4 species because of the extra energy needed by the C_4 species for the secondary carboxylation. At temperatures above 25°C the situation is changed because photorespiration in C_3 species lowers the quantum efficiency for net CO_2 fixation. In C_4 species, on the other hand, the CO_2 pump has made it possible to counteract the encroachment of the oxygenase reaction at high temperatures; the levels of photorespiratory enzymes are also lower in C_3 species and hence the quantum efficiency of the net fixation of CO_2 falls very little at elevated temperatures.

However, we must superimpose a multitude of other environmental interactions onto these considerations before we can provide any assessment of the adaptive value of any of these attributes. The following examples are offered merely to illustrate the importance of such considerations and to emphasise the danger of premature generalizations in respect of the adaptive significance of particular metabolic attributes.

3.8.1 Distribution of C_4 species in North America and Central Europe

Teeri & Stowe (1976) carried out a stepwise multiple regression analysis of the distribution of C_3 and C_4 grasses in relation to a number of environmental variables in 27 geographical zones of North America. The proportion of C_4 grasses was highly correlated with the July minimum temperature but not with any other factors. When this study was extended to the C_4 dicotyledons over the same region (Stowe & Teeri, 1978), distribution correlated with rates of summer pan evaporation and dryness ratio of the habitat and not with minimum July temperature. It seems probable that, in the dicotyledons, the total energy load is greater than for the grasses because the leaves are generally broader, and this brings the greater water use efficiency of the C_4 syndrome into sharper focus in the complex of selective factors. Further experimental work is required in order to test these speculations. Doliner & Joliffe (1979) carried out a multivariate analysis of the distribution and C_3 and C_4 species in relation to habitat parameters in California and Central Europe; they emphasise the need for a more detailed knowledge of the presence of C_4 species but concluded that the frequency of C_4 species correlated only with high temperatures and aridity of the habitat.

3.8.2 Distribution of C_3 and C_4 shrubs in Utah

Caldwell, White, Moore & Camp (1977) studied the balance between *Ceratoides lanata* (C_3) and *Atriplex conferta* (C_4), two shrubs of similar habit which form mixed communities in the Great Basin of Utah. The two species appear to be in equilibrium. When their growth, productivity, carbon balance and water consumption are compared on an annual basis, the two species are seen to perform in a similar manner. When these parameters are investigated in more detail on a monthly basis, it is found that there are seasonal differences between performance of the two species which are compensatory when integrated over the whole year; *Ceratoides* shows greater growth in the cooler periods of spring and later summer than does *Atriplex*, but slows down in the hottest summer months at which time the growth of *Atriplex* is increased. It is of interest that studies of quaternary deposits in this region suggest that the relative proportions of the two species may have varied over the past 20 000 years in relation to fluctuations in the climate. If substantiated these findings could be particularly instructive as they seem to suggest that even when a C_4 species is present and functioning in a given area, and if the assumption is made that it possesses the intrinsic capacity to adapt to these conditions, this does not occur: it appears rather that the C_4 mechanism is at a positive disadvantage. Two further examples will serve to take us deeper into this question; these are cases of the altitudinal distribution of C_4 grasses in Kenya and the case of *Spartina* × *townsendii* in Europe.

3.8.3 C_4 species in Kenya

Tieszan, Seryinska, Imbamba & Troughton (1979) studied the distribution of C_3 and C_4 grasses along an altitude and moisture gradient in the vicinity of Mount Kenya. They found close correlations between increasing numbers of C_3 species and decreasing numbers of C_4 species with increasing altitude; a number of exceptions to the correlation were related to features of micro-habitat and local rainfall factors. One may suppose that in this region the C_4 syndrome had been available in the flora over a long period of time and could have taken over completely had there not been positive disadvantages to it at the higher altitudes. Much of the early work on C_4 metabolism was carried out on maize and other tropical grasses, from which there has emerged a tendency to suppose that there are intrinsic features of C_4 metabolism which render it unsuited to efficient functioning under temperate conditions. Reference has already been made to the lower quantum requirement of C_3 species below 25°C. In the mist zones at higher altitude in Kenya this could place them at an advantage; even here however the total daily irradiance will often exceed that of the habitats in which *Spartina* × *townsendii* flourishes in Europe, which we shall consider next. The metabolic work on tropical grasses also raised the problem of whether there are inherent limitations to C_4 photosynthesis at low temperatures. As an example we may take the case of pyruvate-inorganic phosphate dikinase, an enzyme which is of central importance in C_4 species for the regeneration of PEP in the mesophyll cells. The enzyme was discovered by Hatch & Slack (1968) and has been studied extensively (Sugiyama & Boku, 1976); it comprises four identical subunits which are probably hydrophobically bonded since they appear to dissociate readily at low temperature. In maize, for example, which will not photosynthesise below 10° C, in-vitro preparations of the dikinase show increasing rates of inactivation as the temperature is lowered below 10° C; at 0° C in the presence of 2 mM Mg, $t_{1/2}$ for inactivation is about 5 min (Hatch, 1979). This inactivation may, however, be virtually prevented by the presence of physiologically feasible concentrations of enzyme substrate (2 mM pyruvate) or product (0.1 mM PEP). One might suppose that cryo-stabilizing concentrations of such metabolites could be achieved in the course of evolution. Temperature does not appear to be a barrier to ecological success in the case of *Spartina* × *anglica*, and we may draw from this example further pointers to the complexity of environmental influences which may contribute to the fitness of an organism.

3.8.4 C_4 PHOTOSYNTHESIS IN *SPARTINA ANGLICA*

Spartina anglica is a vigorous allotetraploid which arose from a hybridisation between the native *S. maritima* and *S. alterniflora*, an introduction from North America. Details of the cytology and attendant nomenclature of the species are complicated and need not concern us here. Suffice it to say that the native parent *S. maritima* has remained a scarce species, of relatively weak growth, which rarely sets seed in Britain, whilst the other parent, *S. alterniflora*, has never gained a foothold. *Spartina anglica*, however, produces fertile seed, probably in part by apomixis; and now occupies over 12 000 hectares around the coast of Britain, often spanning salt marshes from the coastal to the landward side. It has Kranz anatomy and the gas-exchange characteristics of a C_4 species (Long, Incoll & Woolhouse, 1975; Long & Woolhouse 1978 a and b), as do both parent species. The productivity of the species is relatively high, of the order 40 tonne ha^{-1} a^{-1}; of which an exceptionally large amount, over

50%, is invested in below-ground parts, most notably a massive rhizome system. Clearly, from the lack of fitness displayed by the parent species of *S. anglica* it must be supposed that the key to its fitness does not rest primarily with the C_4 photosynthetic system. The reasons for the lack of fitness in *S. maritima* and *S. alterniflora* in Britain are not known; the lack of vigour in the former suggests that it may be a poor competitor particularly on account of its greater light requirement than its C_3 neighbours; for *S. alterniflora* conditions are probably too far below the normal temperature requirements of the species. *S. anglica* may have inherited a measure of temperature adaptation from *S. maritima* in that it is able to photosynthesise at similar rates to neighbouring C_3 species even at 5–10°C (Long & Woolhouse, 1978a; Long & Woolhouse, 1979; Long & Incoll, 1979).

There is still some suggestion of imperfect adaptation to temperature, on the other hand, when the seasonal pattern of growth is considered (Woolhouse, 1980). As noted the vigour and productivity of the species is high. The huge below-ground system must impose a severe respiratory demand upon the plant, but against this the rhizome provides the abundant air spaces needed for oxygen transport in the anaerobic estuarine muds, whilst the stout anchor roots enable the plant to establish and maintain itself in the seaward side of the marsh. Indeed, in places where wave action is not too vigorous it may be seen that the species is in essence creating its own habitat as it goes along, as the stout stems and rhizomes slow the movement of water and encourage silting. Thus the burden of the root system may be offset by the gaining of a new niche in which, there being no other species, the light-sensitive C_4 mechanism is spared the burden of competition from other species. Support for this view is provided by experiments in an area where *Phragmites communis* was invading *Spartina* marsh in Poole Harbour (Ranwell, 1972). The *Spartina* became etiolated, then ceased to flower, then failed to produce tillers and finally died over 25 m landward of the seaward limit of *Phragmites.* When the stems of *Phragmites* were cleared from a square metre of ground in the landward extinction zone of *Spartina*, the *Spartina* recommenced growth, tillering and the production of flowers.

An additional burden to the carbon balance of *Spartina* is imposed by the salinity of its habitat which may have an osmotic potential of the order -4MPa (-40 bars). In these circumstances energy is required for the secretion of salt through specialised glands and for the production of a compensating organic osmoticum within the tissues. The energy requirements of the salt pump are difficult to measure and estimates are not available; O'Leary (unpublished data) has estimated the energy expenditure of providing the organic osmoticum at the equivalent of 1.2 tonne $ha^{-1} a^{-1}$, which would be some 3% of the total biomass production. This provision of osmoticum could be another factor pushing the species towards an unfavourable carbon balance and heightening its sensitivity to shade; on the other hand −4 MPa (−40 bars) is a substantial opposing osmotic potential, which raises the question of whether, in these circumstances, the greater water use efficiency of the C_4 mechanism may be advantageous even though the energy loads and hence leaf temperatures may not be high in these temperate coastal environments. Finally, we may return to the CO_2 compensation point of the C_4 syndrome. In lower reaches of a salt marsh *S. anglica* may be submerged for a substantial proportion of the daylight period; it has been shown that under experimental conditions the species may survive for more than four months totally submerged in clear sea water. Under these conditions the massive subterranean root and rhizome system, with large air spaces adapted to support O_2 diffusion to the root tips, could function in reverse to permit passage of CO_2 to the leaves. In these circumstances the low CO_2 compensation point associated with the C_4 mechanism could be an important factor in maintaining the gradient of CO_2 in the direction of the leaves: quantitative estimates of this internal recycling of CO_2 should be sought.

Many of the factors relating to the light, temperature, carbon balance and water relations of the C_4 syndrome which may affect *S. anglica* in the salt-marsh environment clearly require further work in order that their relative significance can be placed on a firm quantitative basis. I have judged it opportune to consider this example in some detail however in order to illustrate the intricacy of constraining factors which may weigh in opposing directions upon various aspects of a particular metabolic pathway.

3.8.5. Ecological aspects of CAM

We have seen that CAM metabolism may be a fixed attribute in some species but appears to be inducible by water stress or salinity in others. CAM species are widely distributed on a global scale, they may occur as epiphytes in tropical rainforests, and as characteristic species of environments such as walls, rock ledges and other periodically dry habitats in almost all of the major climatic zones. CAM species reach their greatest abundance in arid, semi-desert regions where evapotranspiration rates are often extremely high. The productivity of CAM species is generally low, the limits being set presumably by the capacity for organic acid storage in the dark fixation period. There have been an abundance of laboratory studies of CAM species with respect to their water relations, gas exchange characteristics and metabolic regulation; which have provided a rich ground for speculations concerning their ecological significance (Osmond, 1978); the paucity of good measurements of the energy, carbon and water balance of these species under field conditions stands in marked contrast.

3.9 CONCLUDING REMARKS

Many physiologists will be appalled by the inadequacies of this article; amongst the gross omissions, they might point for example to the total failure to explore the anatomical and biochemical aspects of adaptations to shade in the photosynthetic apparatus (for a review see Boardman, 1977). Even worse perhaps they would consider the exclusion of detailed treatments of temperature adaptations of photosynthesis (for a review see Berry & Björkman, 1980). Whole chapters could be written concerning photo and thermo adaptive features of the photosynthetic membranes and the enzymes of carbon assimilation. Such matters are of undoubted importance for a proper understanding of the adaptive responses of a given species or of particular genotypes, not least in the economic context of improving crop production. For present purposes, however, we have defined as our central concern the energy costs of adaptive responses in photosynthesis and, as we have seen, we begin to be able to make a few stumbling pronouncements concerning the overall energetics of the broader aspects of the main mechanisms. But of the *energy costs* of maintaining the enzymes needed to catalyse these different mechanisms, or of making and maintaining the light-capturing and energy-transducing systems needed in different environments, we know nothing. We are in total ignorance of the extent to which such costs bear upon the selective advantage of a particular photosynthetic mechanism. In these circumstances the author has preferred the guidance of Wittgenstein: 'Whereof one cannot speak, thereof one must be silent', *Tractatus Logico-Philosophicus*.

Chapter 4

Maximizing Net Energy Returns from Foraging

COLIN R. TOWNSEND AND ROGER N. HUGHES

4.1 INTRODUCTION

Optimal foraging theory is a way of conceptualizing resource utilization by animals which adopts the *a priori* reasoning described in Chapter 1. Foraging has associated costs (in searching, handling food, etc.) as well as the benefit of energy gained. The basic assumption is that the foraging behaviour which endows an animal with the greatest fitness (optimal foraging) is that which maximizes the net rate of energy gain, and we take this as our phenotypic measure of fitness. The *energy maximization premise* can be applied to all aspects of foraging behaviour but it explicitly ignores the almost universal trade-offs between time or energy invested in foraging and that available for other biological activities such as obtaining a sufficient supply of an essential nutrient or avoiding enemies. Optimal foraging theory therefore predicts optimal behaviour when the energy maximization premise is applicable, that is, given that the rate of energy gain from food is the only limiting factor.

For any given spatial distribution of food there will be a theoretical optimal allocation of search effort which maximizes the overall rate of encounters between forager and food. Similarly, for any particular array of potential food types, differing in abundance, energy content and ease of capture, there will be an optimal choice of diet which maximizes the net rate of energy return to a forager. It is not particularly difficult to discover the optimal strategies by means of quite simple mathematical models. (Even for non-mathematicians careful perusal of the models presented in this chapter should repay the effort. However, the significant predictions which are generated by the mathematics will also be presented in words.) The crucial question is whether real predators are able to behave like their hypothetical optimal counterparts. Thus, having derived some predictions, the results of experiments to test whether predators conform to them will be discussed. Often the optimal solution will appear to require sophisticated computational powers on the part of the animal and attention will be focussed on the means by which the solution may be accomplished or approximated.

4.2 OPTIMAL PATCH USE

Mobile animals, while foraging, usually encounter areas of habitat differing in the density or quality of food present. These areas may have naturally discrete boundaries, such as flowers visited by sunbirds and tree canopies gleaned by warblers, or they may merge with one another, as do different parts of a cockle bed searched by oystercatchers. Assuming that an animal can perceive the area over which it forages approximately as a patchwork of different food densities, then the energy maximization premise requires that the animal should aportion its foraging time between patches so as to maximize the net rate of energy gain. Development of this idea has led to what Charnov (1976) calls the *marginal value theorem.* The problem is to maximize E/T where E is the net energy gain during a foraging period of length T. The time taken to travel between patches is assumed to have an average value T_r, during which feeding is not possible. The average time (T) to exploit a single patch, including the time taken to get to the patch from the previous one, is:

$$T = T_r + \sum P_i . T_{si}$$

where P_i is the proportion of patches which are of productivity i and T_{si} is the time spent searching in these patches. The average net energy gain (E) from a patch is:

$$E = \sum P_i . E_{ti} - T_r . E_r$$

where E_{ti} is the net energy gain after searching for T_{si} time units in a patch of productivity i and E_r is the locomotory energy cost per unit travel time.

The average net energy gain per unit time (E/T) will be:

$$\frac{E}{T} = \frac{\sum P_i . E_{ti} - T_r . E_r}{T_r + \sum P_i . T_{si}}$$

The optimal time (T_{si}) for searching each type of patch is found by setting $\mathrm{d}(E/T)/\mathrm{d}T_{si}$ to zero for all patch types simultaneously and solving the equations.

The result is that

$$\frac{\mathrm{d}E_{ti}}{\mathrm{d}T_{si}} = \frac{E}{T} \quad \text{for all patch types.}$$

This says that *the animal should leave a patch when the instantaneous net rate of energy intake* ($\mathrm{d}E_{ti}/\mathrm{d}T_{si}$) *falls to a value equal to the average net rate of energy intake for the habitat as a whole* (E/T). This is the marginal value theorem. It follows that animals should ignore patches where the instantaneous net rate of energy intake would be less than the average for the habitat (although in most cases animals will necessarily 'waste' some time visiting such patches before ascertaining that they are not in the optimal set). All patches more productive than the average should be used, but the less productive the patch the sooner it will be depleted to the average productivity. This result is illustrated graphically in Fig. 4.1(a). The predicted consequence of this is that *there will be a concentration of foraging activity to progressively fewer patches until all*

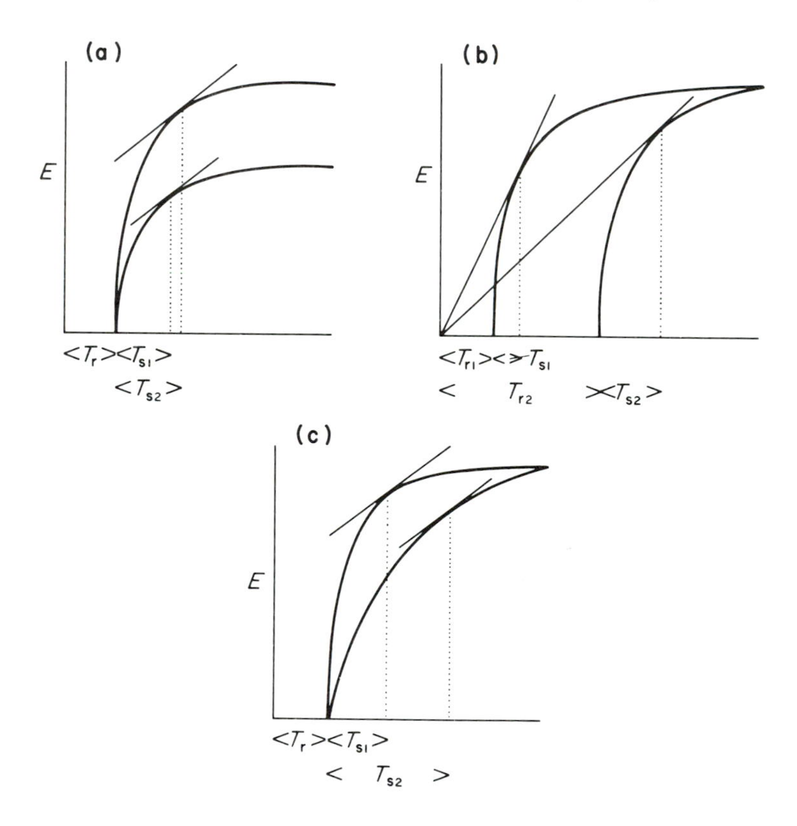

Fig. 4.1 Optimal patch use—the marginal value theorem. The optimal time to spend foraging in a patch depends on: (a) the productivity of the patch; (b) the cost of travel between patches; (c) the searching efficiency within a patch. The ordinates represent the cumulative net energy gain from the patch. The abscissae represent time. T_r is the average travel time between patches, T_s is the optimal time to spend foraging in a particular patch. The slopes of the tangents represent the average net rate of energy gain in the habitat.

patches are depleted to the average productivity of the habitat.

A test of the first prediction has been provided by a combination of field observation and laboratory experiments on the freshwater net-spinning caddis larva, *Plectrocnemia conspersa*, which feeds very largely on chironomid and stonefly larvae (Hildrew & Townsend, 1980; Townsend & Hildrew, 1979, 1980). In an experimental stream it was found that fifth instar predators had a giving-up time (time from last prey capture until leaving the patch, i.e. quitting the net) of 1.7 days. This corresponds to a 'marginal' capture rate of about 0.6 items per day (average-sized chironomid prey). The productivity of their natural environment averaged over the whole year was found to be 0.7 items consumed per day (chironomids and stoneflies combined). Expressed in terms of biomass, the marginal capture rate in the experiment was 69 μg per day, the average consumption in the environment was 65 μg per day. These results lend support to the marginal value theorem.

The second prediction was tested in the laboratory by Hubbard & Cook (1978) who studied the behaviour of the parasitoid *Nemeritis canescens* foraging in an arena containing patches of its host *Ephestia cantella* which were concealed in plastic petri dishes covered with sawdust. The patches contained varying numbers of hosts and one question Hubbard and Cook addressed was whether the terminal rates of encounter with healthy hosts were the same in each patch type. They could judge the rate of encounter by noting the frequency of a characteristic 'cocking' movement of the body which occurs after oviposition. Their result is shown in Fig. 4.2(a) and conforms to the prediction that all patches should be depressed to a common level of host abundance with the richest patches being depleted to the greatest extent. Hubbard and Cook were also able to show that as exploitation proceeded the amount of time which was spent on the patch of highest density declined and the patches next in rank were added into the optimal set in the way predicted by their model (Fig. 4.2(b)).

The predictions and experiments considered so far have dealt with the optimal spatial allocation of search effort in a habitat with a given array of patches and a particular value for average rate of net energy intake (E/T). A further prediction can readily be generated if habitats with different average profitabilities (i.e. different values of E/T) are to be compared. *An animal should remain for longer in a patch containing a given number of prey in a habitat which as a whole is less profitable.* Krebs, Ryan & Charnov (1974) working with tits feeding on mealworms in artificial pine cones produced evidence which was consistent with this prediction. In a poorer habitat the tits demonstrated a longer giving-up time. However, in the case of the net-spinning caddis larva

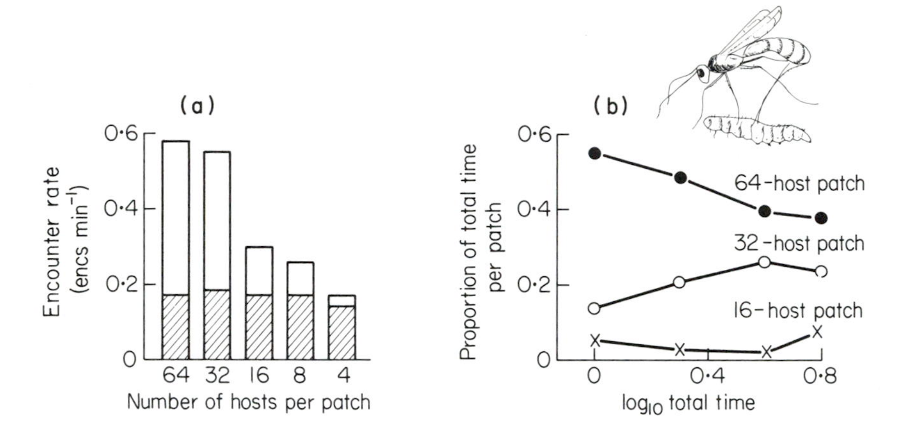

Fig. 4.2 (a) The estimated terminal encounter rate of *Nemeritis canescens* with healthy hosts (filled areas) compared with the initial rates (open areas). (b) The effect of total time on the proportion of total time spent per patch for the top three ranking patches. From Hubbard & Cook (1978).

studied by Townsend & Hildrew (1980) the giving-up time appeared to be fixed at a value appropriate to the average capture rate through the year, producing markedly sub-optimal foraging during the short summer period of peak prey abundance.

Different average profitabilities of patches in various habitats will not only be due to differences in food density but also to differences in time spent obtaining food. An ingenious method of testing the optimal patch use model was devised by Cowie (1977) who saw that the predicted outcome of increasing average travel time (T_r) between patches would be that the animal should remain feeding longer in each patch (because of the decrease in the average net rate of energy gain for the habitat, Fig. 4.1(b)). Cowie tested this prediction by allowing great tits (*Parus major*) to forage in two 'habitats', one where food (bits of mealworm concealed in sawdust) was available in small cups hung from artificial trees in an aviary and the other, representing a habitat with a greater average inter-patch travel time, where identical cups were covered with cardboard lids. The time spent removing a lid to give access to the 'patch' of food was regarded as analogous to travel cost. As predicted, the birds spent longer feeding from cups with lids, even though similar amounts of food were available in each case.

A second factor which involves extra time costs in foraging and thus a lower average profitability for the habitat in special cases is *setting-up time*

(Hildrew & Townsend, 1980). Animals which need to make an investment in a patch before they can begin to exploit it 'lose' time analogous to travel time between patches. Good examples in the terrestrial environment are provided by web-spinning spiders and in freshwater by certain net-building, predatory caddis larvae. Note that Cowie's experiment corresponds just as closely to a test of the influence of greater setting-up time as to a test of the influence of greater travel time.

Finally, it should be noted that the availability of food in patches will depend not only on differences in food density but also on differences in search efficiency as influenced by physical features of the patches. For a given food density, a lower searching efficiency reduces the rate of encounters with food, thus lowering the effective productivity of the patch (Fig. 4.1(c)) and this must be taken into account when calculating the optimal searching time per patch. Experimental demonstrations of the effect of greater environmental heterogeneity in reducing searching efficiency and hence feeding rate have been made by Hildrew & Townsend (1977) and Nelson (1979).

4.2.1 Appraisal

In many cases there is no doubt that foragers have evolved behaviours which increase their foraging efficiency in a patchy environment. More time is spent in prey-rich patches as evidenced by reports of aggregative responses both by invertebrate and vertebrate consumers (e.g. Hassell & May, 1974). But how far can the behaviour be described as optimal, yielding the maximum possible rate of net energy income? This is what optimization models are designed to explore and there is evidence that a variety of foragers, including great tits, an insect parasitoid and a caddis larva, approach the predicted optimum allocation of search effort.

It is of interest that the caddis, *P. conspersa*, appears to have a fixed giving-up time appropriate to average conditions through the year and so forages sub-optimally during times of highest prey density. It is, perhaps, not surprising that the sophisticated gathering of information (perfect knowledge of the environment is assumed), computation of parameters in the optimization equation and calculation of the optimal solution, which the theory apparently demands, should be beyond the powers of such an insect. However, its behaviour does approach the optimum and for tits and others animals the fit with the model is even better. How is this accomplished? In the case of great tits some learning is apparently involved and possibly the bird can store information about the last few inter-catch intervals and about the profitability of the habitat as a whole, although no direct evidence of such a capability has been presented. In most cases, it is probably by using rather simple 'rules of thumb' that foragers approach the optimum. For example, Ollason (1980) has developed an interesting model which indicates that a forager may closely approximate to optimality simply by virtue of an ability to remember and by leaving each patch if it is not feeding as fast as it remembers doing. Alternatively, a forager's search effort is often concentrated on rich patches because of *area-restricted search* (increased rate of turning and/or turning back from patch boundaries) in response to prey capture (Pyke, Pulliam & Charnov, 1974; Waage, 1979), and in other cases by means of a constant giving-up time (like a clock that is reset after each capture; if the giving-up time runs out without a prey capture the patch is quitted) (Hassell & May, 1974). Waage (1979) has argued that the parasitoid *Nemeritis canescens* may achieve near optimal foraging for hosts in habitats with differing profitabilities by means of a simple, innate response in which low oviposition rates over several days enhance responsiveness to stimuli associated with host patches and thus increase time spent in patches in a poorer environment.

4.3 OPTIMAL DIETS

When faced with a choice of alternative food items, the energy maximization premise says that the diets which confer the greatest fitness (optimal diets) are those which maximize the net rate of energy intake. Using this basic assumption, it is possible to build models which predict the optimal diets of diverse animals ranging from macrophagous carnivores and certain herbivores (principally seed, pollen and nectar feeders) to microphagous filter and deposit feeders. The approach unless modified, will usually not be appropriate to browsing and grazing herbivores for whom mineral nutrients are often the limiting factor (Belovsky, 1978; White, 1978) and in some cases it is argued that herbivores select food plants to avoid over-ingestion of any one plant toxin (Freeland & Janzen, 1974). No appropriate feeding models have yet been developed for true parasites or sap-sucking insects.

Because of their very different feeding mechanisms, macrophages and microphages require different theoretical developments of the energy maximization premise.

4.3.1 Macrophages

Suppose an animal, searching for food, comes across several different kinds of food item, say different kinds of seed, different species of insect prey, or different sizes of some common prey organism. Is it possible to predict which of the encountered food items the animal should select to eat and which, if any, it should reject? The prediction is possible if we make a simplifying assumption and then apply the energy maximization premise. The simplifying assumption is that the animal can rank all food items according to their *energy value*. The energy value is the net amount of energy the animal is able to assimilate from the food item (gross energy extracted minus energy costs of handling and digesting the food) divided by the time taken to handle the food item, i.e. the net energy yield per unit handling time. The energy maximization premise says the animal should accept those food items which, if included in the diet, would maximize the net gain of assimilated energy per unit foraging time and reject those items which would lower the net rate of energy gain.

Therefore it is necessary to calculate the net rate of energy gain for any given set of food items comprising the diet and then to find out whether this rate of energy gain is increased or decreased when other food types are added to the diet. As a first approximation, let the average net energy gain per unit foraging time (E/T) equal the average total net energy gain during the foraging bout (E) divided by the total length of the foraging bout (T). The average total net energy gain is the sum of the average numbers of each food type eaten (N_i) multiplied by their appropriate net energy yields (E_i). The average number of each food type eaten is equal to the rate of encounter with each food type (λ_i) multiplied by the time spent searching for food, so that $N_i = T_s\lambda_i$. The average total energy gain will be $T_s\Sigma\lambda_i E_i$. Searching for food and handling food are usually mutually exclusive events, so that the average length of the foraging bout is equal to the average time spent searching for food (T_s) plus the average total time spent handling food. The latter is the sum of the average numbers of each food type eaten (N_i) multiplied by their appropriate handling times (H_i). As before, $N_i = T_s\lambda_i$ so that the average total time spent handling food is $T_s\Sigma\lambda_i H_i$ and the average length of the foraging bout is $T_s + T_s\Sigma\lambda_i H_i$. We can now write the average energy gain per unit foraging time as:

$$\frac{E}{T} = \frac{T_s\Sigma\lambda_i E_i}{T_s + T_s\Sigma\lambda_i H_i} = \frac{\Sigma\lambda_i E_i}{1 + \Sigma\lambda_i H_i}$$

Suppose the diet contains just two food types where the energy value of the first exceeds the second, so that $\frac{E_1}{H_1} > \frac{E_2}{H_2}$. The average net energy gain per unit foraging time when the diet includes both types is given by:

$$\frac{\lambda_1 E_1 + \lambda_2 E_2}{1 + \lambda_1 H_1 + \lambda_2 H_2}$$

and by:

$$\frac{\lambda_1 E_1}{1 + \lambda_1 H_1}$$

for a pure diet of food type 1
and by:

$$\frac{\lambda_2 E_2}{1 + \lambda_2 H_2}$$

for a pure diet of food type 2.

It is now possible to find out whether the animal should feed only on food type 1, or on food type 2, or include both in its diet. For it to be optimal to specialize of food type 1, the average net rate of energy gain from the pure diet must exceed that from the mixed diet, so that:

$$\frac{\lambda_1 E_1}{1 + \lambda_1 H_1} > \frac{\lambda_1 E_1 + \lambda_2 E_2}{1 + \lambda_1 H_1 + \lambda_2 H_2}$$

This inequality can be rearranged to the form

$$\frac{E_2}{H_2} < \lambda_1 \left[E_1 - \frac{E_2 H_1}{H_2} \right] \tag{4.1}$$

Note that the inequality shown in equation 4.1 contains λ_1, the rate of encounter with food type 1, but does not refer to the encounter rate with the second food type. Whether or not it is worthwhile feeding purely on the first food type is therefore seen to depend on the difference in energy value between the two food types and on the encounter rate with the first (more valuable) food type, but to be independent of the encounter rate with the second (less valuable) food type.

By similar argument, specialization on the second food type would be worthwhile if:

$$\frac{E_1}{H_1} < \lambda_2 \left[E_2 - \frac{E_1 H_2}{H_1} \right] \tag{4.2}$$

However, this inequality is always false because $E_1/H_1 > E_2/H_2$, so that it is never worthwhile to feed solely on the second (poorer) food type if both food types are encountered while foraging.

The inequalities in equations 4.1 and 4.2 yield the following predictions:

1 *the more valuable food type should always be accepted when encountered*;

2 *the less valuable food type should be rejected if the encounter rate with the more valuable food type is sufficiently high.* The critical value of this encounter rate will decrease as the difference in energy values of the two food types increases;
3 *the diet should expand to include the poorer food type if the encounter rate with the more valuable food type falls below the critical value. Conversely the diet should contract to exclude the poorer food type if the encounter rate with the more valuable food type rises above the critical value.*

These arguments and predictions can be applied to any number of food types if they can be ranked in order of energy value. The general form of the inequality equation 4.1 for many food types is:

$$\frac{E_m}{H_m} < \sum_{j=1}^{m-1} \lambda_j \left[E_j - \frac{E_m H_j}{H_m} \right]$$

The general predictions are therefore that *the most valuable food type should always be accepted when encountered, but as the encounter rate with it falls, the diet should expand to include the next most valuable food type. As the encounter rates with the more valuable food types continue to decline, the diet should expand further to include sequentially the progressively poorer food types.* How does the feeding behaviour of real animals compare with these theoretical predictions?

Elner & Hughes (1978) tested the predictions with shore crabs (*Carcinus maenas*) fed with mussels (*Mytilus edulis*). Mussels are a common natural food of shore crabs and their energy values are easily measured by dividing the total energy content of the flesh by the time the crabs take to crack open the shells to eat the flesh. The energy values of mussels for a given size of crab form a peaked curve (Fig. 4.3). Elner & Hughes chose three mussel sizes of known energy value for their experiments, as indicated in Fig. 4.3(a). Crabs were presented with the mussels scattered haphazardly over the floor of the aquarium. As they were eaten, mussels were replaced by others of a similar size in order to keep prey density constant. The ratio of mussels in each size class was kept constant and was chosen so that all size classes would occupy the same area of floor space in each feeding trial, thereby reducing any potential bias caused by larger prey being easier to detect. Three feeding trials were made using different total densities of mussels. In the first trial, the number of rank 1 mussels was sufficient to satiate the crabs. In the second trial the number of rank 1 plus rank 2 mussels was sufficient to satiate the crabs, while in the third trial the total number of mussels was far below that necessary to satiate the

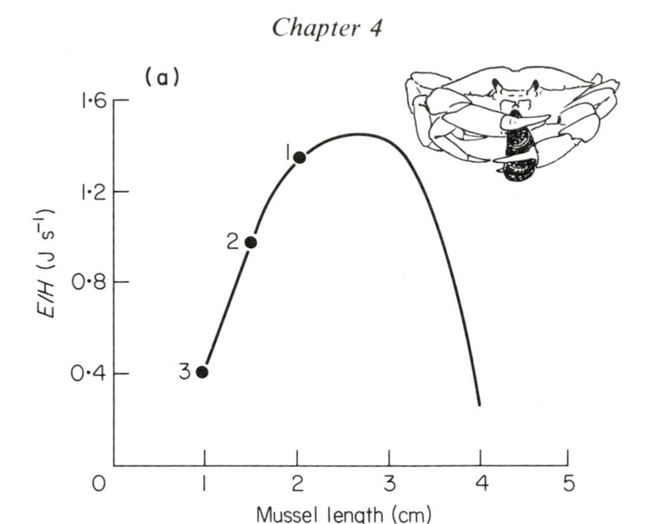

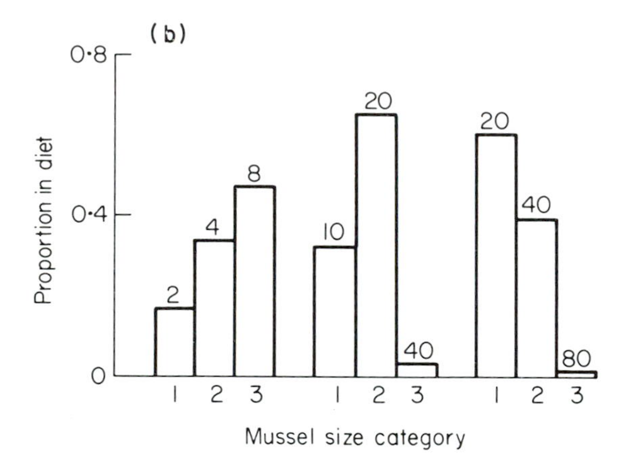

Fig. 4.3 (a) Construction of a prey value curve for mussels (*Mytilus edulis*) eaten by male shore crabs (*Carcinus maenas*). Energy value (E/H) versus shell length. The numbers refer to the mussels used in the optimal diet experiments. (b) Diet of shore crabs fed on three sizes of mussels where the energy value of $1 > 2 > 3$. The numbers of each size class offered, shown over the histograms, were chosen to equalize the floor area covered by each. From Elner & Hughes (1978).

crabs. The crabs showed a clear preference for the rank 1 but also took many of the rank 2 mussels in the first trial (Fig. 4.3(b)). As total prey density decreased in the second trial, the number of rank 2 mussels eaten increased. In the third trial at very low prey density, the rank 3 mussels, which were almost totally rejected in the other trials, were eaten in abundance. There was therefore only partial agreement with the predictions of the optimal diet model. The diet expanded as the preferred mussel sizes became scarcer, but the least valuable mussels were never totally excluded from the diet, even when the most valuable mussels were sufficiently plentiful to satiate the crabs. Careful behavioural analysis showed that all encountered mussels were handled for 1–2 s before being accepted or rejected. In accordance with the theory, rank 1 mussels were always accepted. Lower ranking mussels were rejected if encountered immediately after consumption of a rank 1 mussel, but would finally be accepted after a sequence of encounters with equal ranking mussels, the necessary length of the sequence increasing as the rank decreased.

Ringler (1979) observed brown trout (*Salmo trutta*) feeding on drifting food items in an experimental stream. Trout detect their food visually and handling times are very short and essentially constant for different sizes of food item. This contrasts with the shore crab which uses tactile and olfactory stimuli to detect and evaluate prey, and where handling time is long and varies with mussel size. Ringler selected three categories of prey with contrasting energy values; brine shrimps with a food energy content of 15 J, small crickets and mealworms (103 and 105 J, respectively) and, finally, large crickets and mealworms (240 and 230 J, respectively). All prey items had been frozen for storage and thawed out before use. Food was liberated into the flowing water of the experimental chamber from a 'hopper' at a rate of 10 food items per min during a 5-minute observation period each day for 6 days. Each item was approached from downstream and briefly fixated before being sucked into the mouth. After recording the identity of each food item drifting over the fish, Ringler was able to calculate the daily energy intake for fish feeding randomly among the drifting food items and for fish feeding optimally. The optimal diet was calculated on the basis of the energy maximization premise whereby the largest food items would always be included in the diet, and if the total number of captures exceeded the number of large food items available, then the next largest items would also be taken, and so on. On the first day of the experiment, the diet differed little from that predicted by a model of random feeding (Fig. 4.4). By the sixth day, however, due to selective feeding on larger items, the fish had more than tripled the average energy intake although they never quite reached the predicted optimum. The small departures of the stabilized diet from the optimum were due to trout continuing to take small numbers of brine shrimp.

Krebs, Erichson, Webber & Charnov (1977) tested the predictions of the basic optimal diet model by presenting great tits (*Parus major*) with large and

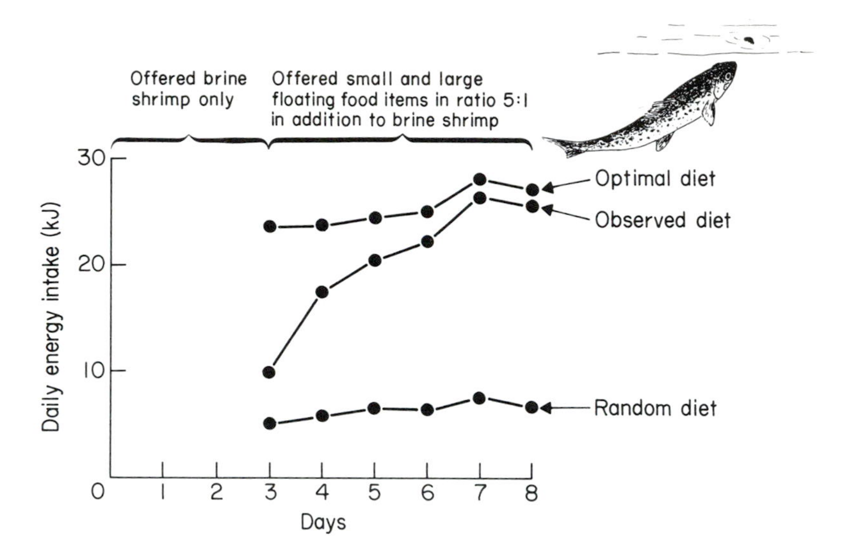

Fig. 4.4 The observed mean diet of trout during a 6-day experimental feeding programme, together with theoretical diets assuming random feeding (items taken according to their abundance in the drift) and assuming optimal foraging (items taken in accordance with their ranked size). From Ringler (1979).

small pieces of mealworm larvae placed on a moving 5 cm wide conveyor belt. Like the trout, the great tit is a visual predator and the appearance of food items for about 0.5 s as they moved across a 'window' in the floor of the cage was analogous to the drift of food items across the visual field of the trout. The large pieces of mealworm consisted of 8 body segments with a mean total weight of 69 mg and the small pieces were made up of 4 segments with a mean total weight of 33 mg. Both large and small pieces of mealworm could be picked up and swallowed by the great tits with equal ease, the handling time being very short. But in order to increase even further the difference in energy value between the two food types, small strips of white plastic tape were stuck lengthways along the top of the small pieces of mealworm. This made the two food types appear to be the same size, but greatly increased the handling time for the 4-segment pieces because the great tits had to hold them under one foot and peel off the tape before swallowing them. Encounter rates with the two food types were controlled by adjusting the densities of the 4- and 8-segment pieces of mealworm placed on the conveyor belt. At low encounter rates with both food types, the better and poorer items were eaten in similar proportions (Fig. 4.5). At high encounter rates with both food types, the great tits showed

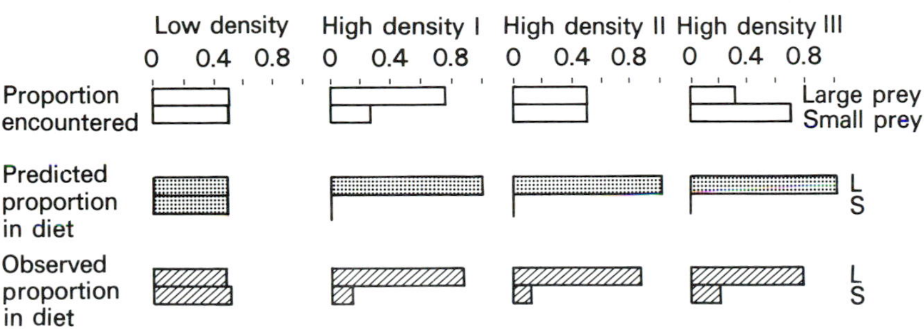

Fig. 4.5 Observed and predicted diets of great tits feeding on large and small pieces of mealworm. From Krebs (1978).

an overwhelming preference for the better food items but still ate a small proportion of the poorer items. Even when the encounter rate with the better food type was made much greater than that with the poorer food type, the great tits still took a small proportion of the poorer items. Clearly, the great tits were able to evaluate and recognize the two food types even though these did not differ in apparent size. The observed diets were close to those predicted by the basic optimal diet model, but differed in the small number of sub-optimal food items still included in the diet when optimal items were plentiful.

4.3.2 Appraisal

A crab, a fish and a bird seemed to choose diets close to those predicted by the basic optimal diet model, suggesting that the theory has general applicability. But how can we account for the small departures from expectation shown in the diets of these and other animals? It is worthwhile to re-examine some basic assumptions of the optimal diet model.

It is assumed that animals can rank a variety of food items. This ability requires first that the animal can associate specific recognition stimuli with the appropriate energy values. Such a perception may often be hindered by extraneous factors such as the confusing effects of prey movement, turbidity of water, the effect of background in promoting crypsis and so on, thereby causing mis-identification and sub-optimal feeding from time to time. Further, the correlation between specific recognition stimuli and the actual dietary value will often be imperfect so that it will become worthwhile for animals to occasionally 'sample' food items as they are encountered (Zach & Falls, 1978; Hughes & Elner, 1979). Finally, sampling, although sub-optimal in the short term, may be required if rank order is not constant because of the appearance of new food types (Heinrich, 1979) or because of intrinsic changes in value of existing ones (Westoby, 1974). (Note that the need for a balance between sampling and exploitation has also been argued in the case of choice of foraging area—Krebs, Kacelnik & Taylor, 1978.) Value may not be constant if an animal learns to recognize or handle food items more efficiently as a result of experience, where learned ability is reinforced by each encounter but attenuated by the absence of encounters. When handling times, and hence energy values, are made functions of specific encounter rates in the optimal diet model, the predicted diet may change from specialization on one food type, through a generalized diet on both types, to a specialized diet on the second diet, or it may switch from a pure diet on one to a pure diet on the other, depending on the encounter rates and learning efficiency (Hughes, 1979). (The energy maximization premise can therefore be used to predict the phenomenon of switching diets, forming a gratifying link between optimal foraging theory and the *searching image* concept of behavioural ecologists and the concept of *apostatic selection* used by ecological geneticists.)

The potential for this kind of behaviour seems to exist. Fig. 4.6(a) shows the increased efficiency with which dogwhelks (*Thais lapillus*) handle mussels after successive meals and Fig. 4.6(b) gives analogous results for bumblebees (*Bombus vagans*) feeding from a complex, zygomorphic flower (Morgan, 1972; Heinrich, 1979, respectively).

Second the ranking assumption requires that animals can 'memorize' a reference ranking at least for the duration of a foraging bout. If animals need to learn the ranks of food items from experience it is likely that the learned information will tend to be forgotten without subsequent reinforcement. We might expect such foragers to sample different food types from time to time in order to calibrate their memorized reference ranking. In other cases it is quite likely that the reference ranking is innate. For example, bluegill sunfish (*Lepomis macrochirus*) appear to approximate to optimal diet predictions when feeding on cladocerans by using the simple criterion of taking prey items which appear largest. When prey are rare, small prey will frequently appear largest by virtue of proximity to the fish. When prey are common, the fish are highly selective because large prey 'appear' largest most frequently (O'Brien,

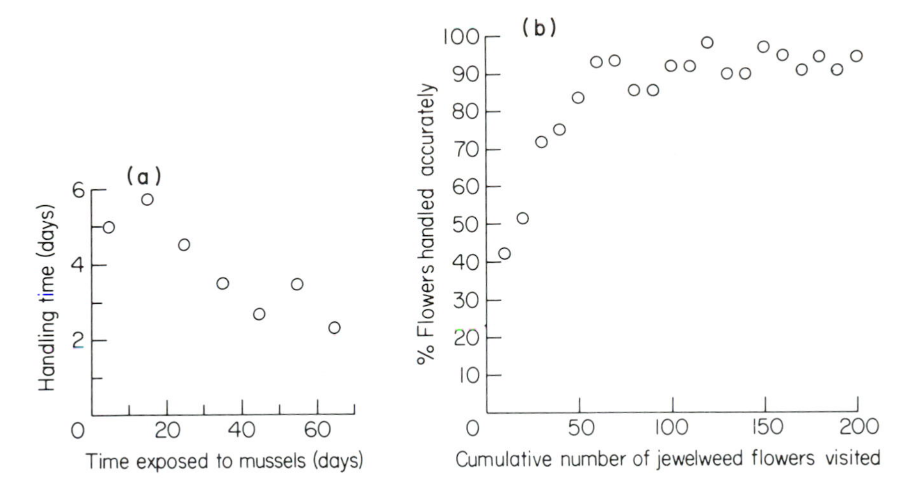

Fig. 4.6 (a) The time taken for dogwhelks (*Thais lapillus*) to drill and eat mussels (*Mytilus edulis*) plotted as a function of the length of time exposed to a diet of mussels. From Morgan (1972). (b) Handling accuracy by bumblebees (*Bombus vagans*) of the complex, zygomorphic jewelweed flower (*Impatiens biflora*) as a function of consecutively visited flowers in workers initiating their foraging careers. From Heinrich (1979).

Slade & Vinyard, 1976). This kind of behaviour will never achieve the predicted optimum exactly because a proportion of small prey will occur close to the predator and appear largest even when density of larger items is very high.

The optimal diet model also assumes that animals can recognize the value of food items the instant they are perceived. This may often be true for animals using vision to detect food, especially where there is development of search images for specific food items. However, in other cases animals may take an appreciable time to evaluate an encountered food item before accepting or rejecting it. This may apply particularly for animals using touch or olfaction to recognize food items. Incorporation of *recognition time* into the optimal diet model modifies the inequality in equation 4.1 thus:

$$\frac{E_2}{H_2} < \lambda_1 \left[E_1 - \frac{E_2 H_1}{H_2} \right] - \frac{\lambda_2 E_2 R_2}{H_2}$$

The new (last) term allows for the time taken in recognizing food items that are subsequently rejected. The rate of encounter (λ_2) with the second food type now appears in the inequality so that *the optimal diet depends on the encounter rates with both food types if recognition times are sufficiently long*. This can be expressed intuitively by saying that sub-optimal foods should be eaten when they are so abundant that the time wasted recognizing items which should otherwise not be eaten would cause the net rate of energy gain for a pure diet to fall below that from a mixed diet. Recognition times are likely to be very short (small fractions of a second) for birds, yet Krebs (1978) thought that significant, but unmeasured, recognition times may have contributed to the persistence of great tits in taking the sub-optimal food items when optimal food items were plentiful.

Finally, the optimal diet model ignores any possible limits to the time available for foraging. There will be a trade-off between fitness gained from time invested in foraging and from time invested in other activities, so that an animal is expected to minimize the time spent foraging per unit energy gain. In itself minimizing T/E amounts to maximizing E/T so that a *time minimizer*, in this sense, is the same as an *energy maximizer* as far as optimal foraging theory is concerned. However, if searching for food has to be curtailed at some arbitrary point in time (e.g. in response to tides in the case of intertidal gastropods) or if it is advantageous to restrict individual foraging bouts to time intervals so short that not many food items can be found and handled in each (e.g. where prolonged lapses in territorial defence would run the risk of intrusion by a competitor, or where prolonged foraging bouts would expose the animal to a high cumulative risk of mortality), then the optimal diet may be broader than that appropriate to unlimited foraging time. Whether this is so will depend on how short individual foraging bouts are, relative to the mean time to encounter specific food items ($1/\lambda_i$) and to the handling times involved. Similarly, if the cumulative probability of encountering an intrinsically highly valuable food item becomes very small as the time available for searching dwindles towards the end of a foraging bout, then the average energy gain during the time remaining may be increased by eating all encountered items that can be handled in time, even if they would not be worth eating if time available for foraging were unlimited. Such a change in diet selection would require the animal to 'know' how much time it had left for foraging during a particular bout. Menge (1974) noted that as foraging proceeded during a low-tide period, the Californian dogwhelk (*Acanthina punctulata*) became less selective and ate more barnacles (*Balanus* and *Chthamalus* spp.) than the more valuable periwinkles (*Littorina* spp.) as compared to earlier in the foraging bout (Fig. 4.7). Periwinkles were much less abundant than barnacles and although feeding exclusively on periwinkles would maximize E/T if search time were unlimited, the probability of finding one during the last half of the foraging bout was too low for it to be worthwhile ignoring barnacles. In some

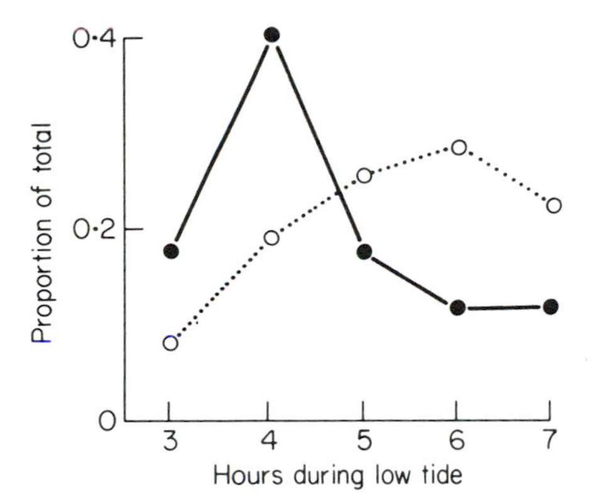

Fig. 4.7 Natural foraging behaviour of the Californian dogwhelk (*Acanthina punctulata*) during low tide. The proportion of first encounters with prey which are periwinkles (—) or barnacles (---) are plotted as a function of time during the low tide period. From Menge (1974).

cases, circumstances might even arise where it would be advantageous to ignore intrinsically valuable food items which have been encountered. This might happen if the handling time were very long, so that if such an item were attacked near the end of a foraging bout of optimal length, either the food item would have to be abandoned before handling was complete, perhaps reducing its value to zero, or the foraging bout would have to be extended, possibly exposing the animal to a high risk of mortality. In general, constraints on foraging time will tend to widen the optimal diet.

4.3.3 MICROPHAGES—FILTER FEEDERS AND DEPOSIT FEEDERS

Phytoplankton, small zooplankton, suspended detritus and bacteria-coated sediment particles comprise an abundant, almost ubiquitous food resource in aquatic habitats. Virtually all water bodies contain their guilds of filter feeders, which utilize the suspended food particles, and deposit feeders, which ingest sediment in order to assimilate the bacteria that colonize detritus particles settled out of suspension, or grow on the film of organic matter adsorbed onto the surfaces of small mineral particles. Except for very small zooplankton, filter feeders and deposit feeders are several orders of magnitude larger than their food particles, which are therefore not processed separately as discrete food items in the sense that macrophages handle each of their food particles separately, but are processed more or less continuously, as if fed into the alimentary canal on a conveyor belt. The interception, sorting and ingestion of food particles occur simultaneously so that it is not meaningful to separate searching time from handling time. Many food particles may be processed simultaneouly so that handling times, which amount to the time spent sorting particles and transferring them from the filter to the mouth, vary little among different particles and do not greatly affect the value of particles as food. Instead, the energy value of food particles is determined by their digestibility and size. The digestibility of suspended food particles varies considerably from easily digested, naked phytoflagellates to the more refractive, cellulose-plated dinoflagellates. Sediment particles are less likely to vary in digestibility because it is the colonizing bacteria rather than the non-living organic matter which is assimilated. Living suspended food particles, such as phytoplankton, increase in their energy value as their volume increases. Non-living sediment particles increase in their energy value as their volume decreases. This is because when sediment particles are packed into the gut of a deposit feeder, the total number of bacteria carried on all the particles is determined by their total surface area. Even though the surface area of an individual particle increases with the size of the particle, the total surface area of many particles packed inside the gut increases as the surface area to volume ratio of the particles increases, i.e. as individual particle size decreases.

Can the energy maximization premise be used to predict the feeding behaviour of filter and deposit feeders given the availability and quality of food particles? This problem was tackled independently with regard to filter feeders by Lehman (1976) and Lam & Frost (1976). Although the two models had somewhat different derivations, their predictions were similar. Whereas with macrophages it is appropriate to emphasize the effects of energy yield per unit handling time of different food items and of food density on the optimal diet while assuming a constant searching speed, with filter and deposit feeders it is more appropriate to emphasize particle size and particle density on particle selection and food processing rate. Filtering rate and sediment ingestion rate are analogous to the searching rate of macrophages.

Consider a filter feeder; such as a planktonic copepod or a mussel attached to the seabed, sifting particles out of suspension. The net rate of energy gain $Q(=E/T)$ equals the rate of energy assimilation (E_a) minus the energy cost of filtering (E_f) minus the cost of rejecting unwanted particles (E_r).

$$Q = E_a - E_f - E_r$$

Let $E_a = \alpha_i D_i E_i F$, where α_i is the proportion of intercepted particles of type i which are accepted for ingestion, D_i is the density of particle type i in the water,

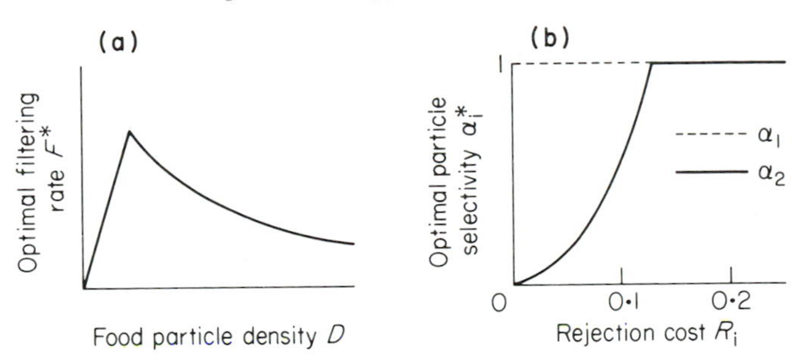

Fig. 4.8 Filter feeders. (a) Optimal filtering rate plotted as a function of food particle density. (b) Optimal particle selectivity plotted as a function of rejection cost. From Lehman (1976).

E_i is the amount of energy extractable from particle type i, F is the filtering rate. Let $E_f = bF^2$, where b is a coefficient of drag. This cost function assumes that the energy expended on the filtering process is proportional to the amount of drag exerted on the moving part of the filter. Lehman assumed that the flow of water over the moving parts would be viscous so that the amount of drag would be proportional to F^2, whereas Lam & Frost assumed turbulent flow with drag proportional to F^3. Predictions remain essentially unaltered by these alternative assumptions.

Let $E_r = (1 - \alpha_i)D_i R_i F$, where $1 - \alpha_i$ is the proportion of encountered particle type i rejected, D_i is the density of particle type i in the water, R_i is the energy cost of rejecting particle type i, F is the filtering rate.

$$Q = \alpha_i D_i E_i F - bF^2 - (1 - \alpha_i) D_i E_i F$$

The problem is to choose values of F and α_i which maximize Q given the densities and energy values of the suspended particles. This is done by setting dQ/dF to zero and solving the resulting equation by iteration on a computer.

Filtering rates (F^*) which maximize Q are plotted as a function of particle density (D_i) in Fig. 4.8(a), where it is assumed for simplicity that there is only one particle type present so that $\alpha_i = 1$. The optimal filtering rate increases rapidly to a maximum as particle density increases from very low values. This maximum corresponds to the particle density at which the gut first becomes fully packed. The decline in optimal filtering rate with further increases in particle density occurs because the gut can be fully packed using progressively lower filtering rates. Once the gut becomes full, the rate of energy intake is limited by the speed of passage through the gut and the rate of digestion. Filtering rates in excess of the gut clearance rate would merely cause a bottleneck and incur filtering costs with no further gain in assimilated energy. Or, if the gut clearance rate were linked to the filtering rate, higher filtering rates would push particles through the gut too fast to be adequately digested.

Particle selectivities (α_i*) which maximize Q at the optimal filtering rate (F*) are plotted as a function of the rejection cost (R_i) in Fig. 4.8(b), where the particle densities (D_i) are held constant. Just two particle types are considered, where type 1 has the greater energy value, which in this case is determined by digestibility rather than size. The optimal diet requires that the more valuable particle should always be accepted ($\alpha_i = 1$). This corresponds to the similar prediction for macrophages. When rejection costs (R_i) are zero, the optimal acceptance rate for the less valuable particle is found to depend solely on the abundance of the more valuable particles, the optimal value of α_2 declining to zero as the abundance of particle type 1 increases. This prediction [not depicted in Fig. 4.8(b)] is similar to the equivalent one for macrophages, except that the optimal value of α_2 for filter feeders may lie between zero and one whereas in the macrophage model it must be either zero or one. Significant particle rejection times are analogous to recognition times in the macrophage model and they have a similar effect on the optimal diet by causing the abundance of both particle types to be important in particle selection. The optimal value of α_2 increases very quickly with rising rejection costs [Fig. 4.8(b)] to the point where selection is no longer profitable ($\alpha_2^* = 1$). With very high rejection costs, even non-digestible particles should be accepted and passed through the gut along with the other particles. Such a situation could not happen with macrophages, since items are eaten individually and a non-digestible item (zero energy value) could never be included in the optimal diet. This difference between the filter feeder and macrophage models is due to the different emphasis placed on searching rate (filtering rate) and handling time, respectively. Otherwise, decreased selectivity with increased selection costs (rejection costs or recognition times) is an analogous prediction in the two models.

With appropriate adjustments to the assumptions, Lehman's filter-feeder model can be applied directly to deposit feeders (Taghon, Self & Jumars, 1978). The model explores the effects of sediment processing rate, digestion rate and particle selectivity on the net rate of energy gain and comes up with predictions similar to the filter-feeding case. Controlling and measuring the availability, digestibility, size and selective uptake of tiny food particles is technically quite difficult, especially when working with sediment. As yet there is little published data adequate for testing the predictions of the filter- and deposit-feeder models, but these predictions pave the way for new experiments, so that critical data can be expected to appear in the future. An encouraging start has been given by Doyle (1979) who found that the amphipod *Corophium volutator* ingested artificial sediments of higher nutritional status more rapidly than less nutritional sediments, as predicted by the energy maximization premise.

4.4 CONCLUSIONS

A foraging animal may need to make decisions at several levels, from the choice of foraging location, through choice of searching method (not considered here but see Krebs, 1978; Pyke, Pulliam & Charnov, 1977; Norberg, 1977) to choice of diet. Not all animals will be faced with all these choices. Many sedentary filter feeders have no choice of foraging location, browsing gastropods have rather stereotyped searching behaviours, and monophagous animals have little choice of diet. However, all animals will face at least one of these choices.

Predictions derived from the energy maximization premise should be regarded as null hypotheses against which to judge the foraging performance of animals. The results presented in this chapter lend support to the models so we can conclude that there is validity in the basic assumption that *the rate of energy gain from food has been a significant limiting factor and has constituted an important selection pressure in the evolution of foraging behaviour.* Of course, it is abundantly clear that efficiency in obtaining food has been selected during evolution as evidenced by the amazing diversity of food-sensing systems, feeding morphologies and behavioural responses which exist. Our intuition is that in most situations where optimal foraging is approximated it will be due to quite simple 'rules of thumb', and in some cases to apparently trivial ones, for example select larger prey, select prey that are easy to catch and handle (slow moving, weak, lacking defensive structures, etc), relax selectivity when hungry, turn more often after a capture, etc. Even with relatively unsophisticated responses such as these, the resulting behaviours may be so near to the optimum that the experimenter cannot tell the difference. They may also be the best that many animals can achieve.

As pointed out earlier, the energy maximization premise is not applicable in all circumstances, notably where essential nutrients are the limiting factors, and the assumption needs to be modified if time constraints on foraging are operating. In addition, it seems likely that sometimes the implicit assumption that environments are composed of repeatable units (MacArthur, 1972), a condition necessary for 'rational' choice, will not hold. In highly stochastic, unpredictable environments it is difficult to envisage how a selective, optimal strategy could have evolved. Finally, it may be true that many foragers exist, and have existed, in environments where prey availability is very low for most of the time. Unselective feeding will have been appropriate in such cases, and predators may not have been subject to sustained pressure selecting for an optimal strategy of the type defined by the optimal foraging models.

Chapter 5

Strategies of Digestion and Defecation

R. M. SIBLY

5.1 INTRODUCTION

In this chapter the *a priori* approach described in Chapter 1 has been applied to animal strategies of digestion. The first step is to identify a satisfactory 'phenotypic measure of fitness' (Chapter 1). For the case of continuous flow digestive systems where survival or reproductive success is critically dependent on the supply of energy 'the rate at which energy is obtained by digestion' is proposed. The next step is to calculate which digestive strategy maximizes the rate at which energy is obtained, and this yields three predictions (section 5.2) with which what is observed in nature or experiment (sections 5.3 and 5.4) can be interpreted. It is not so easy to make unambiguous predictions about digestion in monogastric animals eating meals, and for these a less rigorous discussion (section 5.5.3) is provided.

For ease of reference, terms used in special senses in this chapter are listed here. The reader may wish to skip this section for the time being.

DIGESTIBILITY of a nutrient is:

$$\frac{\text{quantity ingested} - \text{quantity defecated}}{\text{quantity ingested}}$$

where the nutrient may be total food, energy, protein, an amino acid, nitrogen, potassium, etc. In the case of total food the quantities may either be dry weights or energy, and the basis for the digestibility calculation should be specified. Occasionally digestibility is referred to as DIGESTIVE EFFICIENCY when it is necessary to emphasize that this is achieved by an animal using a particular digestive strategy. In the literature digestibility is usually called APPARENT DIGESTIBILITY, the term TRUE DIGESTIBILITY being reserved for quantity absorbed/quantity ingested. Quantity absorbed may differ from (quantity ingested – quantity defecated) if a nutrient is added to the digesta from the body tissues as in the sloughing of cells from the gut lining that occurs especially during the digestion of food rich in fibre.

RETENTION TIME is the time from starting digestion that digesta are retained in the absorptive regions of the gut. Because nutrients are absorbed at different sites, retention time may vary slightly depending on which nutrient is being considered. Retention time is less than THROUGHPUT TIME, which is the time from ingestion of food to defecation of undigested residue. THROUGHPUT RATE is the rate of flow of digesta (g/s) past any specified point in the gut.

Finally, a few of the topics that will not be included: little is said in this chapter about the anatomy of the digestive tract, for which the reader is referred to Morton (1979) for an introduction and Hofmann (1973) for ruminants, or the minute details of its operation for which he/she should consult Davenport (1978, 1977) for monogastric animals, especially man, and Hungate (1966) and Church (1975) for ruminants (see also section 5.5.4). Not much is known about digestive systems which sometimes evacuate the stomach by regurgitating pellets, as in many birds (see Rhoades & Duke, 1975). Coprophagy is not referred to in this chapter, despite some excellent work on rabbits (Brandt & Thacker, 1958; Pickard & Stevens, 1972; Laplace & Lebas, 1975), and other functions of defecation than evacuation of the gut are not mentioned, although an anti-predator function is quite common (see e.g. McDougall & Milne, 1978).

5.2 HOW FOOD SHOULD BE PROCESSED IN CONTINUOUS FLOW DIGESTIVE SYSTEMS

Most prey erect defences that make them harder to digest, because it pays them (usually) to be less profitable food. Cracking these defences is the first objective of digestion, and is achieved by chewing, grinding in a gizzard, chemical or microbial attack. When these defences are breached and the food cells have been separated into their constituents, enzymes break down the bonds that formed, for example, a protein from amino acids. The amino acids are then transported across the gut wall. The process of digestion is complete when all the food has been broken down into components small enough to be absorbed across the gut wall into the blood.

In the digestion of a food item three phases can therefore be distinguished. First, time and effort are invested cracking the food's defences and breaking large molecules into small units, second these small units are rapidly absorbed across the gut wall, and third the rate of absorption declines as the gut completes the digestion of all that it is equipped to process. Arguing in this way we can draw a simple graph to represent the digestive process (Fig. 5.1). The y axis is the net amount of nutrient (e.g. energy) that has been obtained from one gram (wet weight) of food, and this is plotted as a mathematical function of retention time. Net amount obtained is defined as the amount absorbed minus the amount of nutrient invested by the animal in digesting the food. *A priori* such curves must exist for each food and for each nutrient contained therein for every individual animal, irrespective of the trade-offs and constraints that keep these curves the shape they are. Now we can ask how, given the shape of the curve, the food should be processed.

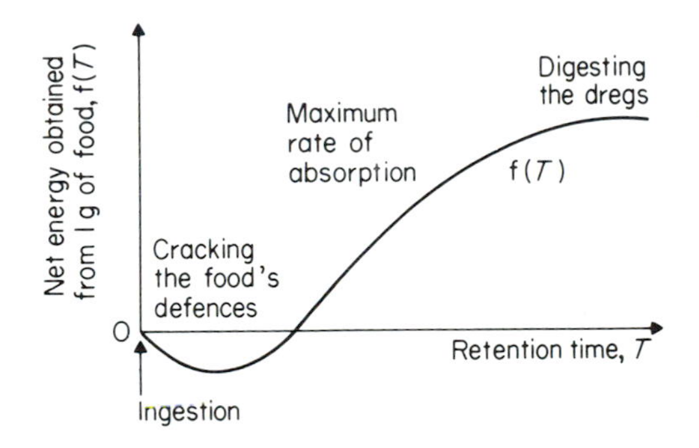

Fig. 5.1 Simple model of digestion. The *net* amount of energy obtained may decline with time while the food's defences are being cracked because energy is provided to crack them but none is absorbed. When the defences are breached energy is rapidly absorbed, but eventually rate of absorption declines when no more food can be digested. $f(T)$ has the same shape as digestibility plotted as a function of retention time.

This question can only be answered by reference to some criterion of performance, which should be a phenotypic measure of fitness. An appropriate measure in this case might be the rate at which energy is obtained. The question of how food should be processed can then be posed more strictly, as, how should food be processed if the rate of obtaining energy is to be maximised? If survival or reproductive success is critically dependent on the supply of some other nutrient, however, the question should be formulated in terms of maximising intake of the limiting nutrient.

Consider a continuous flow system in which food enters the digestive chambers at a steady rate at one end and undigested residue constantly leaves from the other. This is the case for herbivores that graze continuously, such as horses or geese eating grass. Suppose that food items enter the digestive chambers at rate r (g/s) and are retained for time T before the undigested residues are evacuated, at rate $r_{outflow}$. Then if we ignore food leaving by absorption across the gut wall, the weight of material in the digestive chambers is $r.T$.‡ Strictly speaking, $r_{inflow}.T >$ weight of digesta in the digestive chambers $> r_{outflow}.T$.

Let the net amount of energy obtained from one gram of food be a function of time, $f(T)$, as in Fig. 5.1. The amount of food entering the digestive chambers in unit time is r, each gram of which eventually yields a net amount $f(T)$ of energy, so the rate of obtaining energy is $r.f(T)$. Thus for continuous flow digestion:

$$\text{Weight carried} \approx r.T \tag{5.1}$$

$$\text{Net rate of obtaining energy} = r.f(T) \tag{5.2}$$

where r is the rate of ingesting food, T is the time it is retained, and $f(T)$ is the curve in Fig. 5.1. Substituting $r = \frac{\text{weight carried}}{T}$ in equation 5.2 we have:

$$\text{Net rate of obtaining energy} = \text{weight carried.}\frac{f(T)}{T} \tag{5.3}$$

Prediction 1: In order to maximize the rate of obtaining energy the animal should maximize T if r is fixed (from equation 5.2). Thus when rate of ingestion is limiting, digesta should be retained for longer.

T has, however, an upper limit imposed by the size of the digestive chambers. When this limit is attained, Prediction 2 applies.

Prediction 2: For a given weight carried, the optimal strategy is to maximize $\frac{f(T)}{T}$ (from equation 5.3). This holds in particular when gut capacity is limiting. Optimal retention time T^* can be found graphically as in Fig. 5.2.

Prediction 3: Animals eating poorer quality food should have larger digestive chambers, other things being equal. This is because $f(T)$ is lower for a

---

‡ If it is desired to allow for the weight of digesta lost from the gut by absorption across the gut wall, let the weight of digesta absorbed in a short time interval dt after retention for time t be $r.g(t).dt$. Then the weight of digesta carried is $r.T\left(1 - \frac{\int_0^T g(t)\,dt}{T}\right)$. In equation 5.3 $f(T)$ would be replaced by:

$$\frac{f(T)}{1 - \frac{\int_0^T g(t)\,dt}{T}}$$

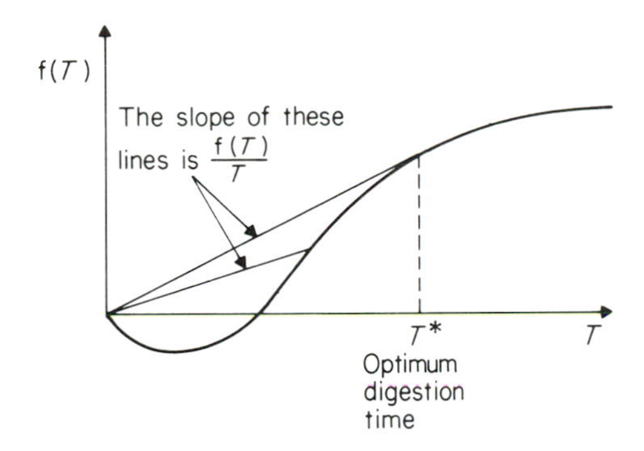

Fig. 5.2 $\frac{f(T)}{T}$ is the slope of the line from the origin to the curve $f(T)$, so $\frac{f(T)}{T}$ is maximized by the steepest line from the origin to the curve (Prediction 2).

poorer quality food, and so the maximum value of $\frac{f(T)}{T}$ is necessarily lower too (as can be seen by consideration of Fig. 5.2). But if the animals' nutritional requirements are the same, and $\frac{f(T)}{T}$ is lower, then the weight carried must be greater (from equation 5.3). We can firmly predict that animals sustaining themselves on poorer food will carry a greater weight of digesta, and should, therefore, have larger guts.

These three predictions apply to animals with continuous flow digestive systems. In summary, if we knew the shape of the curve $f(T)$ for an individual eating a particular food we could predict how long the food should be retained in the digestive chambers. We could also predict the rate of ingesting food, r, necessary to supply the animal's nutritional requirements, and the weight of gut contents $r.T$. This is a predictive theory of how food should be processed, a theory of optimal digestion.

5.2.1 The defences of plants against digestion

Since most of the digestive strategies discussed in this chapter are those of herbivores, it is worth pausing for a moment to consider the digestion of plants. The most important, easily digested nutrients in a plant cell, proteins and soluble carbohydrates, are contained in a structural box composed of cellulose cell walls which vertebrates and some invertebrates lack the enzymes to digest. If the cell contents are to be obtained for further digestion, therefore, either the walls must be torn apart mechanically or they must be broken down by symbiotic micro-organisms contained in fermentation chambers. Thus the strength and resistance to digestion of the cell wall is one of the plants most important defences against being eaten (see Chapter 6). In young, growing tissues, which have to expand and are not required to support weight, the cell wall is thin and fairly flexible, as in the leaves and fruits of grasses and other flowering plants. In older tissues, particularly those that bear weight, the cell wall becomes thickened and hardened by other compounds, notably lignin, the stiffening material found in the stems of tall grass plants and wood.

However, though the plant's digestion-reducing compounds affect the shape of the curve $f(T)$ (Fig. 5.1) and thus the profitability of a food, they do not affect the question of whether animal digestion is optimal given that $f(T)$ is the shape it is.

5.3 CONTINUOUS FLOW DIGESTION: BIRDS EATING PLANTS

In flying animals it is especially important to reduce the weight carried to a minimum. That applies to the contents of the digestive chambers, to food stored prior to digestion, and to the digestive organs themselves. We might even want to formulate another phenotypic measure of fitness based on the weight carried, from which we would expect that, other things being equal,

a bird's digestive strategy should minimize the weight carried.

Of course other things rarely are equal, and as we saw in equation 5.3, it is likely that in birds eating plants the rate of absorption of energy increases with the weight of digesta carried. In this case neither MAXIMIZING ENERGY UPTAKE nor MINIMIZING WEIGHT CARRIED is an adequate phenotypic measure of fitness, since energy uptake can be traded off against weight carried, so the optimal strategy must represent some kind of compromise. We will return to this point later.

Nevertheless we can be sure of Prediction 2 above, that is, given the weight that is actually carried, time of food retention in the digestive chambers should be such as to maximize $\frac{f(T)}{T}$. Furthermore if the quality of food deteriorates as it generally does with the onset of winter then there should be an increase in the weight of digesta carried, and in the size of the digestive chambers themselves (Prediction 3).

5.3.1 SEASONAL CHANGES IN GUT MORPHOLOGY IN BIRDS

It is one of the most striking characteristics of birds forced to eat poor quality foods that their guts change dramatically in length, by as much as 57% in the case of the goose (Table 5.1).

Examining starlings (*Sturnus vulgaris*) caught in the wild Al-Joborae (1980) found that the length of the gut changes month by month in parallel with changes in diet (Fig. 5.3). Thus guts were longer in winter when more plant and less animal material was found in the stomach contents. Seasonal changes in gut length of birds caught in the wild have also been found in great-bearded tits (Spitzer, 1972), reed buntings (Prys Jones, 1977), lesser snow geese

Table 5.1 Wild birds have longer intestines and caeca than captive birds, presumably because they eat poorer foods.

Bird	Body weight (Kg)	Intestine length (cm)			Caecum length (cm)		
		Wild	Captive	Wild/captive	Wild	Captive	Wild/captive
Barnacle geese[1] adults	2	191	122	1.57	29	18	1.61
Red grouse[2] males	0.65	99	72	1.37	144	78	1.85

[1] From Owen (1975); [2] from Moss (1972).

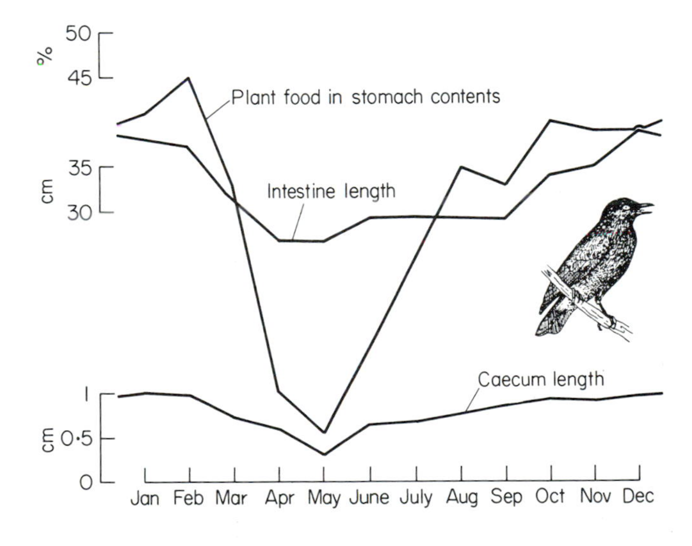

Fig. 5.3 Gut changes parallel changes in diet. Data on starlings from Al-Joborae (1980).

(Ankney, 1977), towhees (Davis, 1961) Californian quail (Levin, 1963), ptarmigan (Moss, 1974) and spruce grouse (Pendergast and Boag, 1973).

To be confident that diet quality affects gut length, we need an experimental demonstration that the guts of adults acclimatized to one diet change when the diet is changed. This has proved difficult with red grouse and geese because the birds could not be satisfactorily maintained on a poor diet in captivity (Moss, 1977; Owen, 1975) but has been possible with Japanese quail (Savory & Gentle, 1976), woodpigeon (Kenward & Sibly, 1977) and starlings (Al-Joborae, 1980) (Table 5.2).

Table 5.2 Experimental evidence that birds on a poorer diet grow longer guts. Acclimatization from a rich to a poor diet reverses that from a poor to a rich diet, so that a particular gut length is characteristic of a particular diet. Acclimatization takes about 25 days: the quail were killed progressively and change in gut size was complete in 3–4 weeks. The woodpigeon were killed after 25 days and the starlings after 14 days.

Bird	Body weight (g)	Poor diet	Rich diet	Intestine length (cm)			Caecum length (cm)		
				Poor	Rich	Poor/Rich	Poor	Rich	Poor/Rich
Quail[1]	125	Artificial	Artificial	51	46	1.1	17.0	14.5	1.2
Woodpigeon[2]	430	Brassica	Grain	220	157	1.4	—	—	—
Starlings[3]	75	Plant	Animal	33	27	1.2	0.8	0.6	1.3

[1] From Savory & Gentle (1976); [2] from Kenward & Sibly (1978); [3] from Al-Joborae (1980).

These results confirm our prediction that a reduction in the quality of a plentiful diet increases the size of the digestive chambers of the birds. The suggestion is that birds increase the rate of obtaining energy but have to pay the penalty of carrying more weight (section 5.3.3). We next consider if there is a limiting rate of obtaining energy that birds do not exceed.

5.3.2 Digestive bottleneck in woodpigeons

To tackle this problem Kenward & Sibly (1977, 1978) maintained 430-g captive woodpigeon (*Columba palumbus*) on a plentiful supply of the cabbage-like tops of Brussels sprout plants, which is a poor food for wild birds. Woodpigeon ingest the leaves rapidly and could obtain their daily requirements in less than 3 h if they could process food fast enough. Theoretically they could store it for digestion later, but in practice the large amounts involved (around 300 g) greatly exceed crop capacity (about 70 g) so this option is not open to woodpigeon. When the time for which food was available was progressively reduced in the laboratory from $10\frac{1}{2}$h/day to $3\frac{1}{2}$h/day the birds did not eat continuously, presumably because a bottleneck in digestion prevented them from processing food fast enough, and so they ate less (140 g/day) and lost weight (8 g/day), even though they increased the amount of food stored in the crop at dusk to the extent that weak individuals were no longer able to fly to roost. The pattern of feeding in the middle of the day (about 20 min/h) did not change, nor did digestive efficiency or throughput rate. We concluded that in this case the gut had probably reached the limit of its digestive capability.

5.3.3 Costs and benefits of longer guts

The obvious day-to-day disadvantages of a longer gut are that it has to be carried around by its owner, and it has to be maintained. Moreover, if it is full to capacity its contents must also be carried. The consequent risks must vary from situation to situation, but it seems very likely that an animal carrying more food takes greater risks because it is slower. Evidence on such a point is hard to obtain, so it is of some interest that hawk attacks on woodpigeon at brassica sites are more successful in the hour before sunset, which is when the pigeons are filling their crops (Kenward, 1978b; Kenward & Sibly, 1978).

Longer, more capacious guts allow a bird to carry a greater weight of digesta, roughly equal to $r.T$, where r is the rate of inflow to the digestive chambers and T is the time that digesta are retained therein (equation 5.1). The advantage of increasing $r.T$ is that the net rate of obtaining energy, $r.\mathrm{f}(T)$ (equation 5.2) can then be increased too. In principle that could either be achieved by increasing r or by increasing T, and it would be interesting to know which actually occurs.

It is difficult to see how one could demonstrate in the field that birds with longer guts can extract more energy/day than birds with shorter guts eating the same food. However, it has been found in two laboratory studies that birds with longer guts process more food per day than birds with shorter guts, while the digestive efficiencies achieved are the same (Savory & Gentle, 1976; Al-Joborae, 1980). Some birds were maintained on a poor diet and others on a rich diet, and it was inferred from other experiments (Table 5.2) that their guts had acclimatized to lengths characteristic of each diet. When given a poor food, 'long guts' and 'short guts' processed it with the same digestive efficiency, although 'long guts' processed more. Thus r increased, but f did not change.

Al-Joborae also measured throughput time (time from ingestion of food item to defecation of the undigested residue) and arrived at an unexpected result. Throughput time was shorter for the 'long gut' birds. The author's explanation for this result is that individual digesta were retained in the small intestine for the same length of time in 'long guts' as in 'short guts' but because throughput rate was higher in 'long gut' birds digesta passed through the other parts of the digestive tract faster. However it is possible that birds with long guts have shorter retention times and that digestion is somehow speeded up, though it is difficult to see how this could be achieved. If segmentation, the mechanical process of splitting up digesta (Davenport, 1977, 1978), occurs faster in the longer gut, then why does it not occur faster in the shorter gut, where it would be even more beneficial?

The advantage of longer guts to starlings is that more food is processed each day, without any change in its digestibility. This is an attractively simple result, and discussion of the complications encountered by other workers will be postponed until later.

By comparison, the advantage of a crop is easy to assess. Food can be stored for digestion in periods when foraging would be unusually hazardous; for example, the crop can be filled towards dusk and digested overnight. The amounts involved can be large, an extreme case being that of the black grouse (*Tetrao tetrix*) in Russia in winter which collects its daily intake in a mere half-hour and is thereby able to spend the rest of the day in its snow-hole (Potapov & Andreev, 1973). Woodpigeon weighing 500 g may accumulate as much as 70 g in the crop at dusk, increasing daily intake by over 30% (Kenward & Sibly, 1977), and white-fronted geese weighing 2.3 kg accumulate as much as 80 g, increasing intake by over 10% (Owen, 1972). Minor complications arise in converting such figures into energetic advantage: the food in the crop may be of lower quality owing to the haste with which it was acquired, but because it is digested overnight and may be retained longer in the digestive tract, more energy may be extracted from it.

The crop may also be used as a temporary food store during the day. When the ability to extract energy from a plentiful supply of poor food is the critical factor affecting an animal's survival, as in the case of woodpigeon eating brassica (above), then we expect the store of food will be replenished before it empties completely so that the gut always operates at maximum capacity. In this case the interval until feeding recommences should depend on the amount stored during a feeding bout. Evidence that these variables are well correlated is presented by Kenward & Sibly (1978).

5.3.4 Geese eating grass

The digestive strategy of geese eating grass is to process enormous quantities without extracting much energy from each gram processed. Food passes through the gut quickly in the order in which it was ingested, and fibre is hardly digested at all. These characteristics greatly simplify the study of digestion in geese.

The amount of grass eaten by geese can be estimated either by multiplying peck rate by peck size, or by working back from faeces production/day to food ingested/day using an estimate of the digestibility of the food. Owen (1972) prefers the second method, and proceeds as follows. White-fronted geese defecate every $3\frac{1}{2}$ min while grazing, producing about 150 droppings in an average grazing day of $8\frac{1}{2}$ h. The weight of a single dropping is 0.735 g dry wt (standard deviation 0.16 g), so faeces production/day is 110 g dry wt. Since the digestibility of grass in winter is about 25 % (Owen, pers. comm. 1979).

$$\text{grass intake} = \frac{\text{faeces production}}{(1 - \text{digestibility})} = \frac{110}{1 - 0.25} \approx 145\,\text{g dry wt}$$

To this we must add the amount of food accumulated in the crop for digestion at the roost overnight, say 15 g dry wt, so food intake/day ≈ 160 g dry wt. Assuming that 24 % grass is dry matter, wet grass intake/day ≈ 670 g.

There is little mixing of grass ingested at different times, at least in wild geese grazing under natural conditions (Owen, 1975). This was inferred by Owen from an examination of the contents of various parts of the guts of geese shot on the feeding grounds. In many cases he found major changes in content between different parts of the gut which presumably reflected the sequence of ingestion in the period before the bird was shot. The changes were not consistent between birds and could not have been caused by differential digestion of different foods. For example, in the guts of a barnacle goose which had been feeding on grass and clover stolons he found: oesophagus, 50 % stolon, gizzard 95 %, first 50 cm of small intestine 96 %, second 50 cm 74 %, third 50 cm 82 % and rectum 35 % stolon. The conclusion was consistent with observations of feeding birds in the field; they alternated their feeding method between grazing and digging for stolons, probably in relation to their success rate in obtaining the latter.

Knowing that there is little mixing of grass in the gut makes it easier to calculate throughput time, since all constituents of the food must be travelling through the gut at the same rate. The throughput time of grass in captive geese is $1\text{–}1\frac{1}{2}$ h (Owen, 1975, on the basis of trials with four species). Owen estimated the throughput time of wild barnacle geese by dividing the gut contents of full birds by the defecation rate. The guts (gizzard and intestine) contained 13.5 g dry wt. (standard deviation 2.8 g) and the defecation rate was 13.48 g/h, giving a throughput time of 1 h. Actually

$$\frac{\text{Gut capacity}}{\text{Defecation rate}} > \text{Throughput time} > \frac{\text{Gut capacity}}{\text{Ingestion rate}}$$

if secretions into the gut are negligible, so 1 h > Throughput time > 45 min.

That fibre is not digested by geese is inferred from a study by Mattocks (1971) in which he was unable to demonstrate any capacity for cellulose digestion in domestic geese, and a study by Marriott & Forbes (1970) specifically designed to mimic geese feeding on pasture. The birds used were captive Cape Barren geese (*Cereopsis novaehollandiae*), an exclusively grazing species in the wild, maintained on a diet of lucerne chaff. When acclimatized to the diet they had low throughput times (1.3 h) and an overall digestive efficiency of 25.8 % (dry matter basis). The digestibilities of the constituents of their food is shown in Table 5.3.

Table 5.3 Composition and digestibility of artificial diet of captive geese. 'Soluble carbohydrate' was assumed to be the residue when the other components had been measured. Note that 'fibre' could be used as a *reference substance*, that is to say, a non-digestible component of the diet with which one can compare the relative amount in diet and faeces of any other component in order to estimate its digestibility. (From Marriott & Forbes, 1970.)

	Soluble carbohydrates	Protein	Fat	Fibre	Ash
Diet (dry wt) (%)	36	19	3	34	8
Digestibility (%)	57	76	26	1	—

Owen considers it likely that digestibility is related to plant characteristics (water and fibre content) and so may increase, for example, in the spring when the grass is growing (Owen, 1976; Owen, Nugent & Davies, 1977). Because geese separate the surfaces of grass leaves during digestion but do not digest the leaf surfaces themselves, Owen used the percentage of leaf fragments that had one but not both leaf surfaces intact as an index to show that the extent of mechanical breakdown of grass leaves varies at different times by as much as 15 %. Owen (1976) considered that these variations were due to changes in the characteristics of the leaves, perhaps age or nutritive value, and not to changes in the digestive capabilities of the birds.

Since grass goes through geese guts fast we might expect geese (other things being equal) to feed as fast as possible at the beginning of the day until their guts are full, then to feed at some lower rate (possibly corresponding to maximum digestion rate if this is limiting as in woodpigeons) for most of the day, reverting to maximum ingestion rate towards dusk in order to accumulate a supply of food for digestion overnight. The actual pattern of feeding has peaks of intake at the beginning and end of the day as predicted (Fig. 5.4) but reaches the dusk peak gradually

5.3.5 RED GROUSE EATING HEATHER

Interest in the digestive strategies of plant-eating birds was stimulated by a comparative study of North American galliforms by Leopold (1953) in which

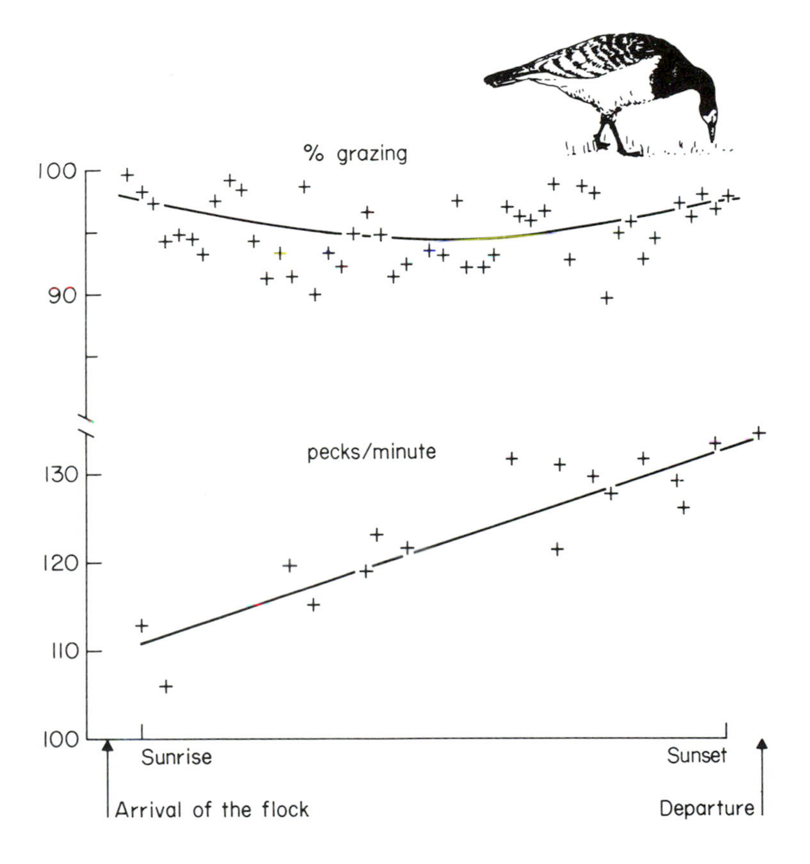

Fig. 5.4 The behaviour of white-fronted geese feeding on grass in mild weather in January in England ($8\frac{1}{2}$ h daylight). (Adapted from Owen, 1972.)

he compared the intestinal morphology of 'browsing' and 'seed-eating' species. By far the greatest differences were in the lengths of caeca (Fig. 5.5).

The digestive strategy of the grouse is very different from that of geese, woodpigeon and starlings. Instead of rapid passage of food through a relatively small gut the grouse retain digesta for long periods in a gut of much greater capacity (Table 5.1), the caeca in particular being much larger. The digestive process is more complex since digesta do not pass through the gut simply in the order in which they were ingested. Instead digesta are separated at the end of the small intestine, one portion entering the caeca through narrow constrictions as a creamy-brown pulp, and the rest continuing to be excreted as woody droppings (Moss & Parkinson, 1972). Moss and Parkinson considered it likely that some if not all of the digestion of cellulose occurs in the caeca (see also Gasaway, 1976; Moss, 1977). The caecal contents are excreted separately as soft glutinous masses (which also occur in geese, though relatively infrequently).

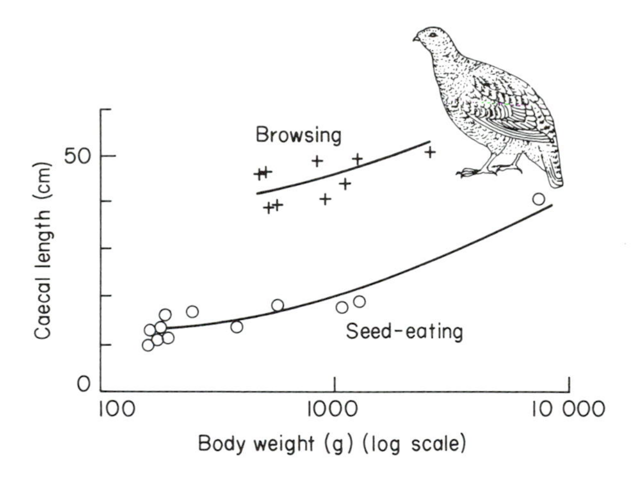

Fig. 5.5 Browsing species have longer caeca than seed-eating species. Comparison of North American galliforms redrawn from Leopold (1953).

Red grouse (*Lagopus lagopus scoticus*) feed largely on heather (*Calluna vulgaris*) throughout the year (Jenkins, Watson & Miller, 1963; Savory, 1978). Heather is a poor food (Table 5.4) and not surprisingly the birds eat large amounts. Daily intake can be readily estimated from peck rate and peck size because grouse, unlike geese, feed mainly from one plant. Wild birds weighing around 600 g eat about 60 g dry wt of heather per day.

Table 5.4 Composition and digestibility of heather diet of red grouse. Heather samples were picked where grouse were seen feeding and corrected according to the known selection exercised by the birds. (From Moss, 1977.)

	Soluble carbohydrates	Nitrogen	Fat	Fibre	
				Cellulose	Lignin
Diet (dry wt) (%)	17	1	8	39	27
Digestibility (dry wt) (%)	87	43	37	38	44

Digestibility of the various components of heather is estimated by using one of the components as a *reference substance* (Moss & Parkinson, 1972). Initially lignin, the woody tissue in plants, was considered, but was found to be digested (see below). In principle any mineral element which is egested/excreted at the same rate that it is ingested could be used as a reference substance for digestibility studies in birds because they excrete their solid nitrogeneous waste along with the faeces, unlike mammals which lose a proportion of each element in their liquid urine. Eventually magnesium was chosen as a reference substance because it is poorly absorbed by herbivores (Moss & Parkinson, 1972) so that even if some was retained, the effect on digestibility estimates would be small.

One of the most surprising results to come out of this line of work was that the digestive efficiency of wild grouse is much higher than that of captive grouse (46 versus 27%, Moss, 1977). Although the reasons for this are not entirely clear, the reader will recall (Table 5.1) that captive grouse have much smaller caeca and intestines than wild birds. It is also worth noting that captive grouse have never been successfully maintained on the diet of heather alone which sustains wild grouse. The efficiency with which wild grouse digest the various components of heather is shown in Table 5.4. Note particularly their astonishing success in digesting lignin and cellulose. Indeed in the digestion of heather wild red grouse are more efficient than captive sheep and red deer (Moss, 1977).

5.4. CONTINUOUS FLOW DIGESTION, MAMMALS EATING PLANTS

5.4.1 East African ruminants

Plant-eating animals in East Africa have a wide range of options as to what they eat and how they digest it. They can achieve high rates of intake and pay a penalty in digestion, or they can be more selective in their foraging and obtain food that is easier to digest. Roughly speaking, monocotyledonous plants (MONOCOTS) are less digestible than dicotyledonous plants (DICOTS). Thus coarse monocot grass forms a poor diet, having tough, fibrous cell walls, but dicot leaves and fruits and seeds form quite a good diet, containing less cell wall and more protein, N-free extracts and water (Hoppe, 1977).

Table 5.5 lists 11 species of East African ruminants arranged in order of body weight together with the proportions of different foods found in their reticulo-rumens. Hoppe suggests that since the dicot diet is richer in soluble components and has less cell wall, the dicot-eaters achieve high absorption rates by processing the food quickly, thus keeping the rumen weight down. Food-retaining structures are little developed in the rumens of dicot selectors (Hofmann, 1973). The monocot eaters, however, employ a different digestive strategy. Their diet is poor in rapidly fermentable components but rich in cellulose and lignin. These are digested, but the process takes several days.

Table 5.5 The reticulo-rumen contents and weights of 11 East African ruminants (see text for details). Heavier reticulo-rumens are associated with poorer, monocot diets. (From Hoppe, 1977.)

	Monocots*	Dicots*	Crude protein†	Body weight (kg)	Reticulo-rumen contents/ body weight (%)	Reticulo-rumen contents/ metabolic weight‡ (%)
Zebu cattle	93	5	3	200	13	50
Wildebeest	94	2	4	200	14	53
Coke's hartebeest	96	1	4	120	11	38
Topi	96	2	6	114	12	38
Haired sheep	92	5	6	25	20	46
Thomson's gazelle	79	20	9	18	12	25
Massai goat	72	25	7	24	20	45
Impala	64	25	7	51	7	19
Grant's gazelle	24	68	7	49	9	23
Bushbuck	10	90	—	27	7	16
Grey duiker	0	100	—	13	10	19
Steenbok	6	94	—	10	7	14
Kirk's dikdik	0	100	17	4	7	11
Suni	6	94	—	4	8	12

* % of total plant parts identified.
† % of dry matter in reticulo-rumen contents.
‡ Metabolic wt = (body weight)$^{0.75}$ is used because metabolic rate is proportional to (body weight)$^{0.75}$. This column indicates the weight of reticulo-rumen contents necessary to supply one watt of power.

Since monocot is a poor diet that can only be digested slowly whereas dicot is richer and can be digested much faster, it would seem that, other things being equal, any animal should prefer a dicot diet. Why do the larger animals (Table 5.5) not eat more dicots? It may be that it simply is not possible for them to obtain dicots fast enough (Bell, 1971). Dicot ingestion rate might be limited by the density of dicots on the ground. Thus larger animals would be unable to eat dicots fast enough to supply their higher metabolic requirements and thus, are forced into slower digestion of a poorer diet.

Certainly the monocot-digesting strategy requires a much larger digestive apparatus to supply energy at a given rate. Since digestion is slower, food must be carried for longer to obtain a given amount of energy from it (T is larger in Equation 5.1) and so the total weight of digesta carried increases. This is dramatically illustrated in the last column of Table 5.5. More poorer food must be carried in the reticulo-rumen to supply the same amount of power to the body.

The same conclusion seems to hold for ruminants in general (Kay, Engelhardt & White, 1980): the reticulo-rumen contents of 10 species of grazers or roughage eaters were on average 13% of bodyweight compared with 9% for 9 species of browsers or concentrate selectors. Within each feeding category, reticulo-rumen contents expressed as a percentage of bodyweight do not vary with bodyweight, which has the interesting consequence that expressed as a percentage of metabolic weight reticulo-rumen contents correlate with bodyweight. This suggests that for each watt of power supplied to the body, the heavier animals carry more rumen contents.

The ruminant strategy is not however the only method ungulates use for digesting high-fibre diets.

5.4.2 Horses versus ruminants

The digestive strategies of mammalian herbivores differ in the site of the fermentation chamber in which are maintained the micro-organisms (bacteria and protozoa) that digest cellulose. In the ARTIODACTYLA (ruminants) fermentation occurs before the food reaches the stomach in the reticulo-rumen, and very high efficiencies of protein utilization are achieved. Dietary protein is largely fermented to ammonia, which is either used directly as a protein source by the symbiotic bacteria or is absorbed through the rumen wall and sent to the liver. Here the ammonia is converted to urea, most of which is returned to the rumen and utilized by symbiotic bacteria. The ingested nitrogen finally becomes available to the animal when it digests the bacteria.

In the PERISSODACTYLA (horses, tapirs and rhinos), on the other hand, fermentation occurs after the stomach, and although the actual process of fermentation appears to be identical to that in artiodactyls it is probable that nitrogen that has escaped absorption in the small intestine cannot thereafter be recovered (Robinson & Slade, 1974). The strategy adopted by the horses is to shorten throughput time at the expense of a reduced efficiency of cellulose digestion (Crampton & Harris, 1969, their Table 20-1). Digesta are processed in 48 h in the horse compared with 70–90 h in the cow, but cellulose digestion is only 70% as efficient (Janis, 1976).

Thus when intake is limited large ruminants outcompete horses, since they extract protein more efficiently from a given quantity of food. However when higher intakes are possible, horses outcompete ruminants by processing far more a little less efficiently (Bell, 1971). The reason why ruminants are unable to do the same is not entirely clear, but evidence that they cannot is discussed below.

5.4.3 Evolutionary history of ruminants and horses

The evolutionary history of the artiodactyls and perissodactyls has been recounted by Janis (1976) on the basis of fossil evidence, including molar morphology, and ecological reasoning as above. The perissodactyls adopted a diet containing cellulose in the Late Palaeocene (Fig. 5.6) whereas even in the Eocene the proto-ruminants had a diet largely free of cellulose. In the Oligocene as the climate became cooler and more arid and growth more seasonal both groups increased in body size and proto-ruminants developed rumination via nitrogen cycling, which improved the use made of protein. The Equidae (horses) had by then adopted a strategy of feeding on the most fibrous of the digestible plants, and so were better able to compete during the Oligocene artiodactyl radiation than were the other perissodactyls.

5.4.4 Digestive bottleneck in large ruminants

The performance of domestic ruminants eating poor but plentiful artificial diets has been studied extensively in sheep, growing cattle and milk cows (see Campling, 1970 and Baile & Forbes, 1974 for reviews on the subject), and an example is given in Fig. 5.7. On the right of Fig. 5.7 it can be seen that intake is relatively low when the food is rich in energy (and low in fibre). As the food deteriorates, i.e. energy content decreases, rate of intake must increase if energy supply to the body is to be maintained (by Equation 5.2). Below an energy

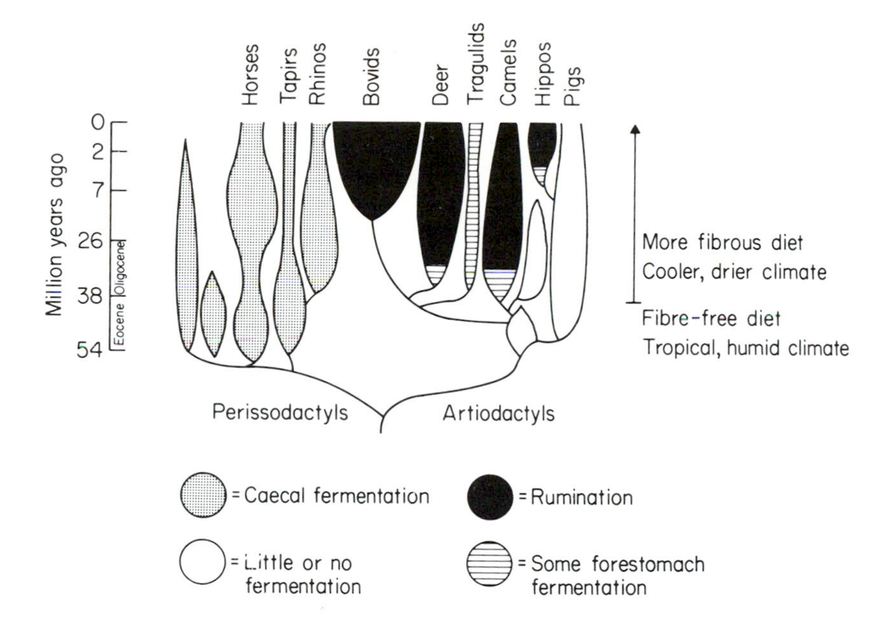

Fig. 5.6 During evolutionary history the perissodactyls, employing fermentation below the stomach, adopted a fibrous diet before the artiodactyls. When the latter evolved fermentation above the stomach to cope with the more fibrous foods of the Oligocene, only the horses competed effectively, eating the most fibrous foods as they had before. (After Janis, 1976.)

content of 590 J/g, however, the rumen is operating at maximum capacity but is no longer able to maintain the energy supply. Whether rate of intake should increase or decrease as food quality deteriorates further depends on the shape of the curve $f(T)$ (Fig. 5.1), a decrease in rate of intake is predicted if optimal retention time increases (by equation 5.1; optimal retention time maximizes $\frac{f(T)}{T}$, see Prediction 2). We cannot evaluate the optimality or otherwise of the observed behaviour without knowing the shapes of the $f(T)$ curves for the different foods, but we can be fairly confident that there is a bottleneck in the digestion of the poorer foods.

The suggestion is that large ruminants faced with high-fibre diets operate their rumens at maximum capacity and process digesta at an optimal rate. Since food cannot be stored we would expect that meals would exactly fill the space made available by recent digestion. Thus meal duration should be correlated with the interval since the last meal. This is true in cows fed on hay (Metz, 1975), and in sheep fed on a high-fibre diet, though not in sheep fed on a

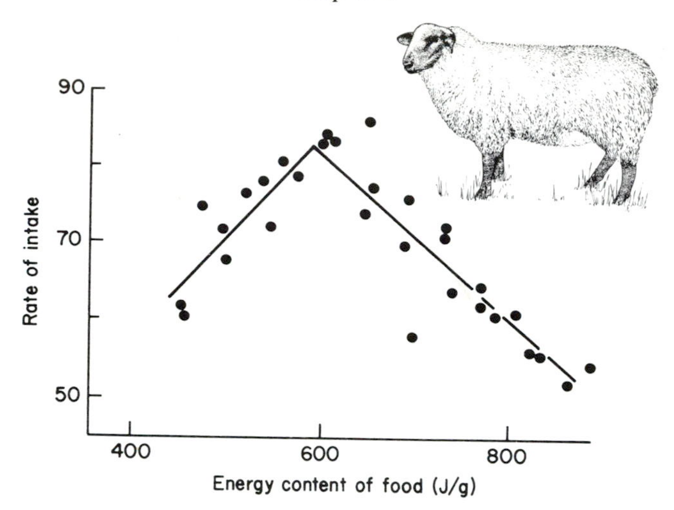

Fig. 5.7 A bottleneck in digestion prevents adult sheep increasing intake to compensate for deterioration of diet when its energy content falls below 590 J/g. Then intake decreases as food deteriorates—but this may be an optimal strategy (see text). Rate of intake measured as dry matter eaten/day and corrected for variation in body weight by dividing by (body weight)$^{0.75}$. (After Dinius & Baumgardt, 1970). Precisely analogous results have been obtained for fowl (Sibbald, Slinger & Ashton, 1960).

more digestible diet (Baile, 1975; Forbes, 1978), as we would expect.

Note that eating to fill the space made available by recent digestion is the reverse of the woodpigeon feeding strategy (section 5.3.3): in cows there is no 'food store', so the digestive chamber is being continually 'topped up' by feeding, but woodpigeon have a food store which can be replenished when empty so that the gut is always kept full.

5.5 THE DIGESTIVE PROCESS

In the preceding sections we have considered the adaptiveness of continuous flow digestion, and because the process operates at a steady rate we have not had to consider changes with time. In studying digestive systems that function intermittently, however, we have necessarily to describe the dynamic operation of the system. This adds to the complexity of the description.

5.5.1 CONTINUOUS FLOW DIGESTION THROUGH ONE COMPARTMENT

For the sake of completeness, we shall begin with the simplest possible case in which digesta flow through one digestive compartment, without any mixing of the food supplied and without selective retention of less digestible particles, as in geese eating grass (section 5.3.4). The remnants of digesta must then be defecated at exactly the rate at which they were ingested, and throughput rate past every part of the system must be the same. If such animals were given marked particles there would be a lag before any were found, and the lag time would approximately equal gut capacity/throughput rate. Thereafter the marked particles would be defecated at the rate at which they were eaten, for exactly the same length of time as that taken to eat them (Fig. 5.8).

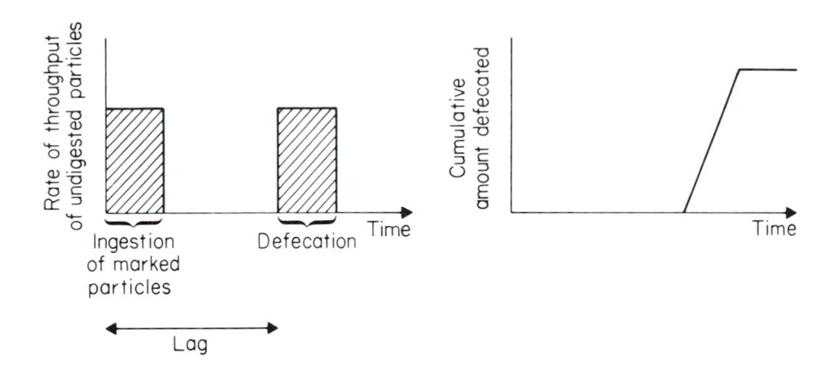

Fig. 5.8 Predicted appearance of undigested particles from a simple one-compartment system where eating is continuous. In curve-fitting to describe data it is preferable to use rate of defecation (left-hand graph), but conventional to use cumulative amount defecated (right hand graph).

5.5.2 Discontinuous flow through one compartment

Slightly more complicated is a one-compartment digestive system operating discontinuously as in monogastric animals eating meals. In such cases the evacuation of the stomach, or the cumulative appearance of markers in the faeces, has almost always been described by a negative exponential curve (Fig. 5.9). The only counter-examples employ a quadratic curve to describe data from man (Hopkins, 1966), rat (Booth, 1978a) and plaice (Jobling & Davies, 1979) (see also six cases referred to in Elliott & Persson, 1978). Diurnal animals which digest continuously during the day may experience a discontinuity at dusk, for example in captive woodpigeon which ate brassica during the day, defecation rate decreased exponentially with time since dusk (unpubl. obs.). There is, however, no change in digestive efficiency at night in white-tailed or rock ptarmigan, probably because they are able to store more at dusk in their larger crops (R. Moss, pers. comm., 1980).

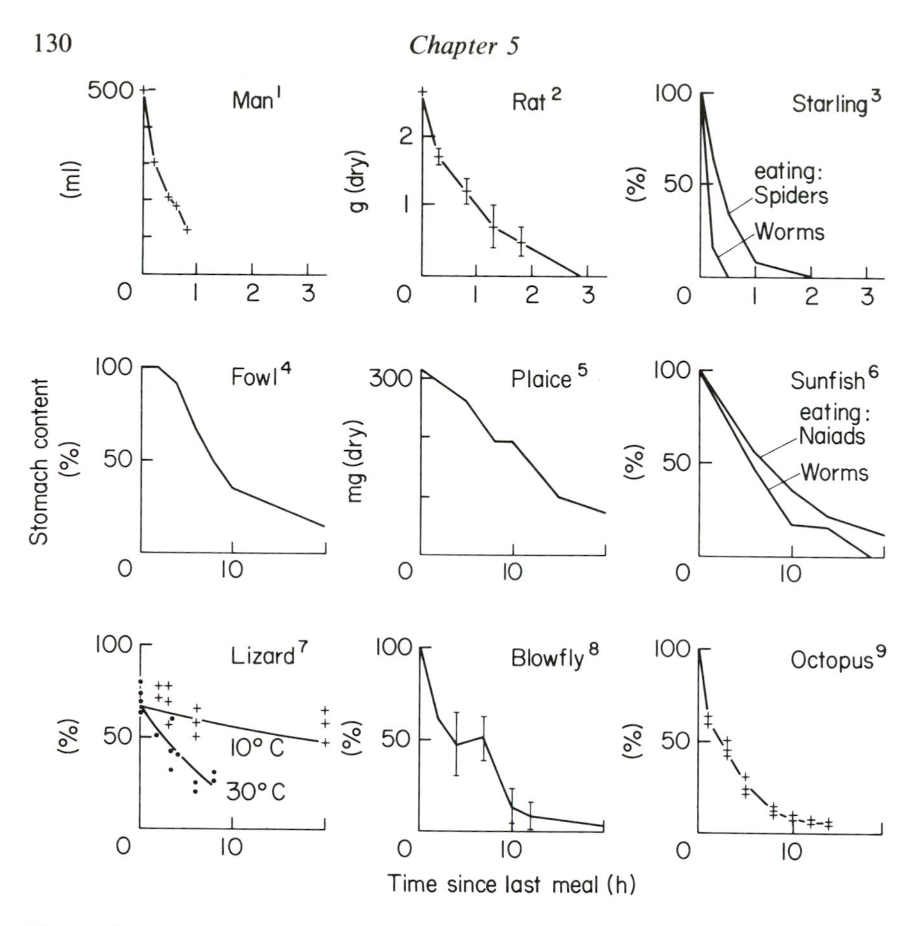

Fig. 5.9 Stomach contents as a function of time in monogastic animals fed meals. 1 Man eating a meal containing 40 % fat, 15 % protein and 45 % carbohydrate. After (Hunt & Knox 1968). 2 Laboratory rats eating chow. (After Booth 1978a). 3 *Sturnus vulgaris* eating Lycosidae and Lumbricidae. (After Coleman 1974). 4 Domestic fowl eating oats. % Markers still inside bird. (After Sibbald 1979). 5 *Pleuronectes platessa* eating minced fish. (After Jobling & Davies, 1979). 6 *Lepomis macrochirus* eating dragonfly naiads and oligochaetes. (After Windell 1967). 7 *Lacerta vivipara* eating mealworms. (After Avery, 1973.) Stomach evacuation is much slower in carnivorous crocodiles, taking up to 13 days (Diefenbach, 1975). 8 Crop contents of *Phormia regina* fed on 0.01 M sucrose. (After Thomson & Holling, 1974.) 9 % Undigested food. *Octopus cyanea* fed crabs (*Cardisoma carrifex*). (After Boucher-Rodoni, 1973.)

5.5.3 WHY DOES THROUGHPUT RATE DECREASE WITH TIME SINCE THE EATING OF A MEAL?

Two physiological hypotheses have been put forward as answers to this question. The first is that 'as enzyme reactions are essentially exponential processes, it is not surprising that gastric evacuation proceeds at an exponential rate' (Elliott & Persson, 1978). A decrease in enzymatic activity with time since ingestion of a meal should, presumably, be mirrored in slower evacuation of digesta from the stomach. However, the author has failed to find good evidence that digestive enzyme reactions proceed in this way.

A second possibility is that distension of the stomach necessarily initiates peristaltic contraction proportionately, the resulting outflow being proportional to the tension in the stomach wall (Hopkins, 1966). The tension in the stomach wall is proportional to its radius, i.e. the square root of stomach contents, if the stomach approximates to an elastic cylinder of fixed length. Hence rate of outflow, $-\mathrm{d}x/\mathrm{d}t$, is proportional to the square root of stomach contents, x. It follows that rate of outflow decreases linearly with time since the eating of the meal, and stomach contents decrease quadratically with time. This hypothesis may seem a little naive in its conception of physiological possibilities, since it supposes that rate of peristaltic contraction is necessarily linearly proportional to the distension of the stomach wall. In only two cases has it been shown that a quadratic model describes stomach evacuation significantly better than an exponential model (man, Hopkins, 1966; rat, Booth, 1978). However, Hunt & Stubbs (1975) have subsequently shown that in man the original volume of meal given is not a determinant of the rate of gastric emptying, whereas nutritive density of the meal is.

Both of these hypotheses about why throughput rate decreases with time are relevant to the question 'Are dynamics of this kind adaptive?' By this we really mean 'Do these dynamics promote survival?'

Does it pay to digest food quickly when there is a lot to be processed, and slowly when little remains?

If there were no cost associated with processing food quickly then the animal should process it all at the maximum possible rate. Then undigested food would not be carried around longer than was absolutely necessary, the digestive process (involving e.g. blood supply to the gut) would take the shortest possible time, and energy would be extracted from the food as quickly as possible. Because the gut does not operate in this way we may be reasonably confident that there is some cost associated with processing food at the maximum possible rate.

Some possible costs (the reader may think of more) are:

1 perhaps less can be extracted from food processed faster;
2 the gut may suffer more wear and tear;

3 the other processing costs (blood supply, enzyme production) might increase;
4 the processing facilities (e.g. enzymes) supplied at the start might be used most efficiently by a diminishing rate of digestion.

Model of a trade-off in which fast processing is eventually sacrificed for some benefit of slower processing

Suppose the instantaneous risk taken by an animal carrying a load x of undigested particles is $\mu_1(x)$, and the instantaneous risk of damaging the gut when digesting at rate $\frac{dx}{dt}$ is $\mu_2\left(\frac{dx}{dt}\right)$, and suppose that a period T is available for digestion. Then the chances of surviving T depend on the decreasing load carried as $e^{-\int_0^T \mu_1(x)dt}$ and the chance that the gut is not injured depends on the digestion rates employed as $e^{-\int_0^T \mu_2(dx/dt)dt}$. If the latter is given weight w we may expect the animal to maximize $e^{-\int_0^T \mu_1 dt} + e^{-w\int_0^T \mu_2 dt}$ i.e. to maximize a composite function of the chance of surviving T and the chance of not injuring the gut. This is approximately equivalent to minimizing $\int_0^T(\mu_1 + w\mu_2)dt$ if the risks are small. If $\mu_1 \propto x^2$ and $\mu_2 \propto \frac{dx^2}{dt}$ then the criterion is minimize $\int_0^T\left(x^2 + w\frac{dx^2}{dt}\right)dt$, which is achieved by an exponential evacuation curve (if T is relatively large) as shown by the data in Fig. 5.9. Quadratic risks $\left(\text{i.e. } \mu_1 \propto x^2, \mu_2 \propto \frac{dx^2}{dt}\right)$ are actually the simplest ones for which exponential digestion is the best strategy.

5.5.4 Continuous flow digestion through two compartments

A further order of complexity is introduced by the addition of another digestive chamber for the fermentation of cellulose, either in front of the stomach (as in ruminants) or behind the stomach (as in the horse and grouse). The cumulative appearance of markers from such systems is always described in practice by the sum of two exponential curves:

$$\text{Cumulative faeces production} = Ae^{k_1(t-\tau)} + Be^{k_2(t-\tau)}$$

where τ is the lag between starting to eat markers and defecation of the first marker, and k_1, k_2, A, and B are constants. Examples are given in Fig. 5.10.

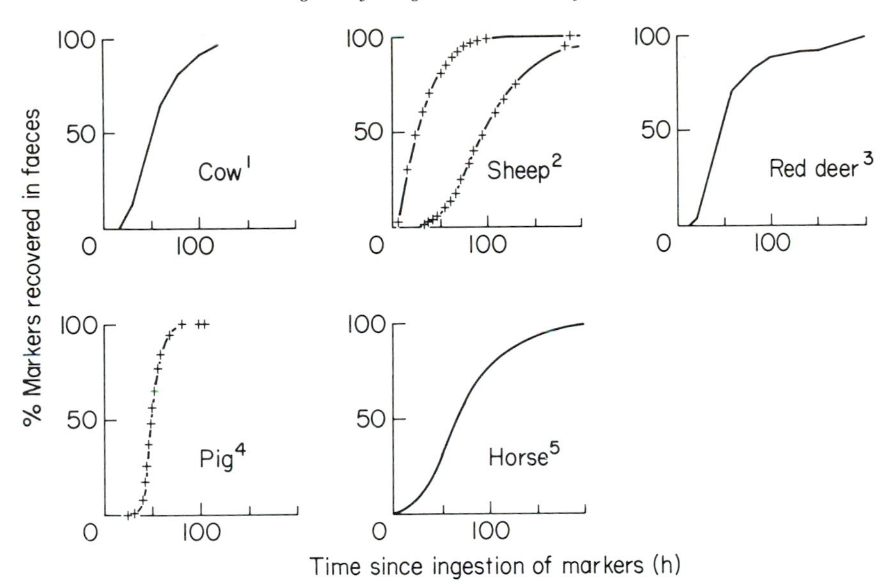

Fig. 5.10 Amount defecated as a function of time. Smooth curves are two-compartment models fitted to the data by least squares. 1 Eating hay. (After Balch, 1950). 2 Eating dried grass (right-hand graph) finely ground (left-hand graph). (After Blaxter, Graham & Wainman, 1956.) (See also Grovum & Williams, 1977; Hartnell & Salter, 1979.) 3 Eating hay. (After Blaxter, Kay, Sharman, Cunningham & Hamilton, 1974.) 4 Eating pig meal. (After Castle & Castle, 1956.) (See also Kidder & Manners, 1978.) 5 Eating hay-grain ration. (After Argenzio, Lowe, Pickard & Stevens, 1974.)

This 'two-compartment model' has enjoyed a rare popularity since its exposition by Blaxter, Graham & Wainman (1956) and Brandt & Thacker (1958), because of its success in describing cumulative faeces production. It was first derived to describe the operation of a system containing two compartments, the outflow of one providing the input to the second. The outflow of a compartment was proportional to its contents. As a simplification, travel between compartments was ignored, except that a constant lag time τ occurred between exit from the second compartment and appearance in the faeces. The interpretation of the 'compartments' is discussed by Balch & Campling (1965), Goldstein & Elwood (1971), Hungate (1966, 1975), Reichl & Baldwin (1975), Grovum & Williams (1973, 1977), Milne, Macrae, Spence & Wilson (1978), Hartnell & Salter (1979), and the assumptions and theory of compartmental analysis is given by Solomon (1960) and Berman (1963). The estimation of throughput time is discussed by Patton & Krause (1972).

A complete description of the gross operation of the reticulo-rumen is not yet possible (Hungate, 1966; Church, 1975), although a number of workers have studied the evacuation of markers from various chambers in the gastro-intestinal tract (e.g. in ponies by Argenzio, Lowe, Pickard & Stevens, 1974; in rabbits by Laplace & Lebas, 1975; Pickard & Stevens, 1972). Argenzio, Lowe, Pickard & Stevens (1974) considered the digestive system of the pony as a five-compartment system, the compartments being the stomach, caecum, ventral, dorsal, and small colons. Markers of various sizes were injected into each compartment, and their movements were charted by killing animals at various time intervals. In the most ambitious application of 'compartmental analysis' so far Mazanov & Nolan (1976) construe nitrogen metabolism in sheep as involving first-order flows of the kind described above between nine compartments. The nitrogen in the microbial population of the rumen is one compartment, and nitrogen flows from it both to the small intestine and to the amino acids in the rumen. Two other compartments are identified in the rumen, viz. ammonia and food input, and four others outside the rumen, viz. the caecum, rectum, body fluids and body tissue. Data used to evaluate the model were obtained by injecting labelled nitrogen into one compartment at a time, and recording its appearance in others. The model fitted the data quite well, and the authors were content with the assumption that underlies compartmental analysis, that outflow from each compartment is proportional to its contents.

5.6 CONCLUSIONS

5.6.1 PRACTICAL DIFFICULTY OF WORKING ON DIGESTION IN WILD ANIMALS

It is often very difficult to see exactly what wild herbivores are eating, especially if they are selective. Without exact information on what goes in, however, the whole study of digestion is prejudiced. Geese (section 5.3.4) are unusually good species to study because everything that goes in produces a natural reference substance in the faeces (leaf surfaces being completely undigested), and faeces are produced so frequently (one every three minutes) that defecation rate can be readily estimated. From the work of Owen (1976) on geese it is clear that

the efficiency with which a herbivore digests a particular plant species may vary (section 5.3.4).

Owen found it difficult to make inferences about digestion in wild geese from observations on captives (see also Harwood, 1975) and this makes work done solely in the laboratory (e.g. Clemens, Stevens & Southworth, 1975) hard to interpret. Digestion in captive grouse (section 5.3.5) is only *half* as efficient as that of wild grouse, even though the captive birds were maintained in large pens on the moors on which they naturally feed, which suggests that

digestive efficiency may vary between captive and wild animals eating the same food.

Thus if inferences are to be made about the digestion of wild animals from studies on captives at least some evidence must be provided that digestion is comparable in the two cases. Some such data are available for ruminants (see e.g. Moran, Norton & Nolan, 1979).

5.6.2 Feeding affected by digestive requirements

Given a particular diet an animal may feed in such a way as to maximize the energy extracted from food, or to reach an optimal compromise between the requirements of digestion and the other functions the animal has to perform. If the animal has a storage chamber (stomach, crop, cheek pouches, or a cache of food) it may fill it periodically so that digestion may proceed at an optimal rate in between. When the ability to extract energy from a plentiful supply of poor food is the critical factor affecting the survival of the animal, then we expect that the store of food will be replenished before it empties completely so that the gut always operates at maximum capacity. As in woodpigeon eating brassica (section 5.3.3) the interval until feeding recommences should depend on the amount stored during a feeding bout. Ruminants, on the other hand, do not have a storage chamber, and ingested food goes straight into the reticulo-rumen. When their ability to extract energy from a plentiful supply of poor food is a critical factor we find that they eat to replace digesta lost by outflow to the intestines, thus keeping the rumen full and operating at maximum efficiency (section 5.4.4).

Given plentiful food some monogastric animals eat at a rate that decreases exponentially with time since the meal began (McCleery, 1977; see also Boucher-Rodoni & Mangold, 1977) and it is tempting to speculate that this is allied in some way to the evacuation of the stomach, which is also described by a negative exponential curve (section 5.5.2). This link is explicitly made in a number of recent physiological control theory models (Booth, 1978b). On the other hand, it may be that as the priority of feeding decreases it pays an animal to allocate an increasing proportion of its time to other activities (Sibly & McFarland, 1976) such as watching out for predators (Milinski & Heller, 1978).

5.6.3 Is animal digestion optimal?

It has been argued that food should be processed to maximize the net rate of extraction of energy, or if another nutrient is limiting, the rate at which it is obtained (section 5.2). This is achieved by retaining digesta longer if food is scarce and rate of intake is limited (Prediction 1). A striking example is provided by the 20–30 kg giant tortoises of Aldabra which retain digesta for 49 days in the late dry season, when food consumption averages 110 g/day dry weight, compared with 6 days in the wet season, when food consumption is 380 g/day (Coe, Bourn & Swingland, 1979). In the invertebrates it is known that both a snail, *Ancylus fluviatilis* (Calow, 1977a), and a woodlouse, *Philoscia muscorum* (Hubbell, Sikora & Paris, 1965), increase retention time when intake is limited, and a similar effect has been found in brown trout *Salmo trutta* (Elliott, 1976). Since it is likely that digestibility increases with retention time (as in Fig. 5.1), digestibility should also vary inversely with feeding rate. A number of examples are cited by Lawton (1970), who classifies this as a type B relationship. Lawton also quotes instances of a type A relationship in which digestibility apparently does not vary with feeding rate; this is only compatible with the author's model if food is *completely* digested at all feeding rates. Definitely not predicted by the author's model is Lawton's type C relationship (digestibility increasing with feeding rate) of which, however, there is only one example (*Carassius*).

If intake is not rate limited then the animal should maximise $\frac{f(T)}{T}$ (Prediction 2), where f is the net amount of energy from one gram of food by digesting it for time T, but no evidence is available to test this directly. Sophisticated models of this type of digestion in filter-feeders (Lehman, 1976) and benthic deposit-feeders (Taghon, Self & Jumars, 1978) have recently been developed (Fig. 5.11 and Chapter 4).

If the available food deteriorates (i.e. lower f) then to maintain the energy supply to the body the animal must increase the weight of digesta carried (Prediction 3). Thus species eating poorer food should have larger guts.

That this is generally true among animals was discovered in the early comparative work in zoology. For example, Marcel de Serres (1813) grouped the Coleoptera into herbivores and carnivores, and remarked that 'généralement, chez les herbivores, les intestins sont plus développés et plus étendus que chez les espèces qui vivent de proie vivante'. This was confirmed in a major study by Bounoure (1919) (Fig. 5.12). Generally in fish the ratio of intestinal length to body length is less than one in carnivores, from one to three

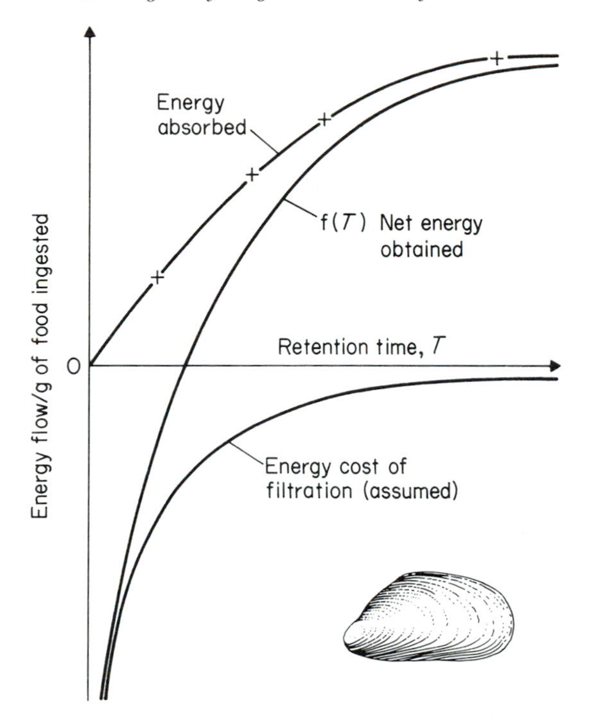

Fig. 5.11 Calculation of f(T) for filter feeders. f(T), the *net* amount of energy obtained from one gram of food retained for time T, is (energy absorbed) – (energy invested in digestion), as in Fig. 5.1. Energy absorbed as a function of retention time is known from experiments (as it is for ruminants Hungate, 1966). The energy invested in digestion in achieving an appropriate filtration rate r is not so readily measured, but is assumed to be an increasing function, say r^2. If the gut is operating at maximum capacity then $T = \frac{\text{gut capacity}}{r}$ (from equation 5.1) and so the energy costs of filtration are $\left(\frac{\text{gut capacity}}{T}\right)^2$. (After Taghon, Self & Jumars, 1978.)

in omnivores, and greater than three in herbivorous species (Jobling, 1978; Braekkan, 1977), and in birds enlarged caeca are associated with increased cellulose in the diet (Farner, 1960, see also Fig. 5.5). Generally in ruminants those that eat better diets have smaller rumens Kay, Engelhardt & White, (1980) and this is true of East African ruminants in particular (Table 5.5).

That some birds at least are capable of increasing (or decreasing) the lengths of their guts in a few weeks was not suspected before the 1970's, and has only been firmly established in the last few years (section 5.3, Table 5.2). Analogous cases in other taxa, though less spectacular, occur in domestic

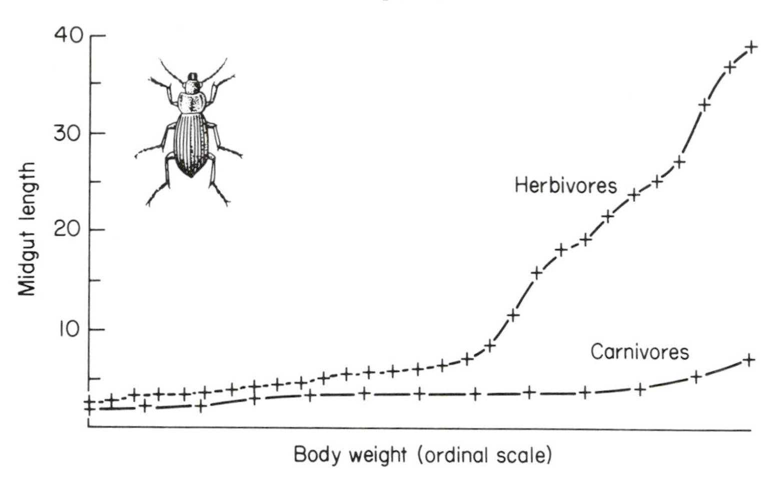

Fig. 5.12 Herbivores have longer guts than carnivores. From a study of Coleoptera by Bounoure (1919). Midgut length was corrected for variations in body size by dividing by $\sqrt[3]{\text{(body weight)}}$.

ruminants (Fell & Weekes, 1975; Warner & Flatt, 1965) and rats (Fabry, 1969). It may be that the cost (i.e. lower survivorship) of carrying around surplus digestive capacity is greater in birds, so that it is worth providing the facility to change digestive capacity when necessary. Obviously there is a great advantage in reducing weight in birds because so much power is saved in flight.

It has been demonstrated in both ruminants and birds that digestive capability affects the amount of energy that can be obtained from food in a day (sections 5.4.4 and 5.3.2).

Turning to large mammals we find that when food is scarce, the large ruminants do better than horses because they are able to extract more energy from a limited supply of food, but when food is plentiful the situation is reversed (section 5.4.2). Quite why each group cannot adopt the strategy of the other when appropriate is unclear. Since the small ruminants (section 5.4.1) can operate with low digestion times it is hard to imagine that the large ruminants cannot. If, however, we accept that the two groups cannot change strategy seasonally, then an interpretation of the evolutionary history of the two groups becomes possible (section 5.4.3).

A functional evaluation of the digestive process is attempted in section 5.5 for three types of digestive system: single compartments operating continuously/discontinuously, and two compartments operating continuously. Digestion in the monogastric stomach eating discrete meals decreases with time since the meal was eaten. This may be functionally related to the shape of the curve f(T) in Fig. 5.1, it may be because of a trade-off between the benefits of extracting energy quickly and the operating costs that entails, or it may (implausibly) be the maladaptive result of an evolutionary inability to produce a digestive system operating in any other way.

Part 3

Partitioning the Resource Input

Introduction

It is not difficult to enumerate the principal metabolic compartments that receive a share of assimilated energy. That which is not used in respiratory metabolism is partitioned between defence, repair, storage, growth and reproduction. In this section, each of these options is discussed in turn, pointing out the benefits and costs of alternative strategies and stressing their ecological, physiological and evolutionary implications.

Energy channelled into defence mechanisms in both animals and plants is unavailable for other metabolic compartments, and this option may therefore reduce the ability to compete with undefended relatives. However, an investment in *defence* may enhance survivorship in the face of disease, parasitism and predation. Chapter 6 provides a thorough account of the physiology of personal defence and of the metabolic costs involved. The author points out the need for field experiments to determine the fitness gains and losses that accrue to organisms with and without defences. Then it will become possible to develop models and test predictions in this area, although a start has already been made.

Within organisms there is a dynamic balance between the generation of damage in molecular, cellular and organ systems and its replacement and *repair*. The benefit of repair is increased lifespan and vitality. Its cost is the energy it consumes and which is therefore diverted from other metabolic compartments. The optimal solution to the partitioning of energy between the options of repair and the rest of metabolism is shown to depend upon the expected lifespan of the organism as influenced by accident, disease and predation. Chapter 7 brings together research findings on all levels of repair and demonstrates the main features of the relationship between repair and life-history, including an illuminating account of senescence.

Slobodkin once made the point that there is a selective advantage to fecundity and not adiposity, implying that spare energy should be converted to offspring production and not storage materials. The cost of *storage* results from the time and energy losses incurred as a result of not investing directly in offspring. Benefits arise in organisms which face the threat of resource scarcity.

Chapter 8 adopts a cost/benefit approach and considers the choice of different molecules, tissues and anatomical sites for energy storage. The relationship between storage and both diapause and migration is discussed, as are the novel functions that storage organs often seem to acquire.

Growth can be considered as the developmental means of achieving a reproductive state. Chapter 9 is concerned with the option between biomass production in growth and respiratory metabolism up to the time of reproduction. The benefits from investing as much energy as possible in growth are that the reproductive state will be reached more quickly, and that the ability to withstand the threats of predation and competition may be enhanced because a large size is reached sooner. However, there are likely to be costs as well as benefits and these are considered under the heading: 'Is growth rate maximized or optimized?'

The final chapter in this section (Chapter 10) is concerned with *reproduction*, the ultimate function of the phenotype. There are obvious gains from transforming as much energy as possible into gametes. However, costs of reproduction arise from its adverse influence on the future survivorship chances of the parent and its subsequent reproductive performance. Optimal investment at any time depends upon the balance of these costs and benefits as they influence the future contribution of the parent to the population. This chapter also considers the relative merits of partitioning the energy made available for reproduction into a large number of small gametes or a small number of large gametes, the investment in parental care and the energetics of hermaphroditism and asexual reproduction.

Chapter 6

Evolutionary Physiology of Personal Defence

DANIEL H. JANZEN

6.1 INTRODUCTION

In general, allocation of resources among defence, repair, storage and growth is moulded by natural selection to maximize the organism's genetic representation in future generations through the medium of sexual reproduction. Here, I emphasize the use of resources for personal defence against predators and parasites. Defence against 'the elements' and against competitors are not directly discussed.

At the outset it must be stated that a predator is a consumer that kills its prey for immediate consumption and that a parasite is a consumer that removes so little of the host's tissue or resource that the host is not directly killed by this act. A predator may therefore be a cobra swallowing a viper, a mouse gnawing up an acorn, a pitcher plant drowning a beetle, or Dutch elm disease killing an elm. A parasite is a mosquito on the back of your hand, a moose browsing a large willow, a black bread mold growing on a mouse's grain cache, or a mycorrhizal fungus on an oak seedling's roots at a time of year when the seedling does not have carbohydrates to spare. Clearly there is a broad grey zone between these two categories, a grey zone that has generated words such as parasitoid (exemplified by an ichneumon wasp larva that takes two months to kill its caterpillar host), hemi-parasite (exemplified by snapdragons that are both tapped into other plants' roots and harvesting their own ions) and mutualism (reciprocal parasitization where for both members the loss is less than the gain and exemplified by bees and flowers).

The array of defence traits is more diverse than is any other array of traits. Mating and sexual isolation traits are of course diverse, but in general each beast or plant is dealing with one acceptable type of organism and *all* the rest are unequivocally unacceptable (except in cases where interspecific introgression is of value to the introgressed). One or a few exclusion rules or traits keep out the unwanted visitors; there isn't much demand for interspecific rape. Not so with the beasts that eat your body and suck your blood. Every species of organism is confronted with easily enough different kinds of predators and parasites to require a multitude of different defences. For example, a seed, hardly more than an embryo with a packed lunch, has at least the following defence traits: synchronous maturation with its sibs and conspecifics; maternal protection through green (immature) fruit hardness and secondary compounds; maternal protection through mature (ripe) fruit; attractiveness to the right animals and unattractiveness to the wrong animals; cryptic colouration of the seed coat; seed-coat hardness and secondary compounds; seed shape and specific gravity that determine its passage rate through a digestive soup; and numerous secondary compounds in the endosperm, cotyledons and embryo (e.g. indigestible starch and other polysaccharides, toxic proteins such as lectins and protease inhibitors, alkaloids and uncommon amino acids, cyanogenic glycosides or cyanolipids, digestion-inhibiting polyphenols and saponins, etc.). A list could be made of several hundred organisms which would eat any given seed if they could get by one or more of the seed's defences; each attacking organism carries a repertoire of largely unique tools for disarming its prey or host, in addition to its more generalized tools.

There are two quite different ways to organize a discussion of defences. I could catalogue known defences and reference examples of each. However, in keeping with the attitude in this book, an attitude usually missing in discussions of defence traits, I will attempt to generalize about the processes in common among personal defence systems, and leave it to the reader to decide if it applies to the particular system he or she is intimately familiar with. As is usual in ecology, the subjects of this essay do not fall in a neat linear hierarchy and no order of importance or 'key position' is intended by my order of discussion. Many of the references cited contain detailed discussions of the mechanics of defences. The interested reader is particularly referred to reviews in Robinson (1969), Harborne (1972, 1978 a and b), van Emden (1973), Zlotkin (1973), Gilbert & Raven (1975), Levin (1976a), Atsatt & O'Dowd (1976), Rugieri (1976), Wallace & Mansell (1976) and Rosenthal & Janzen (1979).

6.2 MULTIPLE USE

Actual defence structures or traits—a spine, fang, raphide, alkaloid, antibody, jump, bite—are very likely to be programmed by more than one gene and to be of several quite different uses to the organism. In like manner, the actual costs and fitness gains attributable to the structure will be intricately apportioned among the different uses. Did snake fangs appear as a response to the value of immobilizing the snake's prey (protection against your prey and enhanced prey-capture ability) or as defence against predators, or both? If we elect the former role, which seems most reasonable, it is still clear that in at least one case, the spitting cobra, one aspect of fang use has been subsequently modified to a purely defensive role (Janzen, 1976c). The cyanogenic glycoside linamarin, which probably protects wild lima bean seeds from a variety of potential predators, is moved by the new seedling into vegetative tissues where it probably serves the same protective role as in the seed but against quite different organisms (Clegg, Conn & Janzen, 1979). However, another seed defence, canavanine, appears to be degraded in part by the seedling as a source of nitrogen in the first several weeks after germination (Rosenthal, 1977). Cellulose, long viewed solely as the master construction material of plants, was probably evolutionarily chosen for the same reason that we construct houses of concrete in areas of high termite activity. There are many structural polysaccharides. It is extremely unlikely that natural selection produced cellulose-rich plants for any other reason than that virtually no higher organism can digest it unaided. Needless to say, the plant 'pays' an enormous cost for this defensive indigestibility; it deprives itself of many repair options and cannot recover the primary carbohydrate from leaves and other plant parts to be shed. However, it is equally striking that plants apparently can enzymatically degrade their own cellulose when producing lateral roots (M. McCully, personal communication) and therefore it appears they can overcome their own defences when the return is high enough. The ultimate outcome of the multiple-use aspect of defence traits is a strong heterogeneity in the world's defence contour. A large viper very rarely uses it fangs for defence and they probably would never have evolved solely for this purpose. However, once evolved for prey capture, viper fangs are present as a horribly intense defence against a rare threat to one of these large reptiles.

6.3 UNTRIED DEFENCE

The function of each defence trait has an active and a passive component. The vast majority of the tannin molecules in oak leaves never see the inside of a herbivore. It is a classic error in herbivory studies to measure herbivore impact only by the amount of materials they eat or by the reduction in fitness caused by herbivory. I suspect that for most plants the biggest cost of herbivory is the cost of the defence chemicals, morphologies and phenologies that keep the bulk of the herbivores away. In other words, to determine the 'size' of the herbivore trophic level, one has to measure not only its biomass and turnover, but also how much resource its presence causes the plants to put into defence, resources that cannot then be used, either directly or indirectly, for reproduction. This principle applies to all trophic levels.

Are untried defences recyclable? The cyanogenic glucosides in a surviving seed are translocated to the new seedling, presumably as part of its defence. The cellulose in a leaf is discarded in the autumn. Snake fangs are used for prey capture between defence bouts. The quills of a porcupine are shed as they age; they are dead tissue from the time they are fully formed on the animal. As a general rule, defence recyclability seems to fall into three categories: structures that are largely for defence and are often made of dead tissue or tissue that is non-degradable to prey and predator alike; structures of living tissues that are usually recyclable and of multiple use; and facultative defences that appear only at the time of the challenge.

Behavioural and physiological traits that appear at the time of the challenge (adrenalin, fever, skunk spray, running, threats, motionlessness, increased liver and kidney metabolism, antibodies, etc.) are well known to zoologists. Analogous traits have been well known in plants since the turn of the century, with respect to phenolics and other defence compounds and chemicals elaborated at the site in response to fungal and microbial attack. Likewise, African observers have often noted that the branches of arid-land *Acacia* respond to browsing by the production of longer thorns, smaller leaves and shorter internodes. However, it was only in the late 1960's that Ryan and his co-workers began to explore the production of protease inhibitors in leaves in response to leaf chewing by insects (Ryan, 1978).

6.4 HOW LARGE IS A DEFENCE?

There is a strong temptation to measure the intensity or size of a defence by its impact on the human that encounters it; it is hard to think of a more irrelevant measure. If a human consumes the products of three ground coffee seeds and gets a mild caffeine kick, coffee is labelled as non-toxic. However, a mouse that tries to eat a coffee seed may be making its entire meal of this alkaloid-rich food and the effect may be as though a human ate a kilogram of ground coffee in one meal. Although consumption of more than 30% rhododendron leaves in the diet causes rickets in deer, rhododendron leaves have been labelled 'non-toxic'; ever see a deer with rickets try to outrun a mountain lion? You may find that you can acclimatize to stinging nettles and acacia-ants; try eating them. A medium-sized parrot can draw blood and sometimes hurt a human slightly;

there is an instance where a parrot broke both wings of an attacking hawk with its bill.

The same caution applies to analyses of why certain parts seem to be defended differently to others. Why should some species of seed be so well defended when the plant makes so many? After all, it only takes one to replace the parent, or so the folklore goes. One of the many answers to that question is that the seed is well defended for the same reason that a young man protested at being sent to Vietnam. The tiny amount of material lost when a shoot apex is eaten is nothing compared to the loss of competitive height status in the time it takes for the plant to replace its shoot tip with an axillary bud. A mutant cedar tree with bubbles of a few milligrams of resin in its seed coats may have a much greater fitness than a mutant that puts massive amounts of resin in its cambial region. The human vascular system can withstand the loss of 100 mosquito stomachfuls of blood in one night, but such an event may give severe anaemia to a mouse. Your defence may be to swat the mosquito or ignore it; a mouse may spend many hours building a mosquito-proof nest or burrow.

6.5 WHY STENOPHAGY?

There seem to be three sorts of phytophagous insects: extreme polyphages which really can feed on many species of plants, host-specific polyphages that have a list of 2–15 or so species of plants (out of hundreds available in their habitat), and monophagous stenophages that feed on only one species of plant (in a habitat rich in plant species) (Janzen, 1980a). There are two puzzles about this defence–parasite interaction.

1 Why are those species that are capable of eating the foliage (or seeds) of many species not capable of feeding on all of them? The dogma is that those which they cannot eat contain some collection of defences which is simply insurmountable to that particular polyphage. However, since each polyphage has a different list of plant species that it can feed on in the *same* habitat (e.g. Rockwood & Glander, 1979) why isn't there one species of herbivore that has the combined defence capabilities of all of them (a sort of mega-army worm; Brattsten *et al.*, 1977)? Part of the answer may lie in a falsehood in the above background. It may well be that the most highly polyphagous species of insects are in fact largely stenophagous or even monophagous as individuals, and therefore the question and its answer move on to the next paragraph (see below). In this view, then, a large herbivore (cow, elephant) that unambiguously eats many species of plants

can be viewed as many herbivores as represented by the various strains of microbes that its gut contains. The large vertebrate herbivore is also, I should add, acting like a human drinking a cup of coffee; it is digesting at a level of toxin dilution at which the damage is less than the gain.

2 If one insect can be stenophagous on one species of presumably well-defended plant, and another species of insect can do the same on another species of plant, why can't there be a third species that possesses both of these quite separate defence degrading systems so that the same individual can feed on either plant species at any time? Furthermore, why can't an insect carry even more quite separate detoxification systems? The answer is not that there is so much resource that there is no selection pressure for such an event to occur. The answer is probably similar to the answer to why a soldier cannot be a warrior, a tank gunner, a radio communications specialist and a demolition expert simultaneously in the same war. Or the answer to why there can be no parrot-like hawk that preys on both mice and large hard tropical seeds.

6.6 ANACHRONISMS AND MISSING ATTACKERS

The study of defences has one particularly difficult aspect. You cannot study the defences of mice against owls in the daytime. Likewise, even if canavanine can be eaten in bulk by an English mouse, it may nevertheless be an effective defence of a Costa Rican canavanine-rich *Canavalia* seed against both the insects and the mice in its native habitat. If a *Cecropia* tree leaves its ants behind when it migrates to an island (Janzen, 1973a), and the *Cecropia* tree does well on the island (or in your garden), we cannot conclude that the *Cecropia* tree was not protected by its ant colony in its original mainland Central American habitats. Experimental demonstration of defence biology requires not only the prey but the predator to be present. This statement applies to both temporal and spatial heterogeneity.

In short, the individual organism cannot know when an attack is going to come, or even if there are any attackers out there. It can only genetically 'know' that there is some probability that an attack will come. In a certain sense, DNA programming may even be viewed as a defence against the costs of being inexperienced and learning by trial and error (W. Hallwachs, personal communication). The problem of missing challengers becomes even more intractable when the defence trait is something also functional in some other context (e.g. wasp stings as nest defence and wasp stings as subduers of prey). Some defences do not get used very often even as the normal state of affairs.

Then there is the complication of the missing evolutionary partner. There may well have been a time when wasps did not protect their nests well, and a number of vertebrates made much of their living preying on the nests. As stinging defences became more sophisticated, these vertebrates evolved other modes of prey capture. The same defences have then also prevented the evolution of new specialists in later millenia because the intermediate stage, as a moderately incompetent wasp nest predator, was too inhospitable. The thick hard nuts of palm fruits may well have evolved in response to mastodon or ground sloth molars 10 000 years ago (Janzen & Martin, 1981), and the contemporary interaction of the nut with nut-crushing peccaries (Kiltie, 1980) and nut-gnawing agoutis (Bradford & Smith, 1977) is pure serendipity.

The opportunity for evolutionary cycles, for anachronisms, for phylogenetic inertia and for uninterpretable natural histories is very great when such a scenario is replicated over millions of animals and plants and their personal defences. I suspect that tropical forests are living museums full of anachronisms left behind by the extinction of one member of a co-evolved pair of species. Even worse for the evolutionary biologist, we are not only dropped into the system knowing nothing of past trends, but nowadays we are often allowed to study only one or two of the members of complexes of defender and attackers. How can one hope to understand the extreme aggressiveness of the African honey bee (Fletcher, 1978) in contemporary habitats with most of the potential honey bee nest predators locally extinct or substantially reduced in density, nest sites greatly reduced in quality and quantity and the genome contaminated by introgression from docile European yard bees.

6.7 CO-EVOLUTION?

The fashion of the day is co-evolution. Every pair of mutualists, parasites and hosts, and predators and prey gets slapped with this label. This is one of those crazy cases where the general theory is probably fine yet often does not apply at all to many of the specific cases. Of course a pair of complex congruent interactants may be co-evolved, but they also very well may not be (Janzen, 1980b). Let me illustrate with a hypothetical example. A hypothetical weevil, *Immunotoxis cyanivorus*, is a euryphagous leaf-eater of the cyanogenic strains of trefoil (*Lotus cyanoambiguous*) in Europe. It has evolved the ability to metabolize cyanogenic glucosides and even use the by-products for normal anabolism. The trefoil has responded evolutionarily by extra-rapid toughening of its new leaves to a hardness the weevil can no longer chew. The weevil exists as a low-density population of individuals constantly searching for tender new trefoil shoot tips, which are widely scattered in time and space. We know they are strongly co-evolved because we watched it happen between 1983 and 2076, following the introduction of the weevil from Australia in 1982 to control trefoil in European pastures. However, in 2041, a tourist from Hamburg carried the weevil to Costa Rica as cocoons in a floral display. Upon emerging, they wandered about in the vicinity of Liberia, Guanacaste Province, trying many of the 900 species of broad-leaved plants native to that region. Not surprisingly, one of them was a legume whose major defence against folivores was cyanogenic glucosides. *Phaseolus lunatus* also bears new leaves for only a short time of each year and the ability of *I. cyanivorus* to search them out allowed it to maintain a breeding population. No subsequent evolution of either weevil or plant occurred with respect to the traits that are directly related to their interaction.

In the current climate of calling everything co-evolution, the Costa Rican entomologist now finding *I. cyanivorus* and describing its interaction with *P. lunatus* would label the system co-evolved, as would a native person watching an introduced hummingbird working African sunbird flowers, watching neotropical syrphid flies that mimic honey bees, or watching an introduced agouti burying palm nuts in Africa. It is a fair guess that most organisms we see today are not being studied in the particular habitat where they evolved the traits they display (hardly surprising when it is clear that most organisms are a collection of traits acquired during evolutionary bouts in a variety of habitats). This means that a great number of congruences are unlikely to be co-evolved. The general process should be that when a species invades a new habitat or geographic region (where the habitat may be the same, but the participants different), it will quickly adjust its foraging and defence to be most congruent with the resources and challenges present. If the fit is good, as it will often be, subsequent evolutionary changes may be at best minor and the apparently highly co-evolved system may have come about through little or no evolutionary change in either partner. When the barn owl, well co-evolved with the mice it hunts, first moved into Australia and started hunting small marsupials, it may not have changed in an evolutionary sense at all. Even if the small prey evolved defences in response to barn-owl hunting pressure, that certainly is not co-evolution. When the first mastodon walked into a neotropical forest, it was instantaneously a pseudo-co-evolved disperser of many species of plants. The elephant that munches spiny acacias for lunch may well have evolved no special traits for this interaction, though the acacia has long been evolving an ever more wicked thorniness.

6.8 COST OF SAFETY CATCHES

Defences add two classes of cost to the resource budget. In addition to the cost of defence production and maintenance are costs of keeping the possessor from being injured. The safety catch of a rifle may be only a small part of the cost of a rifle, but it is largely indispensable. The cost of safety catches is of two sorts. The first is that a defence (or many defences) may be incompatible with another function for a tissue or structure. Secondary compounds, such as lectins, alkaloids, terpenes etc., lacking their individual safety catches (a glucose molecule tacked onto them), are very good defences against organisms that bore in plant tissue, but they cannot be sequestered in bulk in living cells. We, therefore, find the dead heartwood of a tree to be rich in the non-glucoside forms of such compounds but the living sapwood to be very poorly protected. While we cannot prove it, the generally high edibility of animal eggs is probably due to the difficulty of sequestering a really toxic material—and very toxic it would have to be to protect such a nutrient-rich structure—in a liquid medium adjacent to a rapidly developing organism; seeds, on the other hand, which are both not liquid and dormant, can sequester very potent toxins and digestion inhibitors in vacuoles (Orians & Janzen, 1974). Shoot tips of plants (more specifically apical meristems), are generally poor in good chemical defences (though they acquire them quickly after mitosis has ceased). They are thus more like eggs (probably for the same reasons).

The second type is the actual cost of protecting the organism from its own defences. Vacuoles filled with L-dopa cost at least the price of the vacuole itself, the space it occupies in the cell minus the volume of the defensive compounds, and the machinery for transporting the compounds into and out of the vacuoles. The sugar molecule(s) tacked on to inactivate cyanogenic glucosides, alkaloids, cardiac glycosides, lectins, etc. have a cost that may be energetically low in general, but then again at the time of compound construction might be high. They may even have a volume cost at times; seeds, where weight and space are costly commodities, are the only known source of cyanolipids, compounds that are functionally the same as cyanogenic glucosides except when the seedling decides to degrade them for its own energetic uses. A cubic millimetre of lipid will yield much more than a cubic millimetre of sugar. Snakes undoubtedly carry antibodies against their own venom, scabards are as old as knives, and some animals carry such potent chemicals that they mix them up at the time of use, as anyone who aspirates up a bombadier beetle (Carabidae) (Eisner & Dean, 1976) will learn to his chagrin.

Where defence involves a castle, the owner bears a cost of impeded knowledge of the environment and therefore may in fact have to maintain a higher average level of defence intensity than if the owner could facultatively adjust the castle's defences to the challenges as they come along. The seed dispersed via the gut of a large mammal has this problem. It has a hard, thick and liquid-impervious seed coat or nut wall that defends it against molars and digestive fluids during the voyage. However, if it falls below the parent tree without making this voyage, it is sealed in its castle and cannot know the season to germinate. In many species, such seeds do not germinate until a cue reaches them, for example the castle wall becomes scarified and moisture can enter. However, this can be any time during the growing season and the two-week-old seedling may find itself confronted with a northern winter or a tropical dry season. If it is scarified in its voyage through the animal, as its DNA is programmed for it to be, it is in fine shape upon emergence. However, if the mammal is unable to break down its defences, again it sits in its sealed time capsule, and then the accidental scarification that is bound to occur sometime may or may not occur at a good time of year for a seedling to make its start. Once a clam, armadillo, pangolin, or tiger beetle larva has retreated into its armour, it has staked its life on the chance that the attacker cannot broach the walls; there is no opportunity to iteratively tailor a flexible defence to an ever more accurate assessment of the attacker's strength. To make matters worse, the thus-defended animal is deprived of foraging time as well.

6.9 MAMA VERSUS THE KIDS

The parent–offspring conflict is an untouched area of evolutionary physiology, probably because adult humans largely control the worlds of their offspring in such a manner as to raise the fitness or the inclusive fitness of the adult; public study would undermine that control. The offspring is clearly a parasite in any system with parental care, and seed-bearing plants without exception display extensive parental care. Plant zygotes not only grow within the defences of parental tissues (the secondary compounds and physical protection of green fruits) but derive all of their nutrient resources from the parent until weaned at the moment of fruit ripening. Every botany textbook tells us there is a layer of polyploid tissue ($3N$ or better), between the angiosperm zygote and the $2N$ maternal tissues; it is called endosperm and made up of one paternal DNA set and two or more maternal DNA sets. These books offer not a sentence as to the adaptive significance of endosperm. I hypothesise that it is a mediator and controller of physiological squabbles between the parent and the zygote.

Squabbles there should be, and in abundance, as each zygote is in competition with its litter-mates for resources from the parent and somewhere during their development well over half of the zygotes will be rejected (aborted) by the parent in most species of higher plants. The endosperm thus becomes a defensive (and offensive) tissue for both the parent and the offspring. In many plants it atrophies to nothing but cell debris when the final seed crop size has been determined, but in others it goes on to become the site of resource storage in the full-sized seed.

It is not hard to find an analogy in the mammal placenta, though this tissue is purely maternal in origin. Nevertheless it is the first line of defence in both directions. In social animals, numerous social conventions serve the same defensive role; the most familiar are courts, schools, the military, adult chauvinism, marriage codes, wills, etc.

The expenditure of offspring as a type of defence that raises the inclusive fitness of the parent is hardly a new idea or practice; it is not the old or the females that are sent to war in our social systems. Figs pay 50 % or more of their zygotes for pollination services (Janzen, 1979a), allelopathic plants commonly kill their own offspring as well as those of conspecifics and allospecifics, in order to pre-empt the resources in their immediate vicinity, and perennial plants regularly abort large numbers of offspring that they deem unfit or too great a resource drain. What we have failed to explore are the offspring's defences against its parents. In humans, mimicry, addiction to subsidy types, fratricide and sororicide, sexual displays, imprinting, brute force and verbal deception are all behaviours used by offspring to maintain their place in the subsidy line. Surely the same events occur with solitary animals and plants. However, the highly adaptive taboos against exploring such a phenomenon in our own society have not been conducive to their examination.

6.10 WHO ARE THE UNDEFENDED?

As I suspect our ancestors have known for the past five million years, truly hard biotic containers often contain undefended objects. About the only seeds that can be eaten with relative impunity in a tropical rainforest are those that are encased in thick hard fruit (nut) walls, such as many species of palm (coconuts, etc.), *Coula edulis*, *Lithofagus*, Brazil nuts, *Dipteryx panamensis*, *Hippomane mancinella*, etc. The soft-walled edible forest seeds are usually mast-seeding species the world over (bamboo, dipterocarps, niloo, oaks, conifers, beech, pecans, hickory nuts; Janzen, 1971, 1974, 1976a) and even many of these have relatively hard shells. Clams are the turtles and nuts of the invertebrate world unless they happen to have been feeding on toxic small invertebrates. In short, even in sedentary organisms, a very hard or tough castle can act as a substitute for other defences such as stings, toxins and indigestibility.

The organisms that specialize at being truly defenceless appear to live largely in two quite different ways. First, there are those things that live in largely predator- or parasite-free habitats; islands offer the most striking examples. Many species of island plants have been gobbled down to extinction by introduced mainland herbivores like goats and pigs. Galapagos tortoises would not survive one week in Africa and ground-nesting birds are commonly extirpated when cats and/or rats are introduced to oceanic islands. But there are other kinds of islands. Penguins would be 'sitting ducks' for any predator that could get to Antarctica and survive long enough out of water to find them. The tops of tropical mountains and some extremely nutrient-poor terrestrial habitats have vines with above-ground storage tubers, presumably owing to a lack of the sorts of rodents or other climbing herbivores that would gobble them up in a lowland site. When hunting humans hit the North American island, they appear to have encountered a megafauna that was defenceless against tools and weapons, and many species suffered extinction before they could evolve fear of man (Martin, 1973).

Second, there are those things that make their own predator-free habitat by doing something in such concentrated synchrony with their conspecifics that defences are not needed; every available predator has its stomach long since stuffed with food by the time the average individual arrives on the scene. We have predator satiation at the population level in wildebeest calving, 13- and 17-year cicadas, bamboos, northern conifers and Fagaceae, mayflies, termite mating swarms, salmon spawning and passenger pigeon nesting. The ingredients are the physiological ability to synchronize by counting intrinsically (e.g. bamboo; Janzen, 1976a) or by cueing on weather events (e.g. Dipterocarpaceae, conifers and Fagaceae; Janzen, 1971, 1974), to store the resources to breed or otherwise appear in huge numbers at one point at long intervals, to co-occur in large enough numbers that the predators are satiated and to put all the resource into production of individuals and virtually none into individual defences. A periodical cicada, salmon egg, bamboo seed, etc. is essentially defenceless. There is also predator satiation at the level of the individual, such as when a tree produces 100 000 seeds which are on the average located by only 100 weevil females, each of which can only lay 200 eggs, or when on the average a grazing elk is found by only enough mosquitos to lose 10 cm^3 of blood every 24 h. Here, the organism gains more by producing more seeds or more blood than by being chemically or behaviourally defended. The process is extremely common in defence–attacker systems. In short, a defence is not going to arise evolutionarily unless the return is greater than the cost. This is something well reflected in the high prices that housewives pay for unblemished fruit, a fruit type that is very costly in pesticides to produce.

At a more moderate level, it seems reasonable that in many simple defence systems the intensity and diversity of personal defences carried by the prey or host will be proportional to the probability of being found. It has been reasonably argued that plants which carry large crowns of long-lived leaves among many conspecifics on the same terrain for many years are maximally 'apparent' to herbivores and will have to possess a particularly good defence, one that is hardly penetrable even by a specialist. Digestion inhibitors in northern forest tree leaves and in desert evergreens are good examples. At the other end of the scale there are small plants that are scattered among other plants and often occupy a site for only a short time. These escape somewhat in space and time, as do, for example, the very new leaves of trees. Here the challenge is the specialist who searches out a particular species and concentrates evolutionarily on keeping up with its defences. Such plants are expected to be defended by toxic molecules that will stop the occasional generalists, but will be of little use in deterring the specialist (from which escape is largely by playing geographical and temporal hide and seek) (Feeny, 1976; Rhoades & Cates, 1976). However, Freeland points out (personal communication) that the same result can be obtained if we note that the small 'unapparent' plant can ill afford to loose 100 g of leaves to a generalist such as a deer while a large and 'apparent' tree that loses 100 g of leaves to a deer has hardly been touched.

No organism has a defence system that is impenetrable to all attackers. All organisms must escape to some degree in time and space, and aphids more so than elephants. But then again aphids, producing a large number of highly-edible individuals, display yet another kind of defence. In essence, an aphid clone is a highly subdivided camel (Janzen, 1977a); yes, the parts are highly edible, but usually no predator or predator population can find a large proportion of the members of the clone, many of which are rapidly multiplying even as some are being found. The same can be said of grasses, highly evolved to be burned, or grazed by vertebrates (and lawn mowers!). Such organisms have the defence of being able to subdivide and multiply faster than they can be eaten.

6.11 DEFENCE HIERARCHIES

Since many defence traits have multiple functions and origins, and since most organisms have more than one kind of personal defence, it is hard to organize progressions of defences along single axes. Clams are hard on the outside but unprotected on the inside (as are most large tropical nuts), conversely sea cucumbers are soft but well protected inside (as are many large tropical seeds). However, if we construct a scale along the gradient between clams and sea cucumbers, an annoyingly large number of organisms need to be placed in two or more places on the scale. Kentucky coffee bean (*Gymnocladus dioica*) seeds are rock hard and very poisonous inside; aphids are soft and unprotected inside. We are forced to view each organism as sitting at a point in the now famous *n*-dimensional hyperspace; each axis of the hyperspace is a gradient between only two opposite defence traits or allocation behaviours. Just as the *r* and *K* selection dichotomy was clearly inadequate to deal with organisms with complex life cycles; defence repertoires usually cannot be arrayed along a single general-purpose axis.

However, at the level of higher taxa, and often at the interspecific or intergeneric level, the size of the repertoire of chemico-morphological defences is directly proportional to the sedentariness of the individual. In addition to the best-known botanical defensive compound of all, indigestible cellulose, plants offer literally millions of kinds of toxic defence compounds, urticating hairs, fibrous tough covers, stone cells, in striking contrast to the sprinkling of chemical defence analogues offered by the mobile animal world. Where are the really potent chemico-morphological defences among animals? Sessile marine invertebrates (molluscs and soft, exposed organisms like sea cucumbers; see Ruggieri, 1976), social Hymenoptera (see Akre & Davis, 1978), and turtles/armadillos/pangolins/puffer fish/glyptodonts/bird eggs are distinctly non-cursorial. Conversely, if you can run you do not need so much armour. The economic trade-offs are generally clear but specifically unfocused. Does a box turtle make better use of its resources by lumbering about in its castle than does a threatened ctenosaur lizard who runs like hell for its castle? If the ctenosaur is more than about 80 m from its hole in a tree or rocks, it's dead; if the threat is big enough to swallow or crush the turtle, it's dead.

6.12 COST DOES NOT EQUAL EFFECTIVENESS

There is an extremely complex relationship between the cost of a defence and its effectiveness, a complexity that occurs by and large because each organism is defending itself against many different kinds of attacker and because the same defence resources are differentially hard for different organisms to obtain. A few milligrams of caffeine per gram of seed may be as effective a defence against a rodent as 300 mg of tannin per gram of seed, but the caffeine may be worthless as a defence against a caffeine-resistant seed-boring insect. If the plant is in a light-rich and nitrogen-poor habitat, the few milligrams of caffeine may be more difficult to obtain than 300 mg of tannin. But this physiological approach ignores the most serious problem of all in studies of defence costs. True defence costs can only, if ever, be measured by the changes in fitness of the organism (or hopefully some close approximation of fitness like the number of offspring fledged) that occur when the defence trait is present or absent. An enormously expensive defence may be highly selected for if it confers an equally great increase in fitness. The trait called 'a protective ant colony on an ant-acacia' costs at the very least the energy, building blocks and genetic programming and maintenance of Beltian bodies, thorns, extra-floral nectaries and dry-season evergreenness. The gain from this large physiological cost can be eliminated by removing the ants, and the consequence is death for the ant-acacia (Janzen, 1966), a rather high loss in fitness. Yet a very cheap defence may also greatly increase the fitness of an organism. I suspect that the physiological cost of those pigments in an insect's cuticle which gives it such marvellous camouflage is probably a trivial part of the animal's total defence budget. Likewise, the production of cactus spines may be a major part of the growth costs of a cactus shoot tip, but they are dead tissue and once produced bear no maintenance charges. Cactus spines also illustrate yet another aspect of cost/fitness heterogeneity. They may evolutionarily arise rapidly in the face of intense herbivory, but disappear very slowly once that selective pressure is removed simply because they are such a small part of the plant's defence budget. The rate of appearance and disappearance of defence traits in the face of fluctuating selection pressures will depend on both the physiological cost of the trait and the intensity of selection for it.

It is often not appreciated that many kinds of defences bear a cost of preventing the organism from following some kind of resource-gathering behaviour. Selection for downward pointing petioles, few lateral branches, and entire leaves as a morphological way of discouraging vines, cuts down severely on the light-gathering options of a self-supporting plant. But then again, by living this way in the vine- *and* light-rich habitats of early succession in tropical tree falls, it is occupying a habitat where restricted light-gathering options are of minimal consequence. When aculeate Hymenoptera females evolved a sting out of their ovipositor (the egg lubrication gland became the venom gland) they lost the ability to place their eggs deep in host tissue, but they gained the abilities to immobilize their host and protect themselves. Turtles cannot run well but then their suit of armour works well against the majority of predators and apparently turtles rarely eat highly mobile prey. Vomiting is a good solution for getting rid of an unwanted toxin or microbial contaminant, but the meal is lost as well.

6.13 NO DEFENCE STOPS ALL

Defences are not absolute in their effectiveness. The likelihood of their being penetrated in ecological or evolutionary time is a function not only of their imperviousness or toxicity to living organisms, but of the value of what they are protecting. While a few milligrams of tannin per gram wet weight might stop a colobus monkey from eating a leaf, that same amount of defence would not necessarily stop it from eating a seed (McKey *et al.*, 1978). A well-fed tiger may be deterred by the spines of an Indian porcupine, but a starving one may be willing to pay a fair amount of pain and blood to get a porcupine for dinner.

Another aspect of the same problem is offered by the observation that toxicity (and therefore defence) is not an intrinsic trait of any compound (or structure). Canavanine may be a very toxic uncommon amino acid when incorporated in the diets of a variety of insects and vertebrates, yet for a particular beetle, the bruchid *Caryedes brasiliensis*, it is not toxic (though at a cost). *C. brasiliensis* has tRNA which avoids incorporating canavanine in the place of arginine in growing protein chains. Canavanine is even degraded and used in the beetle's protein synthesis (Rosenthal *et al.*, 1976, 1977, 1978). This is just another way of noting that the horns and hooves of bovids do not defend against ticks and mites. The newly-germinated seeds of *Enterolobium cyclocarpum* are such good food for *Liomys salvini*, pocket mice, that they maintain their body weight and even grow fat on a pure diet of them; *Sigmodon hispidus*, another common small seed and leaf-eating rodent in the same tropical deciduous forest habitat, dies within 2–4 days on a pure diet of these seeds (Hallwachs & Janzen, *submitted for publication*).

6.14 DEFENCE VARIANCE

Not only are defences species-specific in their context, but their effectiveness (in terms of fitness outcome) is highly variable among individuals. A fat and well-muscled East African buffalo appears to be little affected by its blood parasites, but if the animal is subjected to a normal dry season it loses weight and the blood parasites become a severe drain (lethal if at high density) (Sinclair, 1975).

Clearly the body weight of the animal itself is a kind of defence, and one that has as real a cost as horns, hide and fleeing behaviour. Again, a zebra's defences are down when she is heavily pregnant, a fact which has led to many a hyaena's full stomach. A plant growing in the shade may have a quite different content of defensive chemicals. Something happens to the defences of an *Enterolobium cyclocarpum* seed when it germinates; agoutis eat the swelling, germinating seed like candy but will refuse the seedling once it has ventured a few centimetres out of the seed coat.

Not only do defences have to be paired off against their protagonists before it makes sense to speak of them, but they also are dosage-dependent. This seems self-evident when speaking of rabbit speed when running away from coyotes, or palm nut hardness when faced with the molars of mastodons, but it also applies to plant poisons. Digitalis, cortisone, cannabinol, morphine, caffeine, theobromine, capsaicin, rotenone and many other familiar botanical drugs are mild perturbers of human neural systems at one dose and lethal at higher doses. When various uncommon amino acids were incorporated into the normal seed diets of larvae of a bruchid beetle, at 1% concentration, 32% had a lethal effect, while at 5% concentration 90% did (Janzen *et al.*, 1977). Viewed from the cost aspect, however, we encounter yet another source of non-linearity. Once the enzyme and substrate chains are set up (evolved) for the production of say 0.1 mg per gram of seed tissue, I suspect that the cost to raise it to 50 mg per gram of seed tissue is much less per milligram than were the factory expenses of the first 0.1 mg. The evolutionary invention of a spine, and the evolutionary lengthening of a spine are acts of quite different physiological, as well as genetic, costs.

6.15 HOSTS AS ISLANDS

Just as parasites compete with each other through the resource budget of their host plant or animal (Janzen, 1973b), they also influence each other through selecting for a defence which then incidentally hurts another parasite (or even predator). There are two components to this. First, several species of parasite may produce in concert a sort of diffuse selection pressure that favours a trait which reduces the intensity of damage from their combined impact. Neither alone might provide enough selection pressure to do this. Ironically, the final outcome might even be the loss of one parasite species from the plant's herbivore load, and this one would not necessarily be the newcomer that caused the significant increase in selection pressure. Second, a major parasite may select for a trait which removes parasites which individually are quite trivial reducers of host fitness. While an example is not at hand, I suspect that this is very common among small foliage-eating insects in tropical forests.

6.16 DEFENCE CONVERGENCE

Convergence in defences is an extremely common phenomenon. There are over 80 genera of small mammals in the world that have (probably independently) evolved spines out of their dorsal hairs (compilation from Walker, 1964), not to mention spiny lobsters, cacti and sea urchins. The urticating spines of *Automeris* silk moth caterpillars and the urticating spines of stinging nettles (*Urtica*) both contain a mixture of acetylcholine and histamine that is injected by hydrostatic pressure through a spine whose top breaks off after puncturing the victim; neither chemical by itself is painful when injected. Eye spots closely resembling the eyes of (presumably threatening) vertebrates have evolved independently on lepidopteran wings many thousands of times. Many secondary compound defences in plants are made by several different biochemical pathways, strongly implying that they are independently (convergently) evolved. All these convergences stem in part from the fact that many different organisms are confronted by the same predators or parasites. Different species of predators and parasites very commonly share vulnerability to defences at the same points. Caffeine in a seed probably acts as effectively upon a mouse as on a boring beetle larva. But convergences are probably also due to the fact that the most economical steps in the evolution of secondary compounds, and other defences, are similar in similar organisms. We find eye spots, and not pictures of mouths full of jagged white teeth on the faces of butterfly wings, because the way lepidopteran wings develop leads easily to patterns of wavy lines, loops and circles with different colours inside and outside the lines. The prominence of polyphenolic digestion inhibitors in plant foliage is a reflection not only of the similarity of digestive processes in herbivores, but the ubiquity of phenolic molecules in plant primary biochemical systems.

An area that requires study in defence biology is recognition of similarity of defence function among very different structures. A straightforward example is fever in a sick adult rabbit and the heat-seeking behaviour of its sick juvenile. Less obvious is that of the ostiole of a fig whose entrance hole is plugged with overlapping scales. The ostiole restricts entry to only a certain size of pollen-bearing wasp and strips the wasp of dirt as it passes through. It is as much a defence against interspecific rape and microbial contamination of ovules as is the stigma and style through which the pollen tube must grow in an ordinary flower (Janzen, 1979b). A colony of social wasps or bees that 'pays' workers to keep out a bird, bear or army ants is no different than a mammal that 'pays' phagocytes, antibodies and leucocytes to keep out microbes. The fungus that produces aflatoxin to protect its grain from the rodent that made the cache (Janzen, 1977b) is no different from the farmer who sprays her apples with parathion to keep out codling moths. Ear wax is probably a powerful antibiotic against the numerous bacteria and fungi that could do so well in that nutrient and moisture-rich incubator; it begs analogy with the intense secondary compound deposition in the walls of holes in the trunks of trees, or the walls of holes to be (Janzen, 1976b).

6.17 WHERE ARE THE DEFENCES?

Categories of defences are certainly not uniformly or randomly distributed over the earth's surface. There appear to be more poisonous snake species per species of snake in Africa than in the neotropics, and I suspect that the reason is that African snake populations have been subject to more intense continuous selection by snake-eating vertebrates than have the neotropical forms (Janzen, 1976c). Such guesses must be made with care, however; nearly 100 % of Australia's snakes are poisonous but for a quite different reason. It appears that the first to arrive were Elapidae, a family containing only poisonous species throughout the world. All but a very few of the spiny mammals mentioned earlier are tropical, while there appear to be just one or two species of skunk in each major habitat type. The Asian rainforest tropics are famous for their terrestrial leeches, parasites that are absent from the African and New World tropics; presumably antibody, mechanical and behavioural defences against their effects are likewise absent from the terrestrial vertebrates of Africa and the New World. We can even predict that terrestrial small rodents would have a very difficult time nesting on the ground in a leech-infested Asian rainforest. Likewise, the northern latitudes are famous for their horrible concentrations of vertebrate-biting mosquitos and other flies; except for certain swamp-rich habitats, many tropical forests are comparatively extremely poor in mosquitos that descend in hordes on large vertebrates. Again, I suspect that the native animals in these sites lack the defences that their northern relatives have. Why is the tapir's tail so short and its hair so thin? Levin (1976b) seems to have located a latitudinal gradient in alkaloid frequency and concentration in plants, and it is clear that the tropics are generally a much greater source of pharmacological drugs than is extra-tropical vegetation (though this may not be much more than a reflection of the much greater species richness of plants in the tropics).

6.18 IN CLOSING

Defences dear to our own way of doing things will long receive attention from humans. On the one hand, the detailed studies of how one organism protects itself from its attackers (e.g. Snyder, 1967; MacConnell, Blum & Fales, 1970; Eisner, van Tassel & Carrel, 1967; Eisner & Shepherd, 1966; Finkelstein, Rubin & Tzeng, 1976; Eisner & Dean, 1976; Eisner & Adams, 1975; Eisner, Kriston & Aneshansley, 1976; Rothschild *et al.*, 1979) are still of great importance. On the other hand, it is clear that we need a much greater understanding of two areas: what do those defences cost in resources and options, and how much is the fitness of the organism that lacks them depressed? For a long time we have been deluded in believing that tightly interacting members of the habitat are likely to be co-evolved, that defence effectiveness is related to defence amount, that the impact of the herbivore array is measured by its size or the amount it eats, etc. A change in attitude is required, instead of more descriptions of yet another alkaloid or its concentration in a tree's foliage. Above all, we need field experiments to tell us the fitness gains and costs that the organism accrues with and without its defences.

Chapter 7

Repair and its Evolution: Survival versus Reproduction

T. B. L. KIRKWOOD

7.1 INTRODUCTION

In its widest sense, the role of biological repair is to sustain the life of an organism and postpone, perhaps indefinitely, the time when it succumbs to an accumulation of random damage. Such damage is an inevitable consequence of the interaction between the organism and its environment. Living systems consist of highly ordered networks of structure and function, and it is natural that such order is intrinsically unstable and tends to give way to disorder. Damage ranges in severity from a change in a single molecule, the result perhaps of thermal noise, to the loss of whole organs and structures. Also, it ranges in frequency from that which occurs almost continuously to that which occurs only in a few individuals within a population. Present-day species are probably all capable of some degree of repair, but there are important differences between them. The most obvious example of an interspecific difference is in the ability to regenerate amputated limbs, which, among vertebrates, is restricted exclusively to certain amphibians (Scadding, 1977). However, there are also important differences at the level of intracellular repair, for example in the excision repair of DNA damaged by ultraviolet light (Hart & Setlow, 1974). It is these differences that raise fundamental questions about the relationship of repair to other ecological factors.

This chapter is an attempt to bring together research on all levels of repair and to relate repair activity collectively to other metabolic processes competing for resources within the organism as a whole. Inevitably, there are gaps, due partly to unequal research effort in different areas and partly to my own biases and limitations. Nevertheless, the overall picture that emerges demonstrates the main features of the relationship between repair and life-history, including, in particular, the crucial association between repair and longevity.

7.2 SOME GENERAL PRINCIPLES

7.2.1 Is repair necessary?

The earliest forms of life were almost certainly simple self-replicating polymers that arose spontaneously. Obviously, they were not initially capable of repairing themselves, and the fact that they survived proves that repair is not an absolute biological necessity. What was, however, essential was that their rate of creating new undamaged copies (births) at least equalled the rate at which existing copies were damaged and lost (deaths). This requirement remains fundamental to any biological system. Further, since natural selection favours any excess of births over deaths (at least in the short term), there is continual selective pressure to increase the birth rate or decrease the death rate. The key role of repair is to lower the death rate by mitigating the impact of damage. This lifts the pressure off the rate of replication and thereby permits a wider range of adaptation.

7.2.2 Pre-conditions for the evolution of repair

When considering the evolution of repair mechanisms, it is particularly important to recognize the existence of constraints which limit what is possible. Before a damaged part can be repaired, two basic conditions must be satisfied:

Condition 1: Information must be available for restoring the part to its original state.
Condition 2: The organism must be able to survive the damage for long enough that repair may occur.

The second of these conditions automatically precludes the repair of some of the more extreme types of damage. For example, animals which rely on mobility to find food or to avoid predators will not long survive the loss of the means of locomotion. Also, the evolution of new repair mechanisms is facilitated if damaged organisms are capable of at least some reproduction without repair, since the selective forces on repair act through the progeny of survivors. Indeed, if damage causes too large a reduction in reproductive fitness it is more likely that selection will favour avoidance of damage, or defence, rather than repair. This must be qualified, however, by the observation that from the point of view of natural selection, damage is only detrimental (and therefore repair only beneficial) to the extent that it reduces fitness.

7.2.3 THE NEED FOR INFORMATION

Access to information which defines normal structure and function is obviously critical, since without it a damaged part cannot be replaced. This information derives, ultimately, from the DNA sequence of the organism's genome, and we can see immediately that certain types of damage to DNA itself cannot be repaired. Also, the molecular machinery for translating DNA into protein may be irreversibly disrupted. As well as being damaged itself, genetic information may be unavailable for repair because it is 'locked away' in an unuseable form. In a multicellular organism, cells and tissues differentiate during development to become specialized in particular functions. Though this process is not fully understood, it is well known that it involves a closely controlled sequence of switching on and off specific genes (Gurdon, 1974). The potential for cells to participate in repair may depend on their ability to de-differentiate by reversing part of this sequence, which may not be possible. Finally, even if de-differentiation can occur, or if a repair pool of undifferentiated cells exists, repair still requires the structural and positional information that would permit an 'action replay' of the developmental programme. While there does not seem to be any overriding constraint to prevent this from being available, there may not always have been sufficient pressure for it to evolve among competing organizational priorities.

7.2.4 COSTS, BENEFITS AND RELATIVE PRIORITIES

The issues in the balance here are familiar ones in everyday life, as well as in the present biological context: damage is detrimental, but repair can only be undertaken at a cost. When is repair worthwhile, and which types of damage should be repaired before others? Before seeking to answer these questions, it is important to be clear about whose interests are at stake and what the units of currency are. In the context of evolution of neo-Darwinian natural selection, costs and benefits relate to genes rather than individuals (see also Chapter 1) and are measured in terms of the expected relative contribution of different genotypes to the gene pool of future generations, i.e. in terms of relative reproductive fitness. This will be defined more carefully later (section 7.5).

If we write D for the average decrease in relative fitness resulting from a particular type of unrepaired damage and R for the cost of repairing it, measured as the decrement in fitness that results from channelling the necessary resources into repair, then repair should never occur unless $D > R$. As well as the immediate cost of repair, R, there is likely to be an 'overhead' cost, C, involved in carrying the potential for repair within each individual. This will include such things as organizational rearrangement, perhaps interfering with growth, as well as the direct metabolic cost of replicating and synthesizing special genes and enzymes. The repair system will only evolve if the net expected benefit over a lifetime exceeds the net cost, i.e. if $F(\mathrm{D}-\mathrm{R})-\mathrm{C}>\mathrm{O}$, where F is the frequency per generation of the particular type of damage. As a first approximation, we may expect the relative priority of different repair systems to be proportional to $F(\mathrm{D}-\mathrm{R})-\mathrm{C}$, so, other things being equal, repair will be most strongly favoured when F is large. In later sections, it will be useful to draw a distinction between damage that occurs frequently, and is therefore predictable, and damage that occurs only rarely. By the above argument, the former is highly likely to be subject to repair, even though individual instances may be relatively minor. The response to this type of damage will be termed 'maintenance' repair. The latter type of damage will often be more serious and may or may not be subject to what will be termed 'emergency' repair.

An alternative to repairing damage is to guard against it by defence (see Chapter 6). This too will involve some cost, P, per lifetime, which may either be a function of F, if damage is actively evaded each time it threatens, or a constant, if the defence strategy is more passive. In the absence of other constraints, defence will be favoured instead of repair if $\mathrm{P}<(F\mathrm{D}+\mathrm{C})$. However, the optimal strategy will often be restricted to a mixture of both defence and repair. In these cases, F itself will be a function of the efficiency of defence.

This analysis goes some way towards illuminating the factors involved in determining the evolutionary and ecological balance between the costs and benefits of repair. In later sections, the links between these factors and an organism's life-history will be explored in more detail. In particular, it becomes important to recognize that D and R are not necessarily constants, as implicitly assumed above, but that they may vary with age. However, before doing this I shall review briefly the main types of repair process and point out the sources of their costs. Present knowledge does not yet permit estimation of F, D, R and C for individual processes, so we cannot explore relative priorities in detail.

7.3 INTRACELLULAR REPAIR

Since all organisms (except viruses) consist of cells, repair processes can be divided into those that occur within cells and those that involve cell division.

The former probably operate in every organism, single- or multi-celled, plant or animal. Furthermore, there are broad similarities in intracellular repair between species as diverse as bacteria and mammals, since all cells have important features in common and are exposed to similar types of damage.

Damage to a cell can be of two main types, that which directly involves the pathways of transfer of genetic information and that which does not. Either can kill the cell, but damage to the informational apparatus is potentially more serious since it can easily become irreparable (see Condition 1 of section 7.2.2), whereas other damage can be diluted by fresh transcription and translation of genetic information. The basis of a cell's function is the replication and translation into protein of the information coded in the polynucleotide (DNA) sequence of the genes. DNA is transcribed into an RNA working copy, which is then translated into protein (Fig. 7.1). Although

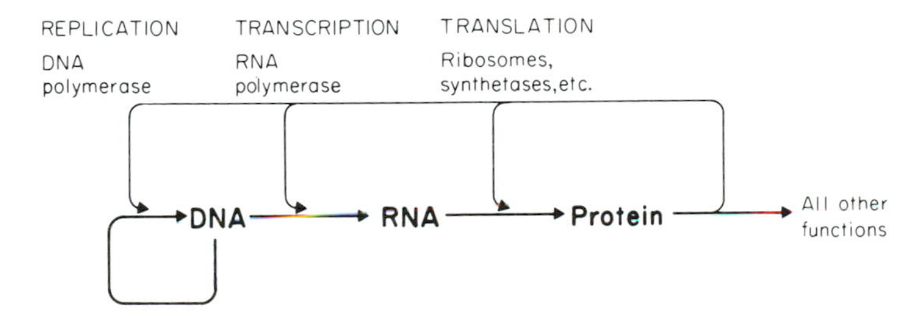

Fig. 7.1 Pathways of transfer of genetic information within a typical cell. The main flow of information is from DNA to protein, but the system as a whole is cyclic since proteins participate in transcription and translation, as well as in DNA replication.

genetic information is carried by DNA alone, its flow within the cell is not purely unidirectional since highly specific proteins (enzymes) are required to replicate, transcribe and translate it, and the system as a whole must remain intact if it is to be of use. Damage which *inactivates* one of its components is not critical if the damaged part can be replaced, but damage which merely *alters activity* can be much more serious. This is because the cyclic nature of the information transfer pathways means that even minor faults can create considerable havoc. For example, a faulty enzyme responsible for a crucial information-transfer step could generate a large number of new faults and some of these might feed back into the cycle. If uncontrolled, this could continue until the cell ceased to function. The most likely form for this type of damage to take is the incorporation of erroneous units (amino acids or nucleotides) during macromolecular synthesis.

Errors in macromolecules have been the basis of two distinct hypotheses concerning progressive loss of cell viability with increasing age. The first was the somatic mutation theory (see Szilard, 1959; Maynard Smith, 1962; Curtis, 1966) involving errors in DNA alone, while the second was the protein error theory (Medvedev, 1962; Orgel, 1963), concerned primarily with feedback of errors in protein synthesis. However, the links between the different levels of information transfer suggest that a general error theory, involving all information-handling processes, is more plausible (Woolhouse, 1969; Lewis & Holliday, 1970; Holliday & Tarrant, 1972; Orgel, 1973; Burnet, 1974; Kirkwood, 1980). Experimental studies of the extent to which informational damage does, in fact, pose a threat to cells are still inconclusive, though there is a growing body of data indicating that the propagation of errors as described above is a widespread phenomenon (Lewis & Holliday, 1970; Holliday & Tarrant, 1972; Branscomb & Galas, 1975; Linn, Kairis & Holliday, 1976; but see Evans, 1977; Gallant & Palmer, 1979 for opposing views). Theoretical studies of the stability of information transfer have identified two main factors through which a cell could control the likelihood of being precipitated onto a pathway of irreversible, and ultimately lethal, error propagation (Hoffman, 1974; Kirkwood & Holliday, 1975; Kirkwood, 1980). These are to keep new errors to a minimum and to destroy old ones as early as possible. Ways of achieving these goals are reviewed briefly below.

Macromolecular 'proof-reading'

The accuracy with which cells synthesize macromolecules is known to exceed the discrimination expected on the grounds of chemical differences between their basic units. This is explained by the operation of additional enzymatic 'proof-reading' steps which reduce the level of errors during synthesis by eliminating mistakes as they arise. Although partly defensive, proof-reading is logically included within the definition of intracellular repair. Generalized 'proof-reading' schemes have been described by Hopfield (1974) and Ninio (1975), in which accuracy can be made arbitrarily high at the expense of rapidly increasing cost. A formal analysis of this law of diminishing return has been made by Bennett (1979), while experimental evidence of a trade-off between accuracy and growth in *Escherichia coli* was found by Galas & Branscomb (1975).

Selective degradation of abnormal protein

At the same time that new proteins are being synthesized, others are being hydrolysed to their component parts. One reason for this seemingly wasteful process is that it permits rapid adaptability to internal and external changes; proteins which are no longer needed are cleared away quickly and do not interfere with new ones. A second reason is that, since proteins cannot themselves be repaired, degradation is the only practicable way of removing damaged or abnormal proteins which might otherwise be harmful to the cell. In fact, there is good evidence that in both bacterial and mammalian cells abnormal proteins are degraded selectively (Goldberg & Dice, 1974; Goldberg & St. John, 1975), and, furthermore, that hydrolysis of abnormal proteins requires metabolic energy (Goldberg, 1972; Bukhari & Zipser, 1973).

DNA repair

DNA is the only molecule within a cell that can truly be repaired. This is because its unique double helix structure permits damage in one strand to be repaired according to the information carried by the other. In view of the primary role of DNA as the carrier of genetic information it is no surprise that a wealth of repair mechanisms has been found to operate in both prokaryote and eukaryote cells (for a recent review, see Hanawalt, Cooper, Ganesan & Smith, 1979). Types of damage that are known to be repaired include missing, incorrect or altered bases, as well as interstrand cross-linkage and single- or double-strand breaks. The two main categories of repair are excision repair and recombination repair. In excision repair, damage is recognized and the lesion, along with adjacent nucleotides, is excised from the strand containing it, to be replaced by fresh synthesis using the intact complementary strand as template. Recombination repair involves strand exchange with intact homologous DNA and is used when excision repair is not possible, for example following a double-strand break or when damage occurs in single-stranded DNA in the region of a replication fork. Precise quantification of the costs of DNA repair is not yet possible, but these include both the cost of repair synthesis itself and the cost of synthesizing repair enzymes and the genes coding for them.

Dilution of damage by turnover of cellular material

In addition to selective scavenging of faulty protein, the natural turnover of cellular material is a potent way of preventing or retarding the accumulation of damage, including toxic metabolic wastes that cannot be transported out of the cell. This is especially true of dividing cells, where 50 % of cellular material must be created afresh between subsequent cell divisions, and lends support to the hypothesis that cellular senescence in higher organisms may, in part, be due to a reduction in the rate of cell division within static tissues (Bidder, 1932; Sheldrake, 1974; Calow, 1978a; Hirsch, 1978). It must be noted, however, that this applies only to structural and functional damage which does not impinge directly on information transfer, since the deleterious effects of damage within the information transfer system may be amplified, rather than diluted, by rapid turnover.

Detoxification

Ingestion of toxic chemicals from the environment may pose a serious threat to life, and organisms have a powerful range of detoxification mechanisms, involving inducible metabolic processes, to render such poisons harmless (Williams, 1959; Parke & Smith, 1977). (In higher animals, systemic toxicity is controlled mainly by the liver, so detoxification is not strictly a mechanism for intracellular repair, but it is included here for convenience.) There is an obvious metabolic cost in maintaining mechanisms for dealing with a wide range of poisons that may be encountered only rarely.

7.4 REPAIR BY CELL DIVISION

In multicellular organisms, damage can occur in which numbers of whole cells are lost or the continuity of a tissue or organ is broken. Repair at this higher physiological level is diverse and is reviewed here only briefly. The basic mode of repair is through cell division, and the immediate metabolic costs relate to the amount of new cellular material that must be generated. Less obvious are the organizational costs associated with such repair. At the very least, any extra genetic information needed to control the process must be carried in the genome and the relevant regulatory enzymes synthesized when necessary.

7.4.1 ANIMALS

Wound healing and organ regrowth

Occasionally, a wound will break the continuity of an organ or a large enough number of cells may be lost, e.g. through toxicity, to stimulate renewed cell division. Most organs and tissues are capable of this type of repair to some degree, and it seems likely that it is regulated by similar homeostatic mechanisms as were responsible for original development, though it is interesting to note that regrowth often slightly exceeds what was there originally, as, for example, in the formation of scar tissue. This wasteful inefficiency might seem to challenge the optimality principle underlying this book, but more probably it reflects the limitations on theoretical options placed by operational constraints. It may be that the homeostatic control responsible for stopping growth works better when neighbouring organs are growing competitively, than when one organ is growing by itself.

Regeneration

An extreme form of damage, but one which does not necessarily result in immediate death, is the loss of a limb or other appendage. Species differ considerably in their ability to regenerate such structures, and the subject of regeneration is complex and fascinating. There is space here only for a few key points, and the reader is referred to Goss (1967) for an excellent general account. First, regenerative capacity declines in inverse proportion to structural complexity; simple organisms have extensive regenerative powers while higher vertebrates (birds, mammals and most reptiles) do not regenerate at all. Second, where it exists, the capacity to regenerate is usually restricted to certain regions, called regeneration territories, and is linked to the presence of special pluripotent cells (i.e. cells with the potential to differentiate into several alternative types), which may arise through partial de-differentiation. Nerve cells also play an important part in determining regenerative potential (Goss, 1967; Elder, 1979).

These observations can be related to the principles discussed in section 7.2. More complex animals may be better at avoiding severe damage (section 7.2.1) or less likely to survive it (section 7.2.2). They may also, through greater specialization, have less ability to make available the information necessary for regeneration (section 7.2.3). In a study of the phylogenic distribution of limb regeneration in amphibians, Scadding (1977) found it to be restricted to certain small salamanders for whom limbs appeared to be least important for survival. These species may be near to the borderline where loss of a limb becomes a sufficiently serious threat to survival that it is not selectively worthwhile to retain regenerative ability.

Cell turnover

Certain types of cell, for example epithelial cells and cells of the immune system, are continually being lost and replaced by cell division. These systems involve both repair and defence, since part of their function is to absorb the impact of extrinsic and intrinsic damage which would otherwise threaten the whole body. Through acting as such a buffer, the cells themselves are highly likely to be damaged and must constantly be replaced. Their turnover also serves to flush out the damage. However, to maintain cell turnover imposes an increased risk of informational damage, including in particular, mutagenic or carcinogenic lesions, on the mitotically active precursor cells. Therefore, it may also be necessary to use special mechanisms to protect these cells (Cairns, 1975).

7.4.2 Plants

At higher physiological levels, repair in plants is organized very differently to that in animals, reflecting fundamental differences in structure. While a certain amount of repair, analogous to wound-healing in animals, is involved in isolating damage by forming protective cambium layers, the main response to damage is to replace rather than repair the damaged part. J. L. Harper (personal communication) has emphasized the critical distinction that all higher plants (together with certain invertebrates) express the genotype by repeated reiteration of modular units of construction, whereas most animals comprise a single but differentiated whole. Modular units may die as a result of damage, competition or environmental change, and their loss may stimulate the production of new units, but it is arguable whether this is true repair. In the present context, repair in whole plants will be restricted to meaning a direct response to random damage. For the most part, it seems that plants keep damage at equilibrium by coupling renewal with defensive organization, as in heavily grazed species which have low growing points or leaves close to the ground (Mahmoud, Grime & Furness, 1975; Grime, 1979).

7.5 REPAIR, LIFE-HISTORIES, AND MEASURES OF FITNESS

The remaining part of this chapter examines the link between repair and the life-history of an organism, with particular attention to the allocation of resources to repair and the prolongation of life. In this context, it is convenient to group repair activities into two classes, maintenance repair and emergency repair, defined in section 7.2.4. In different species, certain repair systems may belong to either of these classes, depending on the frequency with which they are needed. Also, some repair activity may belong to both, especially when different types of damage evoke similar responses. Nevertheless, the distinction is a useful one and will be maintained for the purpose of general discussion. In specific instances, it is relatively easy to decide how a particular repair activity should be classed, so no attempt will be made to do this systematically for the individual processes reviewed above, though we may note that emergency repair will usually be restricted to the higher levels of repair by cell division, for example regeneration.

The life-history of an organism may be defined as the set of co-adapted traits that together determine its age-related pattern of reproduction and mortality. Thus, the life-history is either an *average* for a population or an *expectation* for an individual. Some life-histories are obviously better in a given set of conditions than others, and a considerable body of literature exists in which attempts have been made to identify the ecological determinants of optimal life-histories, including the allocation of energy to different metabolic activities (Gadgil & Bossert, 1970; Taylor, Gourley, Lawrence & Kaplan, 1974; Leon, 1976; for reviews see Stearns, 1976, 1977; Calow, 1977a; Charlesworth, 1980). It is in the context of life-history evolution that we are best able to examine the costs and benefits of repair and the optimization of repair effort. To take our analysis further, we need to relate these factors to some defined measure of fitness.

The measure of fitness most commonly used in previous studies, and that which will be used here, is the intrinsic rate of natural increase, r, in the size of a population with stable age structure. Writing l_x for the probability of surviving to age x and b_x for the average rate of reproduction at age x, r is the unique real root of the equation:

$$\int_0^\infty e^{-rx} l_x b_x = 1$$

(Lotka, 1924; Fisher, 1930). The advantages of r as a measure of fitness are that:

1 it accords with the simple idea of natural selection in which faster multiplying species outcompete slower ones;
2 it is directly related, by the equation above, to quantities l_x and b_x that can be related in turn to repair.

However, there are some disadvantages to using r as a measure of fitness and these should be mentioned, if only to show that they do not invalidate our later conclusions. The first drawback is that species cannot be forever increasing in numbers since density-dependent mortality would soon exact a heavy toll. This realization has generated the alternative concept of K selection (MacArthur & Wilson, 1967; Gadgil & Solbrig, 1972) where the 'carrying capacity' or equilibrium population size, K, is maximized for species under intense competition in stable environments. Since low mortality is one of the characteristics attributed to K selection, as opposed to r selection (which operates primarily in seasonal or colonizing species, where rapid growth in situations of little density-dependent regulation is more important), repair should be mainly K-selected. However, as Stearns (1977) has pointed out, it is difficult to use K itself as a measure of selection since it cannot easily be related to life-history parameters (see also, Charlesworth, 1980, pp. 255–256). Useful as the r–K distinction has proved itself to be in ecological thinking, it is less helpful in the particular context of life-history optimization. Even in strongly K-selected species, an increase in K is attained only through a phase when $r > 0$ and selection then acts through differences in r. Also, even when population size is constant, r may still be at a local maximum, since small changes in the life-history parameters are likely to decrease fitness and result in a negative r. Thus r is a more general measure of fitness and the constraints associated with K selection should be translated into constraints on the variance of l_x and b_x, and, therefore, of r. This is equivalent to finding a local, rather than a global maximum for r. In the present context, we can assume these constraints to exist without needing to specify them, provided we restrict l_x and b_x to forms appropriate for K-selected species. For mathematical and genetical rigour, we should translate r as a measure of selection into u, the long-term probability of genotype survival (Robertson, 1960; Charlesworth & Williamson, 1975). However, I will use r because this is more widely familiar. As one would expect intuitively, both u and r lead to much the same conclusions (Charlesworth & Williamson, 1975).

Using r as the measure of fitness, the next three sections examine optimal maintenance repair strategies for a variety of basic life-history plans. The treatment is largely non-mathematical, since my aim is to describe a general view, but can be related to the mathematical analyses of Taylor. Gourley, Lawrence & Kaplan (1974) and Leon (1976).

7.6 MAINTENANCE REPAIR AND ASEXUAL REPRODUCTION

The importance of repair, and its relation to life-history, are very different in species that reproduce with and without sex. In sexual reproduction, the only materials transmitted to the progeny are the genes and cytoplasmic components of the gametes. Thus, for a new individual to start life without an accumulation of inherited parental damage it is necessary only that such damage is kept out of the germ-line. This principle also applies, of course, to any non-sexual species in which there is a clear distinction between germ-line and somatic cells. For an asexual species without such a distinction, the accumulation of damage may be more serious. Asexual life-histories are very diverse and it would be impossible to consider all types here. I shall review only a few main examples, though it will be relatively simple to extend the arguments to others.

Binary fission

The most simple life-history is that in which the organism divides symmetrically to generate two new and identical individuals. For such an organism it is clear that damage cannot be allowed to accumulate during the lifespan (between successive divisions) or it would lead to rapid extinction. Therefore, repair capacity must at least be adequate to cope with the average rate of damage. Beyond this, extra investment in repair provides a safety margin

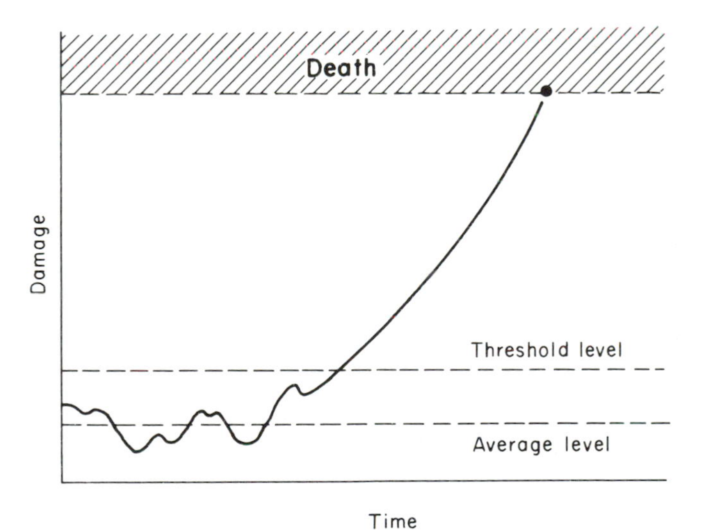

Fig. 7.2 An asexual organism which reproduces by binary fission must maintain the level of damage around a constant average. Above a certain threshold level, the organism is unlikely to be able to prevent damage from accumulating steadily, resulting in death. This accumulation may be rapid or may take place over several generations. The safety margin between the threshold and average levels of damage will depend on the efficiency of repair.

which can accommodate random fluctuation in the level of damage. Death occurs whenever this margin is exceeded (see Fig. 7.2). However, to provide a wide safety margin detracts from the resources available for growth, and the optimal level of repair is that which maximizes r by balancing these effects. For example, writing ρ for the fraction of resources invested in repair, suppose the death rate $m = m_0/\rho$ and the generation time $T = T_0/(1-\rho)$, where m_0 and T_0 are constants. Then it is readily shown that r, which is the root of $2\,e^{-rT}.e^{-mT} = 1$, is maximized for $\rho = m_0 T_0/\log 2$; this must be between 0 and 1 as $m_0 T_0$ must be less than log 2. A similar result can be obtained for other plausible functions $m(\rho)$ and $T(\rho)$.

Asymmetric division

Asymmetry in asexual reproduction can arise in many ways and can involve a single division to form two new individuals or a multiple division to form many. In such cases it may be advantageous to partition damage and/or repair effort asymmetrically (see Fig. 7.3). This may be either an active process, in which the quality of some progeny is raised at the expense of others, or a passive one, the result either of chance or of intrinsic asymmetry in the division process. An example of intrinsic asymmetry is in the asexual division of some species of turbellarian worms into anterior and posterior halves. Each half usually regenerates a whole worm, perhaps from a stock of quasi-embryonic cells (neoblasts) which divide conservatively so as to minimize the incidence of damage, but only the posterior halves can do so indefinitely (Sonneborn, 1930). The finite life of the anterior halves may result from a failure to repair, by renewal, major organs such as the brain (Calow, 1978a).

Sex as repair

It has already been noted that sex has important links with repair, since in sexually reproducing species it is only necessary to keep damage at equilibrium in the germ-line. In species capable of reproducing both with and without sex, sexual reproduction may itself be a repair process of considerable importance. Sex can act as repair in several ways. First, the relatively small number of divisions in germ-cell lineages may help reduce the level of replicative errors. Second, the gametes are often much smaller than the parent or its asexual offspring, so a high level of repair can be invested in them even when the parent cannot repair itself. This 'lifeboat' function of sex is little different, however, to the asexual production of small propagules in response to stress. Third, a more

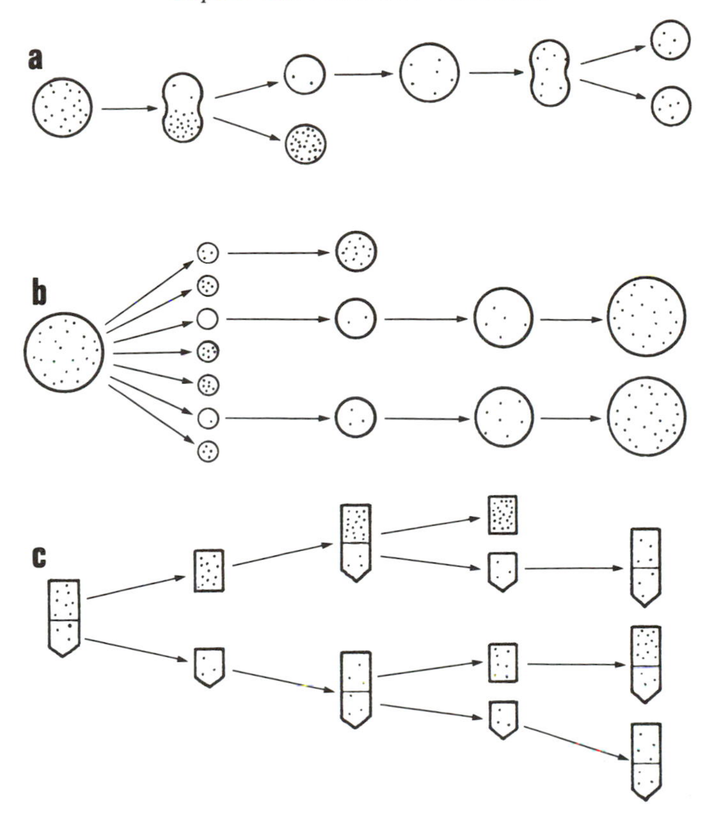

Fig. 7.3 Three alternative ways of partitioning damage (represented by dots) asymmetrically in asexual reproduction. (a) When damage becomes a threat to viability, it may be advantageous periodically to segregate it mainly into one daughter, leaving the other relatively free. (b) Among multiple progeny, damage may be distributed unequally as a result of chance. (c) When division is intrinsically asymmetric, as into anterior and posterior ends of flatworms, only one division product may be able to keep damage at equilibrium; the other loses viability through progressive accumulation of damage.

important role of sex is its potential for informational repair (Muller, 1964; Bremermann, 1979). During a succession of asexual divisions, organisms may accumulate mutations which are neutral in the environment of the time, since the affected genes are not needed, but which may become seriously detrimental if the environment changes. Under these circumstances, sexual recombination can restore viability. Bremermann (1979) has suggested that under suitable conditions the repair benefits of occasional sexual reproduction can outweigh the costs of carrying the mating genes within the genome. It has also been suggested that meiosis may constitute an active repair process in which DNA lesions are removed (Bernstein, 1977; Martin, 1977) and cytoplasmic damage is 'dumped' in short-lived sister cells (Sheldrake, 1974), leaving the gamete itself in prime condition. Finally, the selective effect of competition between gametes may be a potent way to eliminate faults (Cohen, 1977).

Evidence supporting a link between sex and DNA repair is found in recent studies on *Paramecium tetraurelia*, which, if prevented from reproducing sexually, divides only a finite number of times (150–200) before the clone of progeny dies out (Sonneborn, 1954). Smith-Sonneborn (1979) demonstrated that activation of an inducible DNA repair system (photoreactivation of pyrimidine dimers) resulted in an increase in the number of divisions before death. This suggests that DNA damage plays a part in cell death and, therefore, that sex, which rejuvenates the cells, can overcome this damage.

Sex is itself costly, since as well as the metabolic cost of carrying the mating genes and seeking a mate, there is the well-known genetic cost (Maynard Smith, 1978b). Experimental studies have shown that triclad flatworms use more energy to reproduce with sex than without it (Calow, Beveridge & Sibly, 1980), and it will be interesting, therefore, to extend these studies in order to obtain a more complete picture of the costs and benefits of sex in relation to other forms of repair.

7.7 SOMATIC REPAIR IN SEMELPAROUS ORGANISMS

Species that always reproduce sexually are divided into those that reproduce only once in a lifetime and die (semelparous) and those that reproduce repeatedly (iteroparous) (Cole, 1954). In either case, primary interest centres on repair of the soma, since damage must always be kept at long-term equilibrium in the germ-line using any or all of the methods discussed above.

Semelparous reproduction is most likely to evolve in species where the chance of an adult surviving to breed a second time would, in any case, have been small (Charnov & Schaffer, 1973). Since parental care is usually precluded by death, semelparity tends to be associated with large numbers of progeny. Examples of semelparous organisms include the Pacific salmon, octopus, many invertebrates, and annual and biennial plants.

Semelparous life-histories consist of two principal phases, growth and reproduction. During growth, the organism acquires resources. During reproduction, it converts stored resources, together with any it may continue to obtain, into progeny. (In extreme cases, the body of the parent may actually be sacrificed to benefit the young as a source of food.) As in section 7.5, the optimal somatic repair level is determined by balancing growth rate and fecundity (reduced by diverting resources to repair) against the probability of surviving to complete the life-cycle (increased by repair). However, it is also probable that the distinction between the two phases of life will be reflected in the level of somatic repair, since an individual which has embarked on the latter should mobilize all available resources to increasing reproductive output. Somatic repair may not cease altogether during reproduction as this could cause premature death, but it is likely to be considerably reduced, contributing to rapid post-reproductive death (Fig. 7.4). In the octopus,

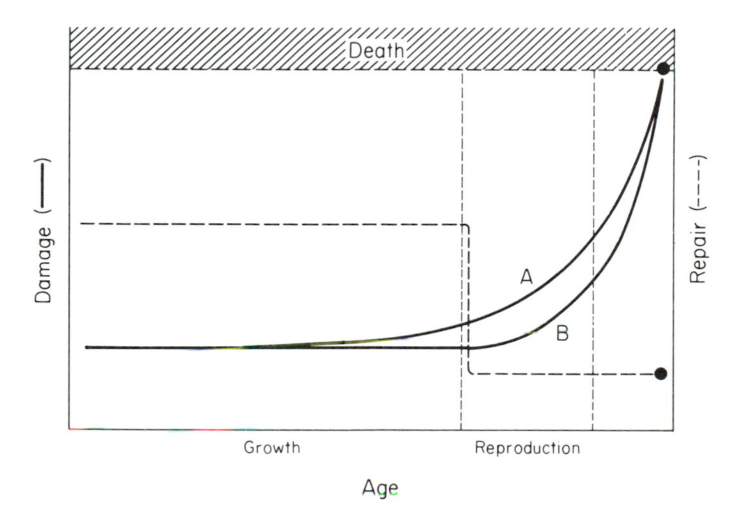

Fig. 7.4 A suggested optimal strategy for somatic repair in semelparous species. During growth, repair remains more-or-less constant, but is reduced at the onset of reproduction. The level of damage in the soma may remain constant during growth (curve B), permitting indefinite pre-reproductive survival, or may increase progressively (curve A). In both cases, damage accumulates rapidly after the reduction in repair, causing early post-reproductive death. The level of damage in the germ-line may be maintained at a constant low level by extra repair, or may be reduced by special repair at the time of reproduction (see text).

Wodinsky (1977) has shown that hormone secretions from the optic gland exert a controlling influence on post-reproductive death, since surgical removal of the gland significantly prolongs survival. These secretions, which, among other things, influence feeding behaviour, may act in part to reduce the level of somatic repair. Also, in semelparous plants, it is well established that hormones regulate senescence and death (Woolhouse, 1974), but it is not clear whether these affect repair.

During the growth phase, the level of somatic repair may be constant, or it may fall progressively in relation to size and maturation (Calow, 1978a) as the need for prolonged survival diminishes. In many semelparous plants and animals, death can be postponed, sometimes indefinitely, by preventing reproduction (see review by Calow, 1979). This extended survival of non-reproducing individuals suggests that repair is more often constant during growth and that the onset of reproduction is itself the trigger for somatic decline. The fact that in some species such individuals do eventually die may be indicative of a decline in repair effort, but more probably reflects insufficient pressure to evolve an appropriately high level of somatic repair.

7.8 SOMATIC REPAIR IN ITEROPAROUS SPECIES

The allocation of resources to somatic repair in iteroparous species is of particular interest since iteroparous lifespans could, in principle, last indefinitely. Reproduction occurs repeatedly during adulthood and is not coincident with completion of the individual life cycle. However, with few, if any, exceptions, each iteroparous species has a characteristic maximum lifespan and individuals which survive the hazards of starvation, predation, accident and disease invariably exhibit a broad spectrum of degenerative change, or senescence, as they approach this limit. This process culminates in certain death, even in highly protected environments. (Note that plants are excluded here since although perennials are capable of repeated sexual reproduction, they may also multiply vegetatively and, therefore, do not have a clear distinction between germ-line and soma.) Why is there a finite limit to lifespan, and what determines the longevities of different species? Is senescence due to inadequate repair? Behind these questions lies an apparent paradox, stated succinctly by Williams (1957), that 'it is remarkable that after a seemingly miraculous feat of morphogenesis a complex metazoan should be unable to perform the much simpler task of merely maintaining what is already formed'. Resolution of the paradox requires either that we explain why it is advantageous to set a limit to lifespan, or that we show it is disadvantageous to invest enough resources in somatic repair to sustain life indefinitely.

A popular argument in favour of the first alternative is that a limited lifespan is beneficial in that it improves a species' long-term chance of survival by promoting the turnover of generations and thereby giving natural selection a better chance to try out new variants in the 'cauldron of changing selective forces' (Woolhouse, 1967). The implication is that senescence is the manifestation of an evolved life-terminating programme. This theory has a long history, being commonly attributed to Weismann (1891) (but see Cremer & Kirkwood, 1981). However, it is now widely regarded as untenable. First, it is unnecessary, and perhaps impossible, to evolve such a programme since natural mortality usually ensures that very few animals survive long enough to senesce (Medawar, 1952). Second, even if this were not the case, the theory faces the difficulty that it assumes priority of group selection (Wynne-Edwards, 1962) over individual selection, which is not normally valid (Maynard Smith, 1964). To prove this, let A and B be two genotypes which differ only in that A has a mechanism to terminate life at some age S, while B does not. Then their respective rates of natural increase are the roots of $\int_0^S e^{-r_A x} l_x b_x dx = 1$ and $\int_0^\infty e^{-r_B x} l_x b_x dx = 1$. Provided $l_x b_x > 0$ for at least some $x > S$, it is obvious that $r_B > r_A$ so that B is fitter than A.

The second alternative, that a high level of somatic repair is disadvantageous, must be considered in the light of a crucial observation on iteroparous life-histories, namely that the force of natural selection declines progressively with age (Medawar, 1952; Williams, 1957; Hamilton, 1966; Charlesworth & Williamson, 1975). The reason for this is simply that since iteroparous species spread their reproductive effort over a considerable portion of their lifespan and since natural selection operates through differences in reproductive fitness, the force of selection falls with age in proportion to the remaining fraction of total expected reproduction. Mutations with late age-specific effects affect only the subsequent reproductive contributions of the small number of individuals surviving to that age, and they are therefore subject to much weaker selection than equivalent mutations with early age-specific effects. For this reason, Medawar (1952) suggested that senescence might, in fact, be nothing more than the result of random accumulation of deleterious mutations with late age-specific effects, since such mutations would encounter little or no negative selection in natural populations (senescence being seen only when accidental mortality is greatly reduced in protected environments). However, there is a logical difficulty with this view since, in the absence of senescence, it is hard to see what would be the timing mechanism to determine 'lateness' at the older end of an iteroparous lifespan (Kirkwood, 1977; Calow, 1978a; Sacher, 1978; Kirkwood & Holliday, 1979). (Note that while it is conceivable that expression of such genes might be cued by normal developmental events, senescence usually occurs well after the end of what is commonly understood as the 'developmental programme'. To postulate the existence during adult life of single, or successive physiological 'clocks' measuring cumulated time, rather than mere periodicity, is unwarranted unless an adaptive role for them, *other than timing the onset of senescence*, can be identified.) A more plausible theory, built on the same foundations, is that senescence is the late deleterious side-effect of genes which confer an early advantage (Williams, 1957). In particular, the 'disposable soma' theory (Kirkwood, 1977; Kirkwood & Holliday, 1979) proposes that senescence results from a strategy of investing only enough resources in somatic repair to ensure the retention of vigour through the normal expectation of life in the wild.

The disposable soma argument can be summarized as follows. Suppose there is a non-senescing species with a given level of natural mortality, say 10 % per year. Then the chance that an individual survives to an age of, say 100 years is only 0.003 %, or 1 in nearly 40 000, which for practical purposes is negligible. If, as seems probable, it costs more to repair the soma well enough for it to last indefinitely than for it to last only 100 years, it would be selectively advantageous to settle for the shorter limit to lifespan and to invest the saving in resources into faster growth or more progeny. By thus trading longevity against reproduction an optimum level of repair for a given ecological niche may be reached. For example, a species subject to high mortality should place relatively high priority on reproduction and low priority on repair of the soma, which will therefore have a short maximum lifespan. On the other hand, a species subject to low mortality may profit by doing the reverse, especially if a protracted period of parental care and learning are necessary. It is worthy of note that, even if accidental mortality were to become very low, it is unlikely that it would ever pay to attain indefinite longevity. Indeed, there may well be a limit beyond which it is not selectively worthwhile to invest any more in repair, even if a significant proportion of the population survives long enough to show signs of senescence. This may already be the case in certain of the longer-lived and more intelligent mammals, especially primates. The main value of the disposable soma theory is not, however, that it explains these well-known correlations between longevity, mortality and fecundity, but that it makes definite predictions which are open to experimental test. In particular, it predicts that longer-lived species should have better somatic repair (Fig. 7.5).

During senescence, there is a general breakdown of all physiological functions (see, for example, Comfort, 1979). This points to the basis of senescence being, at least in part, at a fundamental physiological level within individual cells. This observation is supported by:

1 the finite replicative lifespan of somatic cells in tissue culture (Hayflick & Moorhead, 1961; Hayflick, 1965);

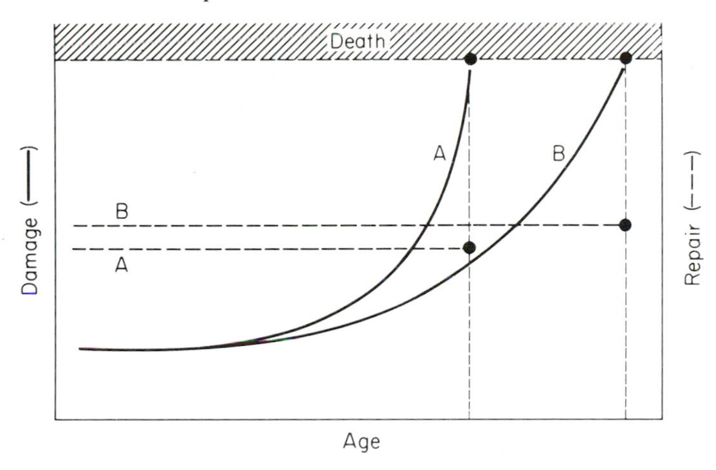

Fig. 7.5 The disposable soma theory predicts an inverse correlation between somatic repair levels and iteroparous longevity. Species A invests less in somatic repair than B and dies from a progressive accumulation of damage (= senescence) at an earlier age. As in semelparous species, the level of damage in the germ-line may be constant, or may be reduced by special repair at reproduction (see text).

2 the inverse correlation of this *in-vitro* lifespan with the age of the cell donor (Martin, Sprague & Epstein, 1974; Schneider & Mitsui, 1976);
3 the suggestion of a positive association of culture lifespan with species' longevity (Hayflick, 1973, 1977; but see also Stanley, Pye & MacGregor, 1975).

Repair at higher levels is, of course, important for survival, as is the design of parts of the body that are not renewed (e.g. teeth), but such systems depend on cell viability for their own function and are less likely to play a primary role in determining longevity. Among the wide variety of theories to explain cellular senescence there are several which directly implicate damage, including attack by free radicals (Harman, 1956), molecular cross-linkage (Bjorksten, 1968), accumulation of toxic wastes (Sheldrake, 1974), amino-acid racemization (DeLong & Poplin, 1977), changes in cell membranes (Zs.-Nagy, 1978), as well as the general error theory (Orgel, 1973; Kirkwood & Holliday, 1979; Kirkwood, 1980). Any such theory must explain not only why somatic cells senesce and die, but also why unicellular organisms, germ cells and cancer cells do not. The error theory can do this easily by postulating different levels of error regulation in different types of cell (Kirkwood, 1977) and may be the strongest candidate for a primary mechanism of cellular senescence, since feedback of errors will progressively impair the cell's ability to repair other types of damage. However, the final death of a cell could be due to any or all of these processes, while another important factor limiting a somatic cell's ability to keep damage at bay may be a reduction in turnover in static tissues (Sheldrake, 1974; Calow, 1977b, 1978; Hirsch, 1978).

To test the disposable soma theory and, if it is correct, to find out which repair processes are the most critical for determining longevity, it is necessary to compare the repair efficiences of different species. Comparative studies by Hart & Setlow (1974) and by Sacher & Hart (1978) have shown a positive correlation of the efficiency of DNA excision repair with species longevity (see also Hart, D'Ambrosio, Ng & Modak, 1979). These results support the theory and highlight the need for further comparative data on other aspects of repair.

The final question in this section is whether an iteroparous organism should maintain a constant level of maintenance repair, or vary it in accordance with events such as maturation and reproduction (Calow, 1978). At present, this question can only be answered in an exploratory way. Consider first the period up to and including the time of maturation, and recall that it is implied in the definition of maintenance repair that it deals with damage which is not normally lethal by itself but which may become so by accumulation. Clearly, any viable option cannot allow damage to reach a dangerous level before maturation, so varying the level of maintenance repair does not directly affect juvenile survival. Indirectly, however, the allocation of resources to repair is important in the extent to which it detracts from other activities. For example, diversion of resources from growth delays maturation and may have a major negative effect on r (Lewontin, 1965). Alternatively, in situations of intense competition, where *rate* of maturation may be much less important, it is essential that scarce resources are used as efficiently as possible to provide adequate defence and minimize the risk of starvation. Thus, across the full range of the r–K spectrum there is an advantage to reducing repair, from the earliest age, to the minimum level that is consistent with optimal longevity. In the case of intracellular repair this corresponds to the time of differentiation of the somatic cells from the germ-line. It is possible the level of repair may vary with juvenile age, though there is no automatic reason why it should. Similarly, there is no *a priori* reason to expect a change at the time of maturation, unless this were to result from re-partitioning resources to allow for reproduction. During adulthood, any variation in repair effort is likely to be inversely related to changes in reproductive effort which can be linked, in turn, to reproductive value (Pianka & Parker, 1975a; Charlesworth, 1980, p. 242). (Reproductive value was defined by Fisher (1930) as expected future reproduction weighted by the probability of survival and taking account of the intrinsic rate of population growth.) For many species, natural mortality is so high that, in the long term, reproductive value may not change very much with age (see Lack, 1954), with corresponding implications for reproductive effort and repair. Short-term fluctuations might arise, however, if seasonal or periodic reproduction generated concomitant reductions in repair effort. Unless such reductions are naturally followed by compensatory increases, this should mean that senescence would occur earlier in the most actively reproducing individuals. On this last point, Comfort (1979) cites evidence that, at least in male rats and humans, the reverse is true and that sexual activity *increases* longevity, though a penetrating study by Calow (1979, and Chapter 10) tentatively concludes that a negative correlation between reproductive output and parental lifespan does indeed exist. For a number of reasons (see Calow, 1979), experimental attempts to resolve this question are complicated by the interaction of many, often poorly quantified factors and the evidence at present remains inconclusive; further careful studies are clearly needed. Late in life, as senescence begins to take its toll on reproductive value, there may be an advantage to reducing the level of repair so as to make extra resources available for a 'last ditch' reproductive effort. This assumes, of course, that parental care is of little importance as death of the parent would automatically be hastened (cf. semelparous reproduction). In practice, these two conditions—a demonstrable effect of senescence on reproductive value and low parental care—may not often be satisfied. Furthermore, the selective pressure to evolve this strategy at the late end of the lifespan would only be weak.

7.9 EMERGENCY REPAIR

The principles governing the use of emergency repair are similar to those for maintenance repair, except that the repair itself demands considerably more time and resources since the damage is usually more serious. We saw in section 7.2 that certain conditions are necessary before the capacity to repair such damage could evolve, and it is assumed here that these are satisfied. The problem is to decide when existing repair potential should be used. In the absence of complications due to strong kin selection or group selection, the overriding principle is simple: repair should be used only when it leads to an increase in residual reproductive value. If reproductive value is zero without repair, then repair should always be attempted. The more interesting case is where reproduction is partially impaired, or where partial repair has already restored some reproductive value. Should the organism aim for complete repair, or should it make the best of the reduced reproductive capacity that remains?

In species which reproduce asexually, serious damage will often be reparable by regenerative renewal similar to that which occurs on division. Thus, provided survival is not jeopardized, repair is almost equivalent to growth and subject to similar controls. When survival is threatened, reproduction may be preferred to repair, especially if the progeny will be unaffected by the damage. In semelparous species, the value of emergency repair depends on when the damage occurs, and in particular on whether reproduction has to be completed within a fixed time or season. Obviously, it is disastrous to put resources, which might otherwise be used for reproduction, into repair that cannot be completed in time. Thus, emergency repair is more likely to occur in younger semelparous individuals than in older ones. To a lesser extent, the same may also be true of some iteroparous species, since reproduction is automatically constrained to occur within the species' maximum lifespan. As senescence approaches, the time required for emergency repair constitutes a larger proportion of the remaining lifespan, and it is less likely to be beneficial. This could be why older amphibians are less able to regenerate limbs than younger ones (Scadding, 1977), and, possibly, why an age-related fall is seen in the efficiency of wound healing (Goodson & Hunt, 1979). However, these phenomena may simply be part of the general decline that occurs during senescence.

7.10 CONCLUSIONS

The relationship between repair, reproduction and longevity is central to both short-term and long-term survival. In the short-term, extra investment in repair increases expectation of life but decreases the resources available for reproduction. In the long-term, genotypes that optimize the balance between repair and reproduction will be more likely to survive.

The optimal allocation of resources to repair depends on life-history and, in particular, on the mode of reproduction. Species which reproduce asexually must keep damage at equilibrium in any part which may be required to generate a new individual. Extra repair effort provides a margin of safety against random increase in the level of damage. In such species, the occasional use of sex may serve as an important repair mechanism. In species which rely entirely on sexual reproduction, damage need only be kept at equilibrium in the germ-line, since damage to the soma is not transmitted directly to the progeny. Therefore, the soma is a 'disposable' vehicle for the germ cells. In semelparous species, the integrity of the soma must be preserved up to, but not necessarily beyond, the act of reproduction; indeed, a suicidal transfer of resources from repair to reproduction may increase fitness. In iteroparous species, a balance must be struck between somatic longevity and reproductive output, taking into account the rate of mortality through causes other than senescence. The optimal repair level should always be less than that which is required to permit indefinite survival. This suggests that senescence is the result of a progressive accumulation of unrepaired damage.

Identification of senescence with the accumulation of damage provides a broad conceptual framework within which a wide variety of basic questions about life maintenance can be studied. For example, instances of rejuvenation, as in the sexual reproduction of *Paramecium* (Sonneborn, 1954) or in the degrowth and regrowth of triclads (Calow, 1977b), may be seen as products of repair. Senescence itself, in its wide variety of manifestations, becomes more comprehensible when viewed in this way, and its evolution is readily explained. However, the link between senescence and damage is not yet proven. Further studies are necessary to confirm:

1 that random damage does accumulate progressively in the soma;
2 that the suggested correlations between somatic repair and longevity are a general phenomenon.

Overall, we do not yet know enough about the costs and benefits of repair to make a detailed comparison of theory with reality. Nevertheless, the general principles that have emerged lead to predictions that can be tested by experiment and indicate areas where more knowledge should be sought. For example, do semelparous organisms reduce repair when reproduction begins? Are there special mechanisms to protect the germ-line? How does repair change with age? What kinds of repair are most important for determining longevity? Some information on these and other points is already available, but a great deal more is obviously needed. Repair is so important to the maintenance of life that there is clear value in a systematic investigation of its biological uses. If any justification other than the drive of scientific curiosity is needed, it may readily be found in the links between inherited repair deficiences and human disease, especially cancer and auto-immunity (Burnet, 1978).

Chapter 8

Storage

CAROLINE M. POND

8.1 INTRODUCTION

Storage, in its strict sense, means the ingestion and deposition in body tissues of any nutrient for which there is no immediate metabolic demand, but which will be utilized later for metabolism, growth, defence or reproduction. Organisms store materials for use during local or seasonal shortages, because they are temporarily unable to gather or synthesize food for themselves, or because the rate of growth or metabolism is temporarily so high that food intake cannot keep up with the demand. Energy-releasing compounds, proteins, minerals, trace nutrients (vitamins), oxygen and water are all stored by one organism or another. Because energy is the theme of this book and because energy storage is widespread among organisms, this chapter will concentrate mainly upon energy storage. However, it is important to remember that in some organisms, such as many green plants, internal parasites and plant-sucking aphids, it is other substances, mainly minerals, which are available only periodically or in small quantities, and hence it is these substances which are stored.

Few organisms have a food supply which is so regular and sufficient that no storage is necessary, and very few organisms are able to store nutrients to meet all possible requirements. There are costs as well as benefits to storage: the additional food material has to be obtained, converted into storage molecules; space must be found for it within the organism's body and it must be supported and/or carried around. All these processes require additional energy expenditure, biochemical machinery and structural tissue. Which nutrients are stored, how much and when storage takes place are integral parts of the organism's strategy for survival. It is often possible to identify ways in which organisms are managing to store as much as possible for as little 'overhead' cost as possible. These ways include appropriate choice of storage molecule, storage tissue, anatomical location of storage tissue and accurate timing of deposition and depletion of the reserve.

When an animal accumulates large quantities of fat it is said to be 'obese'; this word usually implies a pathological condition, i.e. that the handicaps caused by the presence of the storage tissue more than outweigh any advantages obtained from having large energy reserves. Some organisms are adapted to accumulate and maintain large fat stores, and hence for them 'obesity' cannot be said to be pathological; but for many others animals, quite small changes in diet or exercise regime can cause them to deposit larger fat stores than they normally would in the wild. Studies of humans, domestic animals and wild animals in zoos show that this artificially increased obesity does increase morbidity and reduce lifespan, mainly because of increased strain on the heart and vascular system. It is usually impossible to draw a hard line between what is 'normal' and what is 'pathological' obesity without careful study of the species concerned. The actuarial tables, compiled for *Homo sapiens* by life insurance companies, are the only data so far available for any species from which one can predict mortality from body-weight measurements. Fatter people usually have shorter life expectancy, but it is important to remember that, but for the invention of the tin can and the deep freeze, very thin individuals would also be at a disadvantage. Much has been written recently about the physiological and psychological mechanisms in controlling obesity (e.g. Wertheimer, 1965 and papers in the same volume). The purpose of this chapter is to discuss how an organism's 'energy storage policy' is formulated and to explain some of the natural mechanisms by which the ill-effects of obesity are avoided.

8.2 THE CHOICE OF ENERGY-STORAGE MOLECULE

A problem which besets all analysis of storage mechanisms is that it is often very difficult to draw a hard line between what is and what is not a storage molecule. It is widely accepted that the main function of many lipids and carbohydrates is in storage, and some proteinaceous tissues, such as muscle, are degraded and their energy used in metabolism during prolonged starvation, although in well-fed animals their function is clearly not in storage. Conversely some bona fide storage molecules acquire additional functions, as explained in more detail in section 8.7, so the definition of a storage molecule must remain somewhat arbitrary.

A suitable storage molecule must meet the following criteria:

1 it must not have adverse effects upon the cells in which or near which it occurs;
2 the minimum of energy must be used in building up and breaking down the storage material;
3 the tissues to which it will supply energy must be able to metabolize it or its breakdown products;
4 its degradation products must be non-toxic, and if possible metabolically useful in themselves;
5 it must be readily transported from storage site to utilization site;
6 it must of lowest possible mass and smallest possible volume.

No one storage molecule meets all the above criteria. Properties of some common storage molecules are compared in Table 8.1. It is clear that there is considerable variation, as would be expected from such a chemically diverse assemblage of molecules. The fatty-acid components of both triglycerides and waxes vary in their degree of saturation and hence in their energy content and melting point but the alcohol component is always glycerol (Needham, 1965). Since energy is released when the double bond is broken, unsaturated fatty acids have a higher energy content than the same mass of saturated fats. Fats containing more unsaturated fatty acids also remain liquid at lower temperatures than saturated fats, and many poikilotherms have a high proportion of unsaturated fatty acids, notably oleic acid, as a component of their triglyceride energy stores, and fats in exposed sites in homeotherms, including man, contain more unsaturated acids than do fats from internal deposits of the same organism (Schmidt-Nielsen, 1946; West, Burns & Modafferi, 1979). Wild animals usually have a higher proportion of unsaturated fatty acids than their domesticated relations, and hibernating forms such as the bat *Eptesicus fuscus* have the highest proportion (85 %) of unsaturated fatty acids of all (Smalley & Dryer, 1967). In some mammals, such as the pig, and the trout, the composition of fatty acids in storage triglycerides depends to some degree upon their diet (Phillips, 1969; Jeanrenaud, 1965; Shoreland, 1962) but in hens, humans and probably many other species there is much unexplained variation, although the temperature at which the organism is maintained is an important influence on the degree of unsaturation of its storage fats (Fischer, Holland & Weiss, 1962; Hegsted, Jack & Stare, 1962).

The distribution of waxes as storage molecules is even more puzzling; waxes are widespread and abundant in marine organisms, especially swim-

Table 8.1 Some properties of common storage molecules.

Molecular species	Energy content (kJ g^{-1})	Metabolized by all tissues?	Metabolized anaerobically?	How transported in organisms?	Respiratory quotient	Solubility
Carbohydrate						
Disaccharide						
Sucrose		No, but readily broken down to monosaccharides	Yes	In simple solution	1.0	Produces high osmotic pressure
Polysaccharides						
Glycogen	~ 17.6	Yes, degrades to glucose	Yes	In solution	1.0	Associated with 4–5 g $H_2O\ g^{-1}$
Starch	~ 17.6	Yes, when broken down to glucose	Yes	Must be converted to soluble products	1.0	Insoluble
Lipids						
Triglycerides	~ 39.3	No, not brain, red blood cells, some muscle and kidney cells	No	Lipoproteins in insects and vertebrates	~ 0.7	Insoluble and no associated water
Waxes	~ 39.3	Not known, but probably not	No	Not known	0.71	Insoluble and no associated water
Proteins						
Various amino acids	~ 18.0	Used directly by a few insect tissue; broken down to glucose by others	No	Broken down to soluble amino acids	0.74–0.81	Some soluble; all with much associated water

ming forms occurring below 1000 m in depth (Benson & Lee, 1975), and it has been estimated that half of the energy synthesized by phytoplankton becomes waxes at some stage in the cycle. The copepod *Gaussia princeps* has been shown to resorb its waxes in prolonged starvation, although it uses its stores of triglycerides first (Lee & Barnes, 1975). It is not known whether the waxes in other copepods, squids, chaetognaths and the coelocanth are also resorbed in starvation. In calanoid copepods and Odontocete whales all known members of the orders contain waxes in significant quantities (Nevenzel, 1970), but the significance of this fact is obscure. The jojoba (*Simmondsia chinensis*, Buxaceae), a Californian desert bush, is the only flowering plant in which waxes are the energy store in the seeds (Baker, 1970); nothing is known of the functional significance of this anomaly.

The main advantage of lipids as storage molecules is that being hydrophobic compounds they occur 'neat', without associated water, making the density of the tissue around 0.9, much less than any carbohydrate storage molecule. This fact and the absence of associated water mean that more than twice as much energy can be stored per unit weight in the form of lipid than can be stored as carbohydrate (Weis-Fogh, 1967). Nonetheless, water is formed from the metabolism of fats in a larger proportion than from metabolism of carbohydrates (ignoring the contribution of the water almost always associated with carbohydrates), and water thus produced is an essential component of the water balance of flying locusts, desert rodents and many other forms (Schmidt-Nielsen, 1975).

There are two main disadvantages of lipids as energy stores; the fatty-acid components cannot be metabolized anaerobically, and, being relatively large insoluble molecules, lipids cannot readily be transported in body fluids. It is important to note that about 10% by weight of triglycerides is an alcohol, glycerol, which can be converted to carbohydrate and metabolized anaerobically. The glycerol component is also available to tissues such as vertebrate brain and red blood cells which do not normally use triglycerides as a substrate at all. Small organisms with very efficient ventilatory systems, such as flying locusts, may not be limited by the requirement of metabolizing fats aerobically, but it may be a serious disadvantage for larger animals with less efficient respiratory systems, such as vertebrates. The lack of biochemical machinery for anaerobic metabolism of fats may well be the reason for the prevalence of carbohydrate storage molecules in internal parasites (Jennings & Mettrick, 1968; Calow & Jennings, 1974). However, they are also not limited by the need to maintain a small volume or low weight, so there is no selection for the additional biochemical machinery needed to convert carbohydrates to fat; this could account for the occurrence of storage carbohydrates even in Turbellarian parasites of vertebrate lungs (Jennings & Mettrick, 1968).

The limitation of transport within the body also affects larger organisms more than small ones. Both mammals and insects have intermediate lipoproteins, such as phospholipids, which promote transport of fats in the blood, but even so, energy stored as fat cannot be released nearly as rapidly as energy stored as carbohydrate. In many insects, fats are stored in the fat body as triglycerides but appear in the haemolymph as diglycerides, possibly because they are more readily transported and taken up by the muscles in this form (Downer & Matthews, 1976).

Most animal tissues cannot convert lipids back to carbohydrates, although vascular plants and many mico-organisms perform this conversion readily, and of course all three groups can convert carbohydrates to fat in large quantities. Thus once the fat is formed the animal is 'stuck with it'.

We can summarize the pros and cons of storing energy as lipid as follows: from a biochemical point of view a large animal with a high metabolic rate and a restricted respiratory surface is probably worse off using fat instead of carbohydrate. But from a mechanical point of view the animal is much better off, because it can store much more energy for the same weight and volume. The fact that lipids are so widespread as energy-storage molecules suggests that mechanical factors are a more important determinant of 'energy-storage policy' in animals than many recent reviews of the subject would suggest.

Surprisingly few plants store large quantities of fats except in reproductive tissues. Fats are found in high concentrations in some fruits, such as the avocado pear (*Persea gratissima*) and many seeds, (e.g. safflower, *Carthamus tinctorius*) or in both the seed and the fruit, for example the oil palm (*Elaeis guineensis*), but only rarely in the vegetative organs (see section 8.4). One possible explanation has been put forward by Needham (1965): since sugars and starch always occur in association with large quantities of 'bound' water, they act as anti-freeze and anti-desiccant for the water thus stored. Another possibility is that oils, having a higher energy value per unit weight and unit volume, can be more readily sequestered to sites where herbivores cannot reach them. Arctic and polar plants have fewer herbivores and hence can 'afford' to maintain carbohydrate energy stores which would be eaten by herbivores in more temperate climates. A third possibility is that plants cannot translocate an insoluble substance efficiently from storage site to utilization site, particularly at lower temperatures, when most fats would have solidified. Lipoproteins or similar fat-carrier molecules have not been identified in plants, and the nutrients stored as lipid apparently both enter and leave the storage cells as carbohydrate (Salisbury & Ross, 1969). When plants adapt to cold conditions, it is usually low molecular weight sugars and amino acids which increase in the exposed tissues, but sometimes, for example in alfalfa roots, the degree of saturation of the lipids also changes, in the same way as it does in animal tissues exposed to cold.

8.3 THE CHOICE OF ENERGY-STORAGE TISSUE

In some organisms the storage molecules are found in or near the tissues which use them (e.g. the fats in the muscles of eels, salmon etc.), while in others storage molecules are found in tissues apparently specialized for a totally different function, and in a third group there are tissues specialized for storage. The advantages of the first arrangement are clear, since the need for transportation between storage and utilization site is eliminated. The advantages of the second and third arrangements are less clear. The ectoparasitic temnocêphalid worms *Caridinicola indica* and *Monodiscus parvus* (Platyhelminthes; Turbellaria) and the free-living nematode *Pontonema vulgaris* (Nematoda, Enoplida) are typical of several types of lower invertebrates in having glycogen, and in the case of *P. vulgaris* fat as well, scattered in various organs including the intestinal walls, the nerve cords and the body epithelium (Fernando, 1945; Jennings & Colam, 1970). Fat droplets are found in the epithelial cells of the anterior gut lining in the leech, *Hirudo medicinalis* (Bradbury, 1958). This species is an ectoparasite of vertebrates and can flourish if fed as infrequently as once a year. The related species *Glossiphonia complanata* is a carnivore and probably eats much more frequently. Its fat is found in connective tissue cells, which also have an excretory function (Bradbury, 1957).

In many invertebrates, including molluscs and most arthropods, the 'liver' plays an important role in storage (Vague, Boyer, Nicolino & Pinto, 1969). The word 'liver' is in quotation marks, since there are many reasons to doubt that it plays all, or even some, of the functions of the vertebrate liver, and its functions are very different in the various groups. The 'liver' of the scorpion (Arachnida) *Palamnaeus bengalensis* accounts for a fifth of the total body weight, and both proteins and fats are withdrawn from it when the animal is starved in the laboratory (Sinha & Kanungo, 1967) and the hepatopancreas ('liver') of many molluscs and crustaceans shows annual cycles of fat content, implying that one of its functions is as a storage tissue. The principle storage organ of insects is usually called the 'fatbody', but, like other invertebrate 'livers', it is capable of a variety of biochemical activities (Tietz, 1965), many of them unrelated to fat storage. The fatbody should therefore be regarded as another form of invertebrate 'liver' in which fat storage is particularly well developed. There is no doubt that its fat content is depleted during periods in which the insects does not feed, particularly during metamorphosis or the migration of adults. Two-thirds of the body weight of the larvae of the honeybee *Apis mellifera* is fatbody just before metamorphosis (Bishop, 1922) and the fatbody can account for over 50 % of the liveweight of some adult insects just before migration (Beall, 1948).

The mammalian liver contains only about 4–6 % by weight fat and about the same proportion of glycogen. The liver of reptiles (Derickson, 1976) and birds (Mareström, 1966) sometimes contains a much higher proportion of fat. An unusual structure called the 'glycogen body' has been described in the nerve cord of developing chicks (Doyle & Watterson, 1949). It contains 5–10 % of all the glycogen stored in the animal and persists after hatching, but its significance is not known.

True adipose tissue, in which the cells have few metabolic functions unrelated to storage, is well developed only in tetrapod vertebrates, although some fish have small quantities of a similar tissue. Adipose-tissue cells are characterized by being able to undergo enormous changes in volume without ill effect. The individual cells live a long time, possibly for the lifetime of the animal, so the same cells may get filled up and emptied many times. Ninety-eight per cent of the volume of a replete adipose cell may be a single lipid droplet (Wertheimer, 1965), but the remaining 2 % contains enzymes actively involved in the synthesis and breakdown of lipids. Adipose tissue can be by far the most abundant tissue in some animals: just before migration more than half the live weight of the scarlet tanager is adipose tissue (Odum, 1960). The question of whether a constant population of adipose cells expands and contracts as the animal gains or loses weight, or new cells develop, is still not settled. In adult birds, changes in cell volume constitute the chief, possibly only, mechanism (Blem, 1976), but in mammals probably both mechanisms are at work (Young, 1976). As well as their role in fat storage and metabolism, vertebrate adipose tissue and insect fatbody cells also have in common the catylosome, a small cytoplasmic organelle about 1 μm in diameter, which may contain enzymes (Wigglesworth, 1966).

Adipose tissue is found in various sites on the bodies of terrestrial vertebrates and is consistent within a species (Pond, 1978). In poikilotherms it is usually in central abdominal 'fat bodies' and/or around the muscles of the tail. In birds it appears in discrete masses, many of which are under the skin, but which retain their 'organ-like' characteristics. In mammals adipose tissue is found in the fascia between muscles and in the abdomen, and in subcutaneous tissues, where it forms continuous sheets. Some cutaneous glands, particularly the mammary glands, synthesize their own triglycerides (Knittle, 1979); in fact the involvement of the subcutaneous tissues in the metabolism and secretion of fats is a distinctive feature of the Class Mammalia. In other vertebrates, including birds, fat metabolism is mainly in the liver, and the uropygial gland is the only fat-secreting cutaneous gland in birds (Hildebrand, 1974). The question of whether the involvement of mammalian subcutaneous tissues in fat metabolism and secretion arose first in connection with feeding the young on parental secretions, or in connection with the maintenance of fur, has not yet been solved. The present author has argued on other grounds that lactation is an early and fundamental character of mammals (Pond, 1977).

8.4 THE CHOICE OF ANATOMICAL SITE FOR ENERGY STORAGE

For many tissues, such as structural, respiratory or digestive tissues, their location on the body is essential to their function, but storage tissues can perform their functions wherever they are located in the body. There are two major types of constraint acting upon where storage tissues are found in organisms: first, the time that it takes for the store to be liquidated and transported to the site of utilization; and second, the effect of its presence upon the anatomy and physiology of the organism as a whole. When energy is stored in small quantities and for short periods of time, the first constraint is much more important than the second. Organisms which have a high metabolic rate, such as birds and mammals, need a rapidly mobilizable, short-term energy store to meet sudden high energy demands. They maintain two types of energy store in the muscles themselves: a small quantity of creatine phosphate which can be mobilized within 10^{-1} s, and glycogen up to a concentration of 0.5–1.0 %, which can supply energy sufficient for about 200 s of heavy exercise within a few seconds. Further reinforcements of glycogen can be brought to the muscles by the blood from much larger stores (2–8 % by weight of the whole liver, or about 0.16 % of the total live weight in man) but this takes from 1–10 s and in humans the supply is only sufficient for about 10^3 s of exercise (Newsholme & Start, 1973). If energy is still required in excess of food intake, fats are withdrawn from adipose cells at various sites and transported in the blood. However, plasma albumins are required to transport fatty acids in significant quantities in both vertebrates and insects (Allen, 1976), and the adipose cells must be stimulated to break down their triglycerides; both these processes require time and additional biochemical machinery. Fish, such as tuna, salmon and eels, which store oils in the swimming muscles, eliminate this problem since the energy store is at, or close to, its site of utilization.

The effect of the presence of an energy store upon the anatomy and physiology of the organism as a whole is a question which has rarely been addressed. In addition to the cost of obtaining and laying down the energy in the store there is also an inevitable 'maintenance' cost, which may be thought of as having two components. The building and maintenance of the tissues supporting the storage material which one may call the 'larder', and the energetic cost of carrying the energy store (and the 'larder') around if the organism is ambulatory. Almost nothing is known about the cost of building a 'larder' in either plants or animals, except that it is presumably quite high in the case of the camel, which forms additional skin and connective tissue over its hump, and low in the case of those invertebrates which store materials in cells with other functions, since they do not even have to sequester special cells for storage. In higher plants both low-cost and high-cost 'larders' are found. Non-reproductive storage organs include such specially developed structures as corms, tubers and enlarged roots, and since the cellulose, calcium and probably other minerals in the tissue cannot be reclaimed, the cost of the 'larder' must be relatively high. In contrast, some alpine evergreen plants store lipid and protein in their old, shaded-out leaves, and since these would presumably have otherwise been shed, the cost of the 'larder' must be very small in this case (Bliss, 1971). Other plants, for example the Californian buckeye (*Aesculus californica*), store carbohydrates in most of their tissues (Mooney & Hays, 1973).

There is, however, one system in which it may be possible to measure the metabolic cost of building the 'larder' for fat store, and that is the lizard tail, but unfortunately those who have studied the system make no distinction between the energy used to rebuild the tail, and the energy used to replenish the fat store (Vitt, Congdon & Dickson, 1977).

We are all vaguely aware that fat people, fat dogs, etc. run more slowly, jump less well, fatigue more quickly from moderate exercise, have more trouble getting through doors, down foxholes etc. than their thinner relations. But the selective forces acting on the maintenance (as distinct from formation) cost of energy storage, has not hitherto been the subject of experimental or theoretical analysis. Particularly if the storage material remains inside the organism's body for relatively long periods and in substantial quantities, the cost of supporting and carrying around the storage material can be high. It is probably very small in sessile aquatic organisms, such as marine algae and non-swimming coelenterates, but it could be the overriding factor in fast-moving arthropods and vertebrates. When an aquatic animal or plant stores lipids, its volume increases and its density usually decreases. The decrease in density is due to the much lower specific gravity of all fats compared to most other living tissues, and the change has profound effects upon the organism, as will be discussed in more detail in section 8.7.

When a terrestrial animal lays down energy stores, its total mass increases, so more energy must be expended in movement and its muscles and skeletons must be strengthened to support the extra load. Even if the organism does not move, structural tissues must be reinforced if the additional weight is supported above ground. If the storage material is poorly distributed within the body unusual stresses are applied to the organism's support system and additional forces appear during movement, causing animals to break bones and strain muscles much more readily. As with loading trucks, aeroplanes, pack donkeys and racehorses, both the absolute weight increase and its proper distribution on the body can be very important in determining the extent to which the organism is incapacitated when obese. The relationship between an animal's athletic capacity and the abundance and distribution of its storage materials have only recently been the subject of experimental or theoretical analysis. Tucker (1975) mentioned that when a migrating bird utilizes its fat reserves its body profile (as well as body weight) will change, perhaps enough to affect its flight mechanics, but the idea was not elaborated.

In many lizards, fat is stored in small quantities in the liver but in much larger quantities in discrete adipose masses in the abdomen or the tail, or both. The food supply of the desert iguana, *Dipsosaurus dorsalis*, is irregular and unpredictable and it stores relatively large quantities of fat, mostly in the tail which accounts for over half the total length of the body. When resting or walking slowly, the weight of the tail and rear abdomen rests on the ground (see Fig. 8.1), and, like many other lizards, *Dipsosaurus* readily breaks its tail when molested by a predator. It also runs very fast and straight when alarmed, reaching 7.3 m/s under optimum conditions (Belkin, 1961). These circumstances together mean that *Dipsosaurus* is an ideal animal upon which to study the relationship between energy storage and running ability.

In a recent investigation (Pond, unpublished data 1980), I was able to show that the mean running speed drops by 14 % when the lizard is induced to break part of its tail, and replacing the lost mass with artificial weights did not improve the performance. Thus, although the lizard is 5–10 % lighter after tail

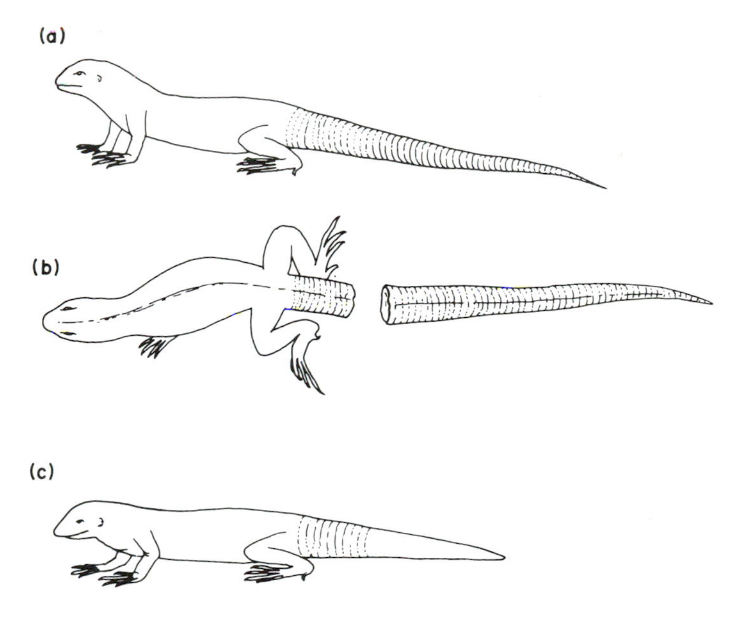

Fig. 8.1 *Dipsosaurus dorsalis*, the desert iguana: (a) intact, standing in the 'alert' posture; (b) with the tail broken off; (c) with the tail fully regenerated, three months later. (Drawn from photographs by the author.)

breakage, its running speed is reduced, not increased. A similar decrease in mean running speed could also be produced by adding weights to the intact tail, so as to increase the body weight by up to 20 %. Both these observations confirm the view that the distribution of mass is more important than its magnitude as regards its effect upon fast running. The question, therefore, arises whether the need for correct distribution of mass between the tail and the rest of the body is an important factor determining the quantity of fat in the intact tail. Again, this question is easily investigated by comparing the change in mean speed of 'escape' running when artificial loads are attached to the lizard at different anatomical sites. A mixture of fine lead shot and soft wax was used because it adheres readily to the skin and is flexible. When the animal's weight was increased by 20 % the drop in running speed was consistently greater when the load was applied to the tail (14–21 % decrease over control) than when it was applied to the anterior thorax (9–11 % decrease) suggesting that if the lizard were 'designed' for fast running alone, its fat stores should be on the thorax. But in fact the lizard spends much less than 1 % of its life engaged in fast running, and a high proportion of the remaining 99 % of the time supporting the anterior part of its body off the ground with its fore legs while the tail just rests on the substrate. A small increase in the weight of the thorax would therefore increase the total energy required to maintain this posture, a cost which may well not be outweighed by the small increase in running speed which apparently could be obtained by transferring some of the tail fat to that site. (Pond, unpublished data). Compared to mammals, the decrease in running speed caused by a 20 % increase in body weight is surprisingly small; a load of 1 kg (0.01 % total weight) is regarded as a significant handicap for thoroughbred racehorses (Anderson, 1942), and few humans can run at even half maximum speed while carrying 20 % of their weight. Other factors which determine the distribution of fat in vertebrates have been critically reviewed elsewhere (Pond, 1978).

Plants differ from animals not only in that they do not move around, but also because their food is made in photosynthetic tissue in daylight. All plants store small quantities of energy, usually as starch grains in the chloroplasts, for 'local' use during darkness. Only some plants have special storage organs, which are usually made of non-photosynthetic tissue, such as corms, tubers and endosperm tissue. Many of the non-reproductive storage organs are underground, probably because they are less subject to freezing when in the soil. Structural tissues are also not required to support the weight of the storage material if it is underground, thus further reducing the cost of maintaining the plant's 'larder'. All storage tissues are potentially attractive to herbivores, and an additional cost to the plant of energy storage is protecting its 'larder' from being eaten. The potato protects its food reserves by infiltrating the starch with an inhibitor of animal digestive enzymes (which is destroyed by boiling). The starchy, swollen roots of Cassava (*Manihot esculenta*) contain hydrocyanic acid in such high concentrations that it is lethal to man and to most other herbivores when raw. A lengthy and dangerous leaching and drying procedure is carried out before the root can be eaten.

Both the quantity and the distribution of storage tissues are under genetic control, and respond, often surprisingly rapidly, to natural and artificial selection. Everyone is familiar with the fact that domesticated forms of onion, apple etc., produce much larger storage organs than their wild relations since it is these storage organs which are harvested and eaten. Since these plants propagate successfully in the wild, the smaller quantities stored must be sufficient for their ordinary needs, but artificial selection has increased it. Some of the more striking differences between otherwise closely related plants are caused by differences in the anatomical arrangement of their storage tissues. For example, *Solanum dulcamara*, the woody nightshade is a wild member of the family Solanaceae, of which several other species have been domesticated in different ways. In the potato *Solanum tuberosum* selection has been for the storage material to accumulate in an underground stem tuber, while in the eggplant (aubergine), *S. melongena*, it forms a massive starchy fruit, while in the closely related tomato *Lycopersicon esculentum* the storage material is more sugary and forms a fruit of very different shape and texture. In the eggplant and tomato, most of the energy is deposited in the fruit, which would normally be shed or removed, and hence is not available to be reclaimed by the parent plant. In contrast the potato is storing food material for vegetative growth the following year.

People who have been buying meat for some years may have noticed that beef joints have got leaner and the fatty tissue of a pork chop is smaller in proportion to the quantity of meat and bone than it was 20–30 years ago. Part of this difference is due to a change in butchering practices, such as killing the animals at an earlier age and cutting the meat differently, but part is also due to the fact that beef cattle, pigs and sheep have been bred to have less fat under the skin and between the muscles than their ancestors of 30 years ago. In Fig. 8.2

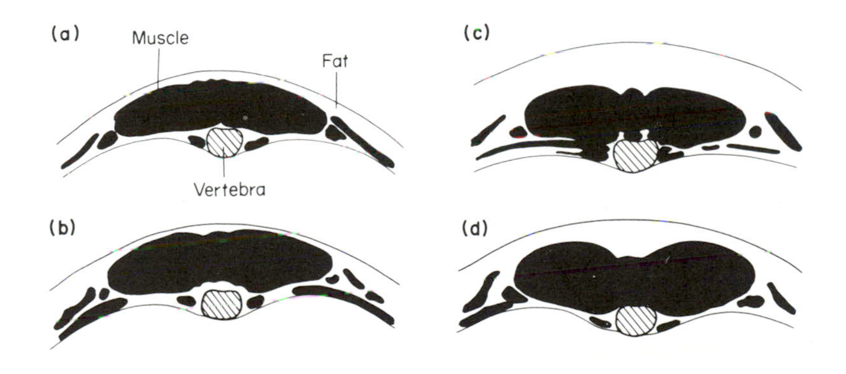

Fig. 8.2 'Lamb chops' from various breeds of sheep of similar age and nutritional status, (a) Blackfaced, (b) Suffolk, (c) Hampshire, (d) Southdown. Muscle is shown in black, bone is hatched, fat is white (redrawn and simplified after Hammond, Mason & Robinson, 1971).

lamb chops taken from sheep of the same age but different breeds are compared; there are clear differences in both the abundance and distribution of fat in relation to the muscles and bone. Fatty tissue also responds very rapidly to selection. The data on Fig. 8.3 show how some features of pig anatomy changed during 41 years (approximately 40 generations) of artificial selection. Fat thickness has changed much more than the thickness of the muscular abdominal wall.

There is also a substantial genetic component in the distribution of fat on the human body, although it is difficult to separate it from the effect of age, nutritional status and infant-feeding practices. But we are all familiar with the fact that related adults of the same age tend to resemble each other in their body build; in particular daughters tend to develop similar figures to their mothers at the same age, in terms of the relative development of the larger fat masses, breasts, buttocks, tummies, etc. As Shattock (1909) first expounded in detail, one of the most dramatic ways in which small, inbred races of Africans differ from each other is in the development of fatty masses on the ankles,

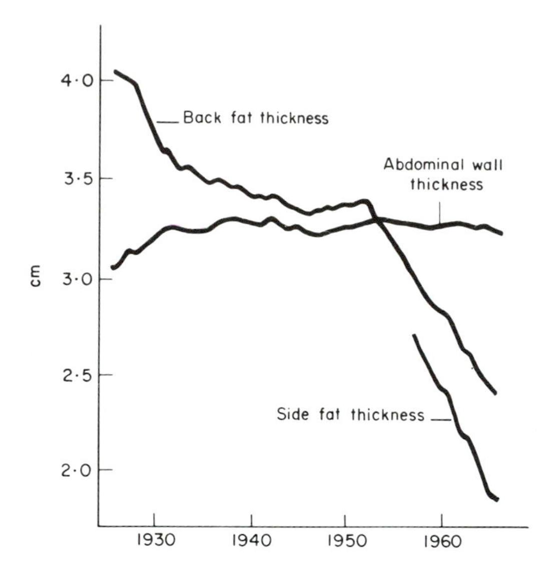

Fig. 8.3 The response of Danish Landrace pigs to selection for less back and side fat and thinner abdominal wall. The latter is mainly muscle and connective tissue (redrawn and simplified after Hammond, Mason & Robinson, 1971).

thighs, buttocks etc. of young adult females. The female figure, or at least what may have been regarded as the ideal female figure, as shown by prehistoric figurines, has also changed greatly over the last 20 000 years (Morris, 1977), as shown in Fig. 8.4.

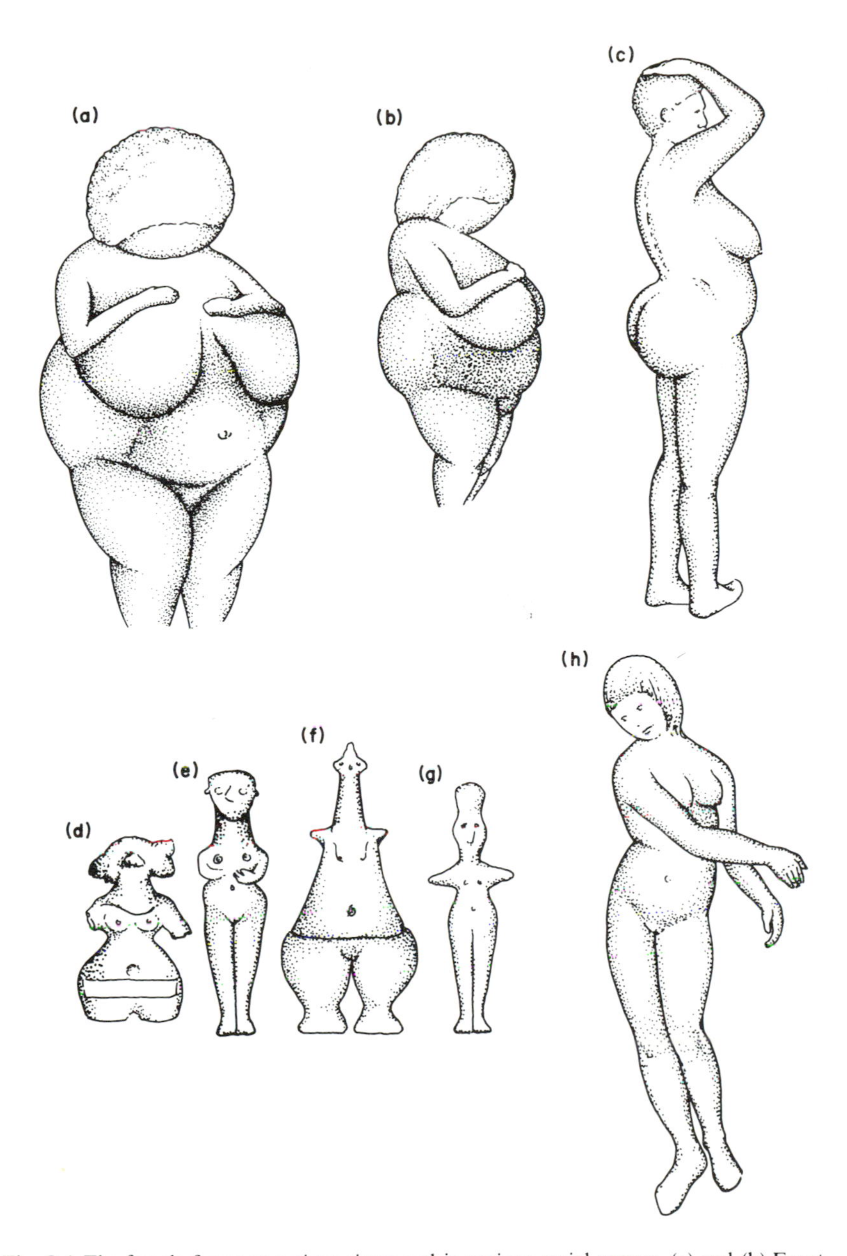

Fig. 8.4 The female figure at various times and in various racial groups. (a) and (b) Front and side view of Old Stone Age figurine from Europe. (c) Hottentot woman. (d) Figurine from Bronze Age India. (e–g) Figurines from various Late Bronze Age in the Eastern Mediterranean. (h) Modern Western European figure. Note that broad hips and thighs are the most consistent feature, but that in European figures lateral spreading is exaggerated, while in African figures the protuberances are dorsal and ventral. ((c–g) redrawn after Morris, 1977).

8.5 FACTORS WHICH DETERMINE THE QUANTITY OF ENERGY STORED

8.5.1 Temporal factors

The time for which an organism can live without food depends upon how much energy it has stored in its body, and upon its metabolic rate. Smaller animals usually have higher metabolic rates (Schmidt-Nielsen, 1975 and see Chapter 2) and hence have greater energy requirements. For example, an energy supply sufficient for 2 days normal activity is 10% of the body weight of a 30 g mouse, but only 1% of a 70 kg man (Pitts & Bullard, 1968); 25% body weight fat will last an elephant 200 days, and a large whale 400 days (Morrison, 1960). But all these calculations completely ignore the energy costs on maintenance, i.e. support and transport, of the storage tissue, which we saw in section 8.4 can be important, especially if storage lasts for a significant fraction of the animal's lifetime.

One way in which organisms reduce the cost of storage is to deposit only the minimum energy reserve, just sufficient to cover the period of starvation. For example, Evans (1969) found that the amount of fat accumulated each day by the yellow bunting *Emberiza citrinella* correlated well with the average night-time temperature for that date, being greater for colder and longer nights. The unicellular alga *Chlamydomonas reinhardi* shows the same phenomenon; as light intensity increases, it starts photosynthesis later in the day and stops earlier, with the result that the amount of photosynthate made is just sufficient to last the night. Very little 'excess' is carried over from day to day (Cohen & Parnas, 1976). Particularly if its journey takes it over water, a migrating bird often cannot stop to feed for many hours at a stretch. The birds fatten very rapidly before migration and their fuel reserves can amount to 50% of the body weight (Odum, 1960). Even so, the normal migration route of many small birds pushes their fuel reserves very close to the limit; if they are delayed or blown off course they frequently die of exhaustion before they are able to reach land and a suitable food source. However, the exact energy requirements are difficult to calculate. One reason is that recent evidence suggests that the migration of some species is carefully timed to make maximum possible use of following winds and other favourable meteorological circumstances (Tucker, 1975). Another reason is that there is a complicated relationship between body weight and energetic cost of flight. Very fat birds have a lot of trouble taking off (Fry, Ash & Ferguson-Lee, 1970), but once in the air, a heavier body can sometimes result in its travelling faster, especially if it spends much of its time gliding (Pennycuick, 1972). Larger birds can migrate longer distances without stopping because they use proportionately less energy for flapping flight and hence can travel further for the same 'payload' of fat. Thus a goose which sets out with 20 % of its body weight as fat can travel 6×10^6 m, while a hummingbird can travel only 0.8×10^6 m with the same proportion of its weight as fat (Tucker, 1975).

A second way that the net cost of energy storage is reduced is by shortening the time for which the energy is accommodated within the body. Since energy comes from the food ingested or synthesized when available, and the purpose of storage is to tide the organism over periods of food scarcity, manipulation of the timing of 'fattening' would seem to be impossible. However, the really dramatic examples of energy storage generally arise when organisms are using large quantities of energy for activities such as migrating, mating, territory defence or rearing young. Although there may be enough food around, they are 'too busy' with these activities to get enough food for their needs. It is in these cases that we can look for ways that organisms regulate the timing of fattening, since the availability of food is not the limiting factor. For example, the white-throated sparrow, *Zonotrichia albicollis*, maintains an almost constant weight throughout the summer in North America, but in the last 10 days before migration its food intake increases by 50 % and its fat content rises to 25 % live weight (Odum, 1960). The bat *Myotis sodalis* in Kentucky also 'leaves it to the last minute' and then gains weight very rapidly just before hibernation begins (Davis, 1970). In many female mammals, the deposition of fat (and other nutrients such as calcium) which will meet the demand of lactation, does not begin to any measurable extent until after gestation has begun (see Pond, 1977). In all these examples the timing of accumulation of energy stores seems to have little to do with fluctuations in food supply, and it is likely that they enable the animal to reduce to a minimum the time for which they have to carry around the extra load. Further evidence for this point of view comes from the observation that many normally solitary species of bird temporarily become gregarious when obese. Living in flocks may afford more protection to birds which incur a greatly increased risk of predation because they are too heavy or unbalanced.

Some male mammals, for example the squirrel monkey, *Saimiri sciureus*, (DuMond & Hutchinson, 1967) and the sealion, *Zalophus californicus*, (Schusterman & Gentry, 1971), deposit fat in anatomically conspicuous sites immediately before the mating season begins, but lose it between seasons, even when there is sufficient food available. The situation implies that there is some cost associated with increased body weight, about 10 % in the case of *Saimiri*, which is not expended unless it is necessary for reproductive success. The fact that the cost is not eliminated even when the animal has plenty of food available suggests that it is not simply a matter of energy for carrying the fat around. Other possibilities for the 'cost' include increased strain on the heart, tendons, joints and bones, reduced ability to get through small spaces and greater conspicuousness.

An alternative to storing energy inside the body is to make 'caches' of food outside the body. Although there are costs associated with accumulating the cache (Andersson & Krebs, 1978), a benefit which is not often mentioned is that the animal avoids the anatomical and physiological stresses of maintaining a fat reserve. Many of the animals which make caches are small carnivorous birds and mammals (Macdonald, 1979), which hunt by active pursuit; so many complicated factors are contributing to the availability of their prey that periods of food shortage are irregular and unpredictable. By making caches rather than becoming fat, the animal avoids the physiological costs, which may turn out to have been unnecessarily incurred, while still protecting itself against starvation. The females of sphecoid wasps of the family Ampulicidae also avoid the need to accumulate fat reserves in their own bodies, and of laying a large yolky egg, by provisioning each nest with a paralysed but not dead insect, upon which the larva feeds when it hatches (Lanham, 1964).

With few exceptions, plant growth is limited more by shortage of minerals, particularly nitrogen and phosphorus, than by energy. When energy-releasing compounds are stored in large quantities by green plants it is usually because the plant will grow, and hence require energy, at a rate greater than it can photosynthesize it. The growth may be of vegetative tissue, such as a shoot after winter dormancy, or reproductive tissue such as a mast seed crop. For example, grasses in northern Utah use 50–60% of their carbohydrate reserves in just one week, when growth begins in spring (Donart, 1969).

8.5.2 The role of individual differences in heredity and experience

One might guess that within any one species, the organisms in the areas with the largest and most constant food supply would have the most storage material. In fact, the limited evidence available suggests that the opposite is true. Comparing different populations of the African elephant, *Loxodonta africana*, Laws, Parker & Johnstone (1975) found that those at Murchison Falls had less fat than the Tsavo population, although the latter lived in an area with less predictable rainfall. Malpas (1977) found a similar relationship between elephants in Ruenzori, which has a more equable climate, and Kabalega, where there is a marked wet and dry season and where the elephants have more fat throughout the year. The indigenous population of the hare *Lepus timidus scotius* have more subcutaneous fat in Scotland, than does the introduced New Zealand population, although the latter lives on better pasture (Flux, 1978). Areas with plentiful food supply also support larger populations of predators, and the increased risk of predation when obese may be the selective force against storing large quantities of fat when the risk of food shortage is low. This notion is supported in the case of camels, which are the only artiodactyls to carry large quantities of fat all year round. Their food supply is scattered and erratic, but since they live in an environment free from large predators there is little selection against increasing their body weight.

Some micro-organisms also have genetically determined differences in the quantity of glycogen stores; Parnas & Cohen (1976) were able to select for increased energy storage in *Streptococcus salivaris* by varying its supply of nutrients. Triclads which experience different regimes of feeding and starvation also differ in the quantity of fat that they store, with less frequently fed individuals storing greater quantities (Calow & Jennings, 1977).

8.5.3 THE AGE OF THE ORGANISM

We saw in sections 8.5.1 and 8.5.2 that the evidence is consistent with the view that organisms do not accumulate energy stores in larger quantities or for longer periods than the circumstances require. However, it is not only the environment which is changing. The age of the organism is also a relevant parameter because it affects the proportion of energy which the organism is allocating to growth and reproduction, and the types of selective pressure acting on an organism may change with age. Storage tissues show more variation in abundance (and sometimes distribution as well) with age than do any other tissues. This circumstance has caused biologists to refer to adipose tissue as a 'late maturing' or 'labile' tissue (Hammond, Mason & Robinson, 1971), implying that the age dependence of the tissue is somehow an inherent property of the storage tissue *per se*. These ideas obscure the fact that the opposing selective forces favouring obesity or leanness change with age.

With the exception of birds and mammals, most newly hatched or newly germinated organisms have relatively large quantities of storage materials. This storage material was accumulated by the parent and deposited in the form of egg yolk, seed etc., to give the developing embryo a 'good start in life'. So large is this maternal 'dowry' that many invertebrates, fish and reptiles are fatter immediately after hatching than at any other time in their lives. The material is consumed as the organism grows and is not replaced until after maturity so most juvenile and young adult organisms have relatively little storage material. In many fish and reptiles, the onset of first reproduction is in part determined by how much energy the animal has been able to store (for example, Telford, 1970; Michael, 1972). In contrast, birds and mammals, particularly mammals, are born with very little fat, often under 2% total weight (Adolph & Heggeness, 1971), and their fat stores increase rapidly during the first few weeks of life, so that they are fattest at the time of fledging or weaning. Human infants have very much more fat, particularly subcutaneous fat, than do other newborn apes and monkeys (Schultz, 1969), but they also fatten during suckling so that they are proportionately fatter at about one year of age than at any other time of their lives. This exceptional arrangement in birds and mammals is due to the fact that the parents both feed and protect the young after birth (Pond, 1977). The young does not need much fat at birth, it can 'rely' upon its parents to feed it, and indeed there is selection for the fetus to be as light as possible when *in utero*. For the same reason, it is usually exempt from the need to move far or fast to catch prey or escape predation and hence there is little selection against becoming very fat. Some nidifugous birds (which hatch well-developed and soon leave the nest) which feed for themselves at an early age have a much more reptilian pattern of fat accretion and depletion; a domestic chicken hatches with 6.6% dry weight fat, which it resorbs during the first four weeks of life (Wilson, 1954). In contrast, in the case of some strongly nidicolous birds (undeveloped at hatching, remain in nest), such as the oil bird (*Steatornis caripensis*), the young have fat reserves sufficient to last them for weeks at fledging and are actually heavier than their parents when they are abandoned (Snow, 1961); young cormorants (*Phalacrocorax auritus*) have only 1% fat at hatching (Dunn, 1975), which increases rapidly during the first month of being fed by the parents.

Most vertebrates, including oviparous reptiles and all birds, lay relatively large eggs, rich in energy-storage materials, mostly fat, to supply the embryo while it is developing. In contrast, mammalian eggs are very small indeed, symmetrical in form with almost no yolk, and show little evidence of ever having had a very yolky egg. The embryo forms a placenta early on in its development and obtains all its nourishment from the mother. It is unclear when in their evolutionary history the ancestors of placental and marsupial mammals first developed the small egg, and in doing so forsook the long established vertebrate 'tradition' of having very yolky eggs. But it is one more way in which the mammalian reproductive strategy means that the acquisition, processing and storage of resources is left to the mother, while the embryo is able to get on with the job of growing and developing (see Pond, 1977 for further details).

Most plants and adult female animals, including birds, build up their storage tissues before reproduction. In lizards, fish and insects, which are among the most thoroughly studied groups, the females accumulate fats and often minerals before mating, and almost all the stores are utilized in making the eggs. In fact, in lizards the provisioning of the eggs seems to be the main function of the bodies in females. Why they do not spend longer developing the eggs and deposit the fats directly in them as they are accumulated, thus eliminating the fat body stage, is not entirely clear.

In their strategy for storage of nutrients in relation to reproduction, mammals are again exceptional. The females do not gain significant amounts of weight before conception, although in humans and deer and probably other species a female which is substantially below normal body weight often does not ovulate (Frisch & McArthur, 1974), and total body fat increases rather than decreases as pregnancy progresses.

Although the question has been much less thoroughly studied, there are also age-related changes in energy storage in males and non-reproducing females. In long-lived organisms with high juvenile mortality due to predation, selection will favour growth to near full size as rapidly as possible (see Chapter 9). Once the animal is large enough that predation pressure is reduced, selection will favour maintaining an energy reserve, especially if the life expectancy of the organism is long compared to the average frequency of periods of food scarcity. One may expect, therefore, that juveniles and sub-adults will have little storage material while big old adults will maintain much larger stores. Large trees probably fit into this pattern, especially if they show mast cropping. While it is usually not possible to tell their age, large and, hence presumably old, snakes, crocodiles and lizards of the bigger species are also frequently very fat, while smaller, presumably younger and still growing specimens of the same species have proportionately less fat (Pond, 1978).

Dipsosaurus dorsalis, the desert iguana, again proves a suitable animal upon which to study this question. As mentioned in section 8.4, when *Dipsosaurus* loses its tail it sheds both the energy store and the 'larder'. The tail then regenerates (see Fig. 8.1), and the new tail differs from the never-broken tail in some interesting ways (Pond, unpublished data). Very few juveniles are found with broken tails, but 45% of adult males and 34% of females in museum collections had broken and regenerated tails. In intact lizards of both sexes, the tail accounts for 47% of the total length of the animal, but only 35%

of the total length in regenerated males, 28 % in regenerated females. There are also differences in tissue composition; the regenerated tail has much less skeletal and muscle tissue and hence more space in it for fat cells to expand. If known lengths of never-broken tail are experimentally amputated, the regenerated structure averaged 56 % of the length removed, but if the tail has previously been broken and allowed to regenerate naturally, and then amputated again, the new tail averages 104 % of the length removed. Its tissue composition is the same as the 'second' tail; that is, a greater proportion of the tail volume occupied by fat, so although the tail as a whole is shorter and lighter, it has more space for fat in it than the never-broken tail. Juvenile lizards probably run a higher risk of predation than their more experienced elders, but since *Dipsosaurus* is known to live at least 5 years, the ability to store fat to tide it over the periods of scarcity that it is likely to meet in that time will be an advantage. The shape and tissue composition differences between the regenerated and never-broken tails can be interpreted adaptively: the lizard regenerates its 'larder' and hence is able to store more fat than it could before the tail was broken, suggesting that the capacity for fat storage is now a 'higher priority' trait than the ability to run very fast or break the tail readily when molested, although the latter abilities may be important to small, young lizards.

Man is unusual in that both sexes show very marked changes in fat quantity and distribution with age. The distribution of fat is similar in infants, but girls retain more fat in all, and most of it is subcutaneous during childhood. In adolescence, the quantity of fat, particularly subcutaneous fat increases, and remains important throughout young adulthood. However, almost all the increase in total body fatness in late middle age is due to an increase in internal fat deposits. In males the relative reduction of subcutaneous fat on the limbs begins much earlier, in teenage, and by early adulthood men are only half as fat as women in proportion to their lean body weight. Beginning in the third decade of life, body fatness increases with age due almost entirely to enlargement of the abdominal deposits, causing the general body shape of men and women to become similar by late middle age (Vague, 1953).

8.6 HIBERNATION

Hibernation, like dormancy in plants, can be viewed as an adaptation to minimize the need for energy storage; the energy requirements may fall to as little as 1 % that of maximum activity, so that the energy stores which the animals are able to maintain in the body last through the winter. Even with this economy, the organism's storage mechanisms are often pushed to the limit of their capacity, and if the winter is prolonged, they run the risk of starvation. This risk is particularly high for small animals, since a greater proportion of their body weight must be storage tissue to survive a given period of starvation (Morrison, 1960). For example, the insectivorous bats *Eptesicus fuscus* and *Myotis lucifugus* in Minnesota deposit fat up to one-third of the body weight, and use it at a rate of 38.5 mg per day while torpid, but if the period of hibernation extends beyond 170 days, due to prolonged cold weather, the bats die of starvation (Beer & Richards, 1956).

Since hibernation may be viewed as a period of starvation predictable both in its length and in the rate at which food reserves will be resorbed, the mechanism of control of deposition and depletion are likely to be well developed. The quantity of fat deposited by ground squirrels (*Citellus* spp.) depends upon the temperature at which they are living just before hibernation begins (Mrosovsky, 1976) and there seems to be an internal 'set point' for body weight as the season progresses (Mrosovsky & Fisher, 1970). Like migrating birds, the timing of fattening is determined by a range of environmental stimuli which act on the endocrine system to maintain tight control. Even in non-hibernating animals, daily feeding schedule and amount of handling etc. affect the rate and quantity of fat deposited in animals as diverse as pigeons, teleost fish (*Fundulus* spp.) and *Anolis* lizards (Meier, Trobec, Haymaker, MacGregor & Russo, 1973).

Hibernation in its true sense is only shown by mammals of below 20 kg adult weight (Young, 1976). Larger mammals such as bears enter periods of torpor in which the body temperature falls to about 34° C, compared to 4–6° C in true hibernators. Nonetheless, there are important differences in the nitrogen metabolism of bears in imposed starvation in summer and natural torpor in winter (Nelson, Jones, Wahner, McGill & Code, 1975). In summer, ketones appear in the bear's urine during starvation, indicating that proteins are being broken down, presumably to release glucose, but no ketones are detected in the urine of torpid bears in winter although they remain much longer without feeding. This suggests either that proteins are exempt from being broken down in winter, or that the urea so formed is being recycled. Different adipose tissue masses are depleted in a definite order in starvation and hibernation and some are not depleted at all (see Pond, 1978), although there is no apparent chemical or cytological difference between undepleted tissues and those from which the fat is withdrawn.

Some mammals, such as the squirrel (*Sciurus* spp.), avoid the problem of winter food shortage by accumulating large food reserves in caches, outside the body. It remains slim and agile and is active throughout the winter.

8.7 ALTERNATIVE FUNCTIONS OF ENERGY STORAGE ORGANS AND TISSUES

We saw in sections 8.2 and 8.3 that storage tissues are bulky and often have a different density to the rest of the organism's body. We saw in section 8.5 that organisms frequently increase their energy stores before particular events in their life history take place, and changes in the quantity and/or distribution of their storage substances are often an easily recognizable manifestation of changes in their hormonal state which have promoted that new event in their life cycle. These circumstances together promote changes in function of the storage organs, sometimes to the extent that its principal role is no longer that of a storage organ. I mentioned in section 8.4 that storage organs are unusual in that they have no immediate function for the organism as a whole at any given time; they are simply a reserve to be broken down at some later time. Nonetheless their presence cannot but make some difference to the shape, density or total mass of the animal or plant, and this difference in itself may acquire functional importance.

One example of change in function of storage tissue is its role in altering the body shape of some higher mammals and birds; the changes in body shape have become an indication of social and/or sexual status, particularly in groups such as the ungulates, carnivores and primates in which vision as well as olfactory and auditory stimuli play a part in individual recognition.

I pointed out in section 8.5.3 that mammals are unusual in that suckling and newly weaned juveniles of both sexes have relatively enormous fat deposits, and that much of this fat is subcutaneous, giving the animal a rotund, smooth outline, podgy cheeks and relatively short, plump limbs. The opposite is true of young adult males; they have much less fat overall, and almost all of it is in internal deposits. The adult males of many higher primates including baboons and man, dogs, wolves and even seals will permit 'cuddly' juveniles of both sexes, which have juvenile fur colour and juvenile fat distribution, to play with them and pull their fur but are less tolerant towards older juveniles which differ in body composition and in coat colour (Jolly, 1972). Adult humans show strong maternal behaviour towards chubby infants of their own (and other) species—much of the advertising for the baby food industry depends upon this fact—while responding less readily to older children with less 'cherubic' figures. It is notable that man differs from most other primates in that infants do not undergo a change in fur (hair) colour at about the age of weaning; a change in fat distribution, particularly a reduction of fat on the distal limbs, seems to have taken its place as an indicator of the change in social status. Similarly, scrawny baby birds are an object of disgust while the English robin owes much of its popularity to its 'plump' breast—notwithstanding the fact that the breast derives its shape from the configuration of the flight muscles, and has very little to do with fatness!

Fatness is not only an indicator of general prosperity, but fatter individuals will also have more reserves to devote to egg development, territory defence or rearing offspring. It is therefore not surprising that the distribution and abundance of fatty tissues have become an indicator of sexual and/or social status in some mammals and birds (see Chapter 13). Superficial fat masses in anatomically prominent positions alter the shape of the body and emphasize differences in skin, plumage and fur colour. Stallions develop a band of fat on the dorsal side of the neck under the mane and in several primates, such as the proboscis monkey (*Nasalis larvatus*), the orang-utan (*Pongo pygmaeus*) and the chimpanzee (*Pan troglodytes*), the adult males develop prominent masses of fatty, fibrous tissue on the face, neck or shoulders (Fig. 8.5). The distribution of fat in humans depends both upon their age and their sex; the sexes are similar in infancy, early childhood and old age, and differences in fat distribution are greatest from early teenage until late middle age (Vague, 1953). Man is also unusual in that the distribution (as well as abundance) of fat is under the control of hormones, particularly sex hormones, and fat can be withdrawn from one part of the body at the same time as being accumulated at another (Skerlj, 1959; Skerlj, Brozek & Hunt, 1953).

The distribution of fat in young adult humans, particularly women, is also such as to exaggerate the principal ways in which the anatomy of *Homo* differs from that of the other great apes: the widening of the pelvis and dorsal rounding of the buttocks, caused by the rotation of sacrum and pelvis and the enormous development of the gluteus maximus muscle for bipedal posture, is emphasized by superficial fat; the waist, made possible by the shorter ilium, increased development of the lumbar vertebrae and the lumbar curvature of the spine, is emphasized in frontal view by lateral deposits on the gluteal fascia and in side view by subcutaneous fat around the navel. The female pelvis is wider than that of the male and is tilted forwards and downwards and hence the femurs are set further apart; these differences are exaggerated by fat masses on the hips and trochanteral region of the thighs (see Fig. 8.4).

There are marked individual differences in the distribution of fat in women which become especially apparent in middle age, and which are at least partly

Fig. 8.5 The role of fat in determining individual differences in orang-utans. (a) and (b) Two different old males showing massive but dissimilar development of the fatty cheekpads. (c) Young male without cheekpads and showing single pectoral fold. (d) Non-reproducing adult female without cheekpads but with massive pectoral fat deposits forming 'breast'. ((a) and (b) redrawn after Schultz, 1968; (c) and (d) drawn from photographs kindly lent by Thomas Ritchie.)

determined by racial origin (Skerlj, Brozek & Hunt 1953). In some isolated racial groups, certain fat masses are enormously developed compared to others, particularly in young women (Shattock, 1909), suggesting that the differences are the result of sexual selection. Some of the more striking examples are shown in Fig. 8.4; similar but less extreme differences can also be detected in white American women (Skerlj, Brozek & Hunt, 1953). There is no reason to suggest that the skeleton and muscles are responsible for more than very minor variations (Oxnard, 1973), so most of the differences are due to the relative development of fat. The most consistent feature is the widening of the pelvis, which, as explained in Fig. 8.6 is the most conspicuous skeletal difference between men and women and between human and ape females. Fisher (1930) pointed out that sexual selection will only become established when there is some advantage to one partner in choosing a mate with a particular character. It is interesting to speculate how far sexual selection for the distribution of fat on the thighs and buttocks was initiated and perpetuated by natural selection for a wider pelvis. *Homo sapiens* has a particularly small pelvic inlet compared to the size of the infant's head, and birth is much more prolonged and difficult than in all other great apes (Schultz, 1969). If the chances of a birth being successful were increased if the women's pelvis was wider, men who chose mates with this character are more likely to leave offspring, thus providing a basis upon which sexual selection can be built up.

Another example of change in function of fatty tissue is its effect on buoyancy in aquatic organisms, from unicellular algae to whales. For most organisms there is only indirect evidence for its contribution to flotation (see Pond, 1978), but in the case of cartilagenous fish there is direct evidence for the involvement of liver lipids in buoyancy control. Many sharks have relatively enormous livers, up to 25% of their total weight (in contrast to about 2% in man and most other vertebrates) and the liver is composed of 75–92% hydrocarbon and lipid by weight. Some hydrocarbons have a specific gravity as little as 0.86, compared to a density of 1.06 to 1.09 for non-fatty fish tissues combined (Alexander, 1970). In at least one species, the spiny dogfish *Squalus acanthias*, the ratio of hydrocarbons (in this case diacyl glycerylethers) to triglycerides could be altered by artificially increasing the specific gravity of the fish by attaching weights to its fins (Malins & Barone, 1970). Thus the composition of its 'storage organ' is affected by factors totally removed from diet or nutritional status. A similar process may occur in pregnant female dogfish (*Acanthias vulgaris*); lipids are redistributed between the muscles and the liver as pregnancy progresses (Hickling, 1930).

The elasmobranchs *Squalus* and *Deania* contain large quantities of alkyldiacylglycerol (specific gravity 0.91) and squalene (sp. gr. 0.86), both of which are rare in teleost fish and mammals. Those lipids are both transported in the serum like triglyceride, but only alkyldiacylglycerol can be broken down for energy metabolism. The only known metabolic function for the hydrocarbon squalene is in the formation of cholesterol and bile salts, (Sargent, Gatten & McIntosh, 1973). Thus some of the lipid is stored as alkyldiacylglycerol which has the advantage of being less dense than triglycerides (average sp. gr. 0.93), but still available as an energy store, while the squalene has lost its function as an energy store. Most elasmobranchs are not wholly dependent on the static buoyancy obtained from low-density tissues such as fats, but can also

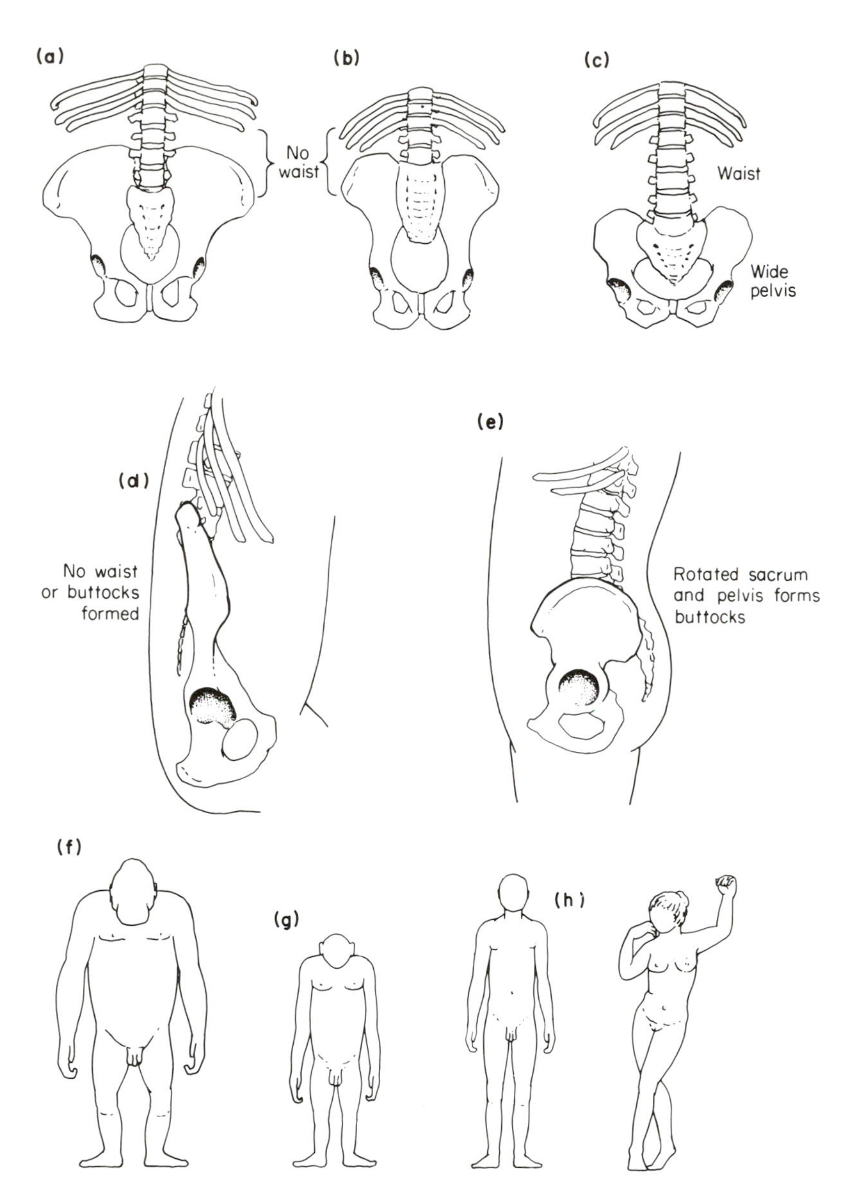

Fig. 8.6 The skeletal basis of the waist, buttocks and hips in humans. Front view of the pelvic girdle, lumbar vertebrae and last three ribs in adult female gorilla (a), chimpanzee (b) and human (c). Ape (d) and human (e) from side. The 'figures' of gorilla (f), chimpanzee (g) and male and female humans (h). Notice that the shortened and broadened ilium and the increased number of lumbar vertebrae form the waist, the rotated pelvis and lumbar flexure of the spine form the buttocks and increased separation of the acetabula form the hips. In apes the 'waist' is convex and the 'buttocks' concave. The skeletal differences are more pronounced in human females than in human males, but the differences are further exaggerated by the disposition of superficial fat in human females. (Mostly redrawn and modified after Schultz, 1968.)

generate hydrodynamic lift. It is not known whether those species which 'commit' large quantities of lipid to very low-density, but metabolically inert substances such as squalene have reduced ability to generate hydrodynamic lift, compared to those whose fat stores enable them to rely on static buoyancy during prosperous periods, but which must presumably revert to hydrodynamic buoyancy mechanisms when emaciated.

8.8 SUMMARY AND CONCLUSIONS

When Slobodkin (1962) wrote 'there is no clear advantage in adiposity', he was probably thinking of the relationship between adiposity and energy put into reproduction. Calow (1979) vaguely alluded to 'ecological and ethological' components of the cost of energy storage. In this short chapter I have tried to expand these ideas and to show that the determinants of how much energy is stored in an organism's body, when and where it is stored, are much more complicated than Slobodkin's concept. In particular, the cost of accumulating, depositing and maintaining an energy reserve cannot be estimated simply by measuring the energy content, as some ecologists have tried to do (Slobodkin, 1961; Calow, 1979). The 'cost' of maintaining an energy reserve is not only measured in joules as the extra work done in support and transportation, but is also measured in terms of the selective disadvantages which the animal incurs because the presence of the energy reserve reduces its mechanical performance. However, unless the energy reserve is transitory, there is a high probability that it will 'acquire' additional functions for the organism, and these sometimes become principal functions. The distinction between a storage organ and a case of reabsorption under extreme starvation of an organ whose main function is not for storage, then becomes an academic one.

Chapter 9

Resource Utilization in Growth

PETER CALOW AND COLIN R. TOWNSEND

9.1 INTRODUCTION

Fitness is ultimately measured in terms of the phenotype's relative success in converting resources to reproductive products. In a completely benign environment it could be imagined that the conversion process would be carried out most effectively by a replicating system consisting only of a small template and an enzyme which catalyses duplication (in a sense, a *replicase* is the minimum phenotype; Calow, 1977a). In real environments, however, there exist numerous risks to survival, and many structures and processes are associated with the replicase system to cushion it from disturbance. Phenotypes are invariably bigger than the minimum conceivable. A consequence of constraints in reproduction is that reproductive products (a single cell in sexual reproduction, multicellular fragments of the parent in asexual reproduction) are generally much smaller and more simple than their parents. Thus, growth can be viewed as the developmental means of achieving the reproductive state. This chapter is concerned with the utilization of resources for growth.

Bigness, *per se*, often carries with it important benefits. Bigger individuals may be at a competitive advantage when it comes to gathering resources, whether these consist of prey items or light quanta, and they may also be less vulnerable to predation. In addition, larger size makes it easier to maintain constancy of body function in the face of variation in the abiotic environment (Chapter 2). In consequence, the absolute increase in size during development is often very substantial.

However, in this chapter we are more concerned with the rate of attaining mature body size. We take as our null hypothesis the maximization principle (Chapter 1) that *production processes of organisms should be adapted to maximize growth and development rates.* This is a plausible hypothesis since the more rapidly an organism grows the sooner it becomes reproductive and begins producing progeny. Even when breeding is limited to a certain time of year a faster growth rate may mean a larger final size and thus a greater reproductive output (Chapter 10). Finally, since small offspring are often very vulnerable to adverse physical conditions, competition and predation, there may be an advantage in quickly getting over this stage.

Rapid development and shortened generation time, greater reproductive output and increased individual survivorship are all positively correlated with fitness. Hence, it is to be expected that any physiological trait which enhances rates of growth and development will have been selected, and two important traits from this point of view are *efficiency* and *rate* of metabolism. Growth rate, for example, is the product of the rate at which resources are made available, and their efficiency of conversion to tissues. We shall explore the physiological limits to these components in the next two sections. We shall then proceed to ask whether there is selective value in the different growth habits which characterize organisms in contrasting environments. There may, of course, be costs as well as benefits associated with rapid growth, and these are considered under the heading: 'Is growth rate maximized or optimized?' Finally, we shall say something about the most appropriate models for metabolism and about the involvement of growth in morphogenesis.

9.2 CONVERSION EFFICIENCIES

9.2.1 General

Resources absorbed from food by heterotrophs, or formed during photosynthesis by autotrophs, are used either as building blocks in the synthesis of tissues or to power the physical, chemical and synthetic work of the organism. The latter takes place via energy carriers like ATP. Hence, the partitioning of input resources occurs as illustrated in Fig. 9.1. The proportion of input resources which is available for work and growth depends, to a considerable extent, on the efficiency with which energy can be transferred from food molecules to the terminal phosphate bonds of ATP. We consider this partial efficiency of the system first before going on to consider the overall efficiency of conversion of input resources to tissues.

9.2.2 Respiration

The mechanisms involved in generating ATP are roughly comparable from one organism to another (Fig. 9.2.). The primitive system operated without oxygen and may have been equivalent to the glycolytic pathway found in many

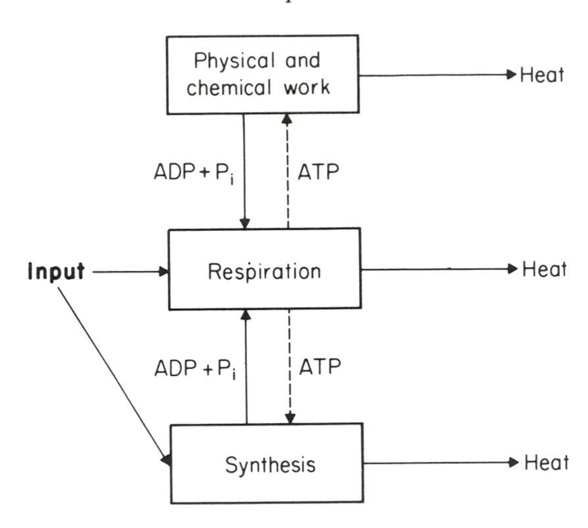

Fig. 9.1 Partition of input energy in organisms.

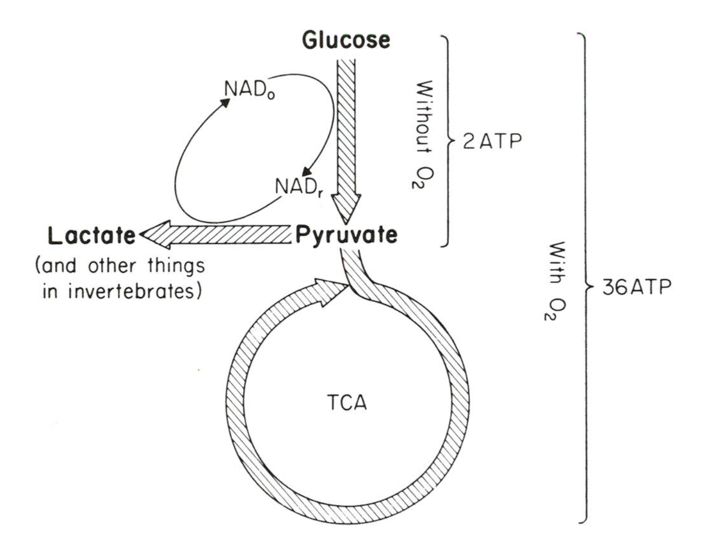

Fig. 9.2 An extremely stylized representation of the major metabolic steps concerned with the generation of ATP.

extant organisms. The details of this part of the process are given in Lehninger (1973) and can be summarized:

$$\text{Glucose} + 2\text{ADP} + 2P_i \rightarrow 2\,\text{Lactate} + 2\text{ATP} + 2\text{H}_2\text{O}$$

Since the standard free energy of glucose is approximately $-2880\,\text{kJ mol}^{-1}$ and each terminal phosphate bond in ATP stores $30.7\,\text{kJ mol}^{-1}$ then the efficiency of conversion amounts to an unimpressive 2% $\left(\text{i.e. } \frac{2 \times 30.7}{2880} \times 100\right)$. In the presence of oxygen the overall balance is:

$$\text{Glucose} + 36\text{ADP} + 36P_i + 6O_2 \rightarrow 6CO_2 + 6H_2O + 36\text{ATP}$$

and the efficiency is, therefore, 38 % $\left(\text{i.e. } \frac{36 \times 30.7}{2880} \times 100\right)$.

With the advent of photosynthesis and the generation of oxygen on this planet metabolism became much more efficient. However, even the efficiency of the oxidative system is not particularly impressive. There are at least two possible reasons for this, both concerned with constraints (Chapter 1). First, the make-up and efficiency of the system we find now may have been fixed by chance happenings at the origin of life. It was presumably just those *initial conditions* which determined that carbon should form the basis of biochemistry, and that phosphate bonds should be used as energy carriers. Second, constraints might be imposed on the efficiency of the system by virtue of what else is required of it. For example, glycolysis and the TCA cycle form the backbone of all aspects of intermediary metabolism, not just those concerned with energy release.

It is interesting to note that obligate anaerobes such as endoparasites and animals which live in habitats experiencing prolonged anoxia (e.g. bivalves and annelids in aquatic sediments) do employ ATP-generating processes which are more efficient than glycolysis. These are discussed by Hochachka & Somero (1973) and a simplified version of how the processes work is given in Fig. 9.3. The reduced NAD (NAD_r) generated in the pathway from glucose to pyruvate is used not to form lactate but to drive an electron transport system in which fumarate, rather than oxygen, is the final electron accepter. ATP is generated from ADP and P_i, using the energy released as electrons are transferred along the chain, and fumarate is transformed to succinate. Fumarate is derived in the first place via oxaloacetate from phosphoenolpyruvate (PEP) in the glycolytic chain. (Oxaloacetate, fumarate and succinate are all intermediaries in the TCA cycle, and this modified anaerobic cycle can be considered as a part of the TCA cycle put in reverse! Some would argue that it was from this kind of beginning that the TCA cycle itself evolved.)

Depending on the exact pathways employed, between 5 and 8 ATP molecules may be formed from the more elaborate anaerobic metabolism,

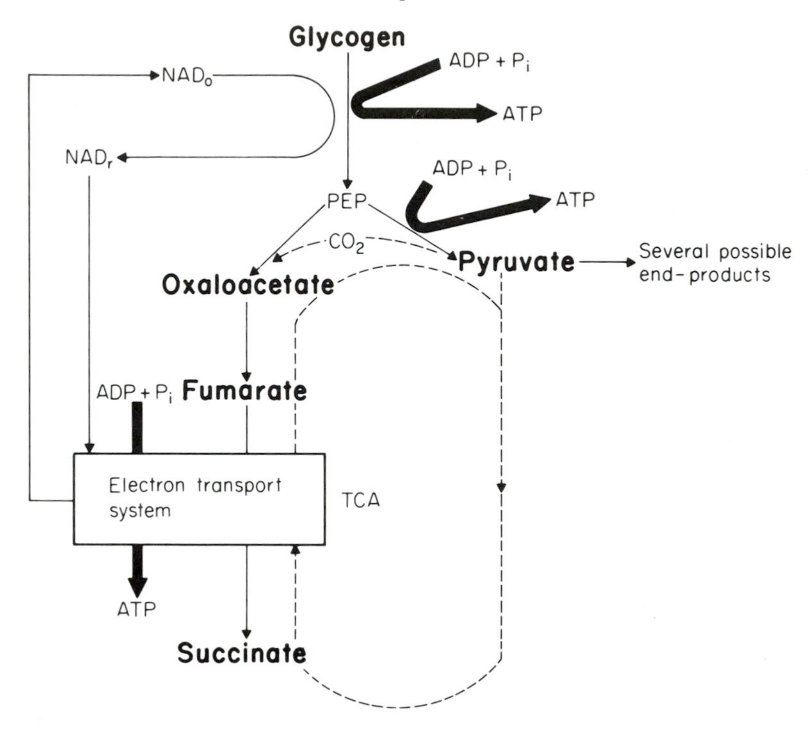

Fig. 9.3 The modified anaerobic pathway used by endoparasites and bivalves living in conditions of hypoxia for long periods.

increasing the efficiency from 2% to between 6 and 8%. Why then is the more efficient system not always employed? One possible answer is that the *more efficient* succinate system yields ATP *at a slower rate* than the lactate system. For example, Hochachka (1976) points out that the succinate system of the oyster heart yields 1 ATP molecule s^{-1} g^{-1} of tissue whereas the lactate system of trout muscle yields 13 ATP molecules s^{-1} g^{-1}. The latter may be more advantageous when sustained output is required over short periods of anoxia, as occurs in overworked muscles, whereas long-term anoxibiosis is likely to favour the more efficient system.

9.2.3 CONVERSION EFFICIENCIES IN HETEROTROPHS

Assuming that aerobic metabolism is taking place and making minimum allowance for the demands of mechanical, chemical and synthetic work in individual animals it is possible to derive a rough estimate of the 'best possible efficiency' for the conversion of nutrients absorbed across the gut walls (A) to tissue (P_g). This figure ($[P_g/A] \times 100$) turns out to be approximately 80% (Calow, 1977c). [Note that P_g/A is not the same as the efficiencies quoted in Chapter 2. The latter are based on population production data which are influenced by mortality and natality as well as growth]. The expectation is that P_g/A values will have been 'pushed' by natural selection towards this theoretical limit.

Figure 9.4 compares actual conversion efficiencies recorded for a variety of

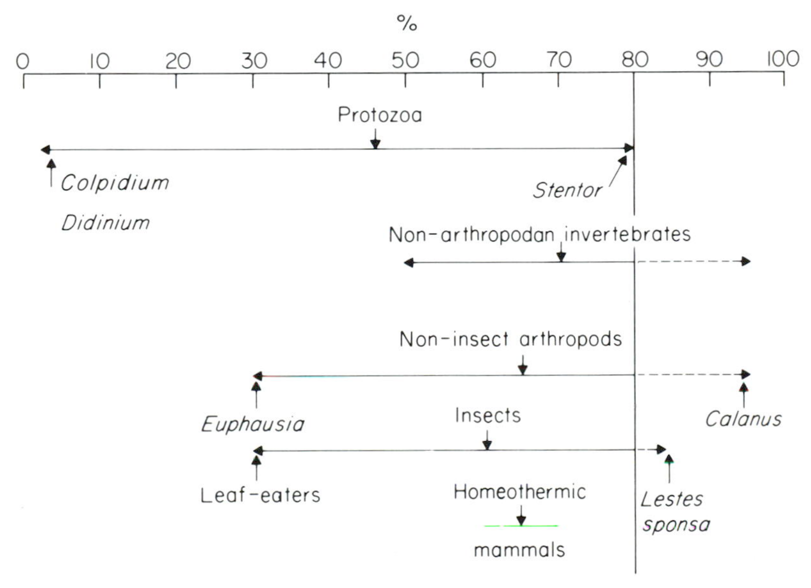

Fig. 9.4 Conversion efficiency ranges in several heterotrophs. Vertical line represents the hypothetical limit.

heterotrophic metazoa with the theoretical prediction. A number of points emerge:

1 the range is wide and cannot be said to lend a great deal of support to the maximization hypothesis;
2 some of the ranges exceed the theoretical limit. This may either be due to errors in measurement or to flaws in the theoretical rationale. In fact, the highest efficiencies are associated mainly with sedentary species; *Stentor*, an attached protozoan, sea anemones in the non-arthropod, invertebrate range, and larval *Lestes sponsa*, a sit-and-wait predatory insect. Planktonic, immature *Calanus* in the non-insect arthropod range is an

'odd one out' in that it moves actively in the plankton but still has a high efficiency. Perhaps this is due to the fact that it is usually surrounded by algal food and does not need to invest much effort in catching it. In general, therefore, the theoretical calculation may have overestimated the cost of locomotion in these species;

3 lowest efficiencies in the protozoan range are associated with very active species; *Colpidium*, *Didinium* and *Tetrahymena*, for example. Hence, in these species more energy may be invested in activity than was assumed in the theoretical estimate. The same must be true of many of the winged insects which invest large amounts of energy in flight (Weis-Fogh, 1954);

4 the discrepancy between the range for homeothermic mammals and the theoretical limit may be explained in terms of the cost of endothermy—the maintenance of high, constant internal temperatures requires the expensive generation of heat.

To summarize, then, the differences between expectations (based on the maximization hypothesis) and actual efficiencies may be explained, at least in part, by the energetic costs of activity and endothermy. However, these energetic costs may bring fitness benefits of their own. For example, endothermy means that metabolism can occur at a high and constant rate irrespective of environmental disturbances and the consequences of this will be discussed below (and see Chapter 2). Similarly, high levels of activity may bring considerable survivorship benefits through increasing the effectiveness of capturing prey and avoiding being killed by predators. It is clear that there are compromises to be made in terms of what proportion of input resources to devote to work as opposed to growth. Our view, and one which has been stressed repeatedly through this book, is that natural selection has acted to *optimize* the form of the compromises so as to *maximize* individual fitness.

9.2.4 Conversion efficiencies in autotrophs

For autotrophic plants and bacteria, Penning de Vries *et al.* (1974) calculated a dry weight conversion efficiency (weight of product/total weight of substrate required to produce it) of 75 % or less, depending on exact conditions. This is less than the maximum efficiency found among heterotrophs. While it is true that autotrophs incur lower costs than animals in physical activity, nevertheless they incur greater costs in reorganizing photosynthates (e.g. glucose) and inorganic molecules (e.g. NH_3, nitrates and sulphates) into fatty acids, for fat synthesis, amino acids, for protein synthesis, and nucleotides, for polynucleotide synthesis. In experiments performed on certain plants a good correspondence was found between the theoretically established relative yield and the actual data derived from experiments. This suggests that metabolism in higher plants may operate close to the maximum efficiency allowed by the theoretical scheme.

9.3 GROWTH RATES

9.3.1 VERTEBRATES

Because of the cost of heat production in endothermy, ectothermic vertebrates are likely to be more efficient converters than endotherms (Calow, 1977b; but cf. Case, 1978). On the other hand, endothermy is likely facilitate relatively constant high rates of feeding. Since growth rate is a product of a rate (How fast are resources made available?) and an efficiency (What proportion of A is utilized in P_g?), it is not apparent on *a-priori* grounds which strategy will produce higher growth rates. It turns out that the maximum growth rates of most endotherms are at least an order of magnitude greater than those of ectotherms (Case, 1978).

Case (1978) has reviewed growth rates of vertebrates. Since these are not independent of size and age he has calculated the maximum rate for each species and used the logarithm of this for interspecific comparison. Plots of $\log_{10}$ maximum growth rate against $\log_{10}$ adult size are shown in Figs 9.5 and 9.6. The growth rates of fishes and reptiles are found to be lower than those of mammals. Precocial birds (hatched in an advanced state and capable of locomotion) grow at a similar rate to mammals, but altricial birds (incapable of locomotion, requiring parental care) grow more rapidly. Not surprisingly, there was much intergroup variation. For example, growth rates in mammals ranged from 66 000 g day^{-1} (a whale) to 0.41 g day^{-1} (a primate). Marsupials also had relatively low growth rates (1–30 g day^{-1}).

9.3.2 INVERTEBRATES

Table 9.1 contains a compilation of maximum recorded growth rates for 26 species of invertebrate ranging through six phyla. Data were taken from studies in which there were no feeding restrictions, temperatures were clearly defined (given as T° C) and, except for the endoparasite, *Hymenolepis diminuta*, and the shelled snails, size was measured directly as fresh weight. The sizes of the adults (g fresh weight) expressed as logarithms are denoted as LAW. Most

Table 9.1 Adult sizes and growth rates of invertebrates.

Species	$T°C$	LAW	LGR_T	LGR_{10}	LGR_{38}	Code in Fig. 9.7	Source
Platyhelminthes							
Hymenolepis diminuta	§1	−0.92	−1.22	—	−1.22	(P)	Hopkins *et al.* (1972)
Dugesia gonocephala	c.10	−1.30	−3.44	−3.44	−2.39	P	Abeloos (1930)
Nematoda							
Plectus palustris	20	−5.52	−6.40	−6.87	−5.82	N	Schiemer *et al.* (1980)
Annelida							
Lumbricus terrestris	§2	1.00	−1.86	−1.86	−0.81	O	Satchell (1967)
Dendrobaena subrubicunda	10	−0.30	−2.52	−2.52	−1.47	O	Michon (1954)
Erpobdella octoculata	10	−0.89	−2.61	−2.61	−1.56	L	Calow (Unpubl. obs.)
Erpobdella testacea	10	−1.09	−2.41	−2.41	−1.36	L	Calow (Unpubl. obs.)
Mollusca							
Planorbis contortus	10	−2.60	−4.25	−4.25	−3.20	M	Calow (Unpubl. obs.)
Ancylus fluviatilis	10	−1.70	−3.70	−3.70	−2.65	M	Calow (Unpubl. obs.)
Arion rufus	20	1.00	−0.96	−1.43	−0.38	M	Abeloos (1944)
Arion subfuscus	20	0.48	−1.37	−1.84	−0.79	M	Abeloos (1944)
Arion hortensis	20	0.70	−1.15	−1.62	−0.57	M	Abeloos (1944)
Arion intermedius	20	0.00	−2.52	−2.99	−1.94	M	Abeloos (1944)
Arion circumscriptus	20	−0.30	−2.75	−3.22	−2.17	M	Abeloos (1944)
Octopus cyanea	26	3.60	1.73	1.26	2.31	(M)	Wells (1978)

Table 9.1 (*contd.*)

Species	$T°$ C	LAW	LGR_T	LGR_{10}	LGR_{38}	Code in Fig. 9.7	Source
Echinodermata							
Strongylocentrotus drobachiensis	§3	2.3	−0.54	−0.54	0.51	E	Vadas (1977)
Arthropoda/Insecta							
Dixipus morosus	18	0	−1.60	−1.99	−0.94	I	Teissier (1931)
Galleria mellonella	32	−0.70	−1.62	−2.49	−1.45	I	Teissier (1931)
Bombyx mori	32	0.64	−0.62	−1.49	−0.44	I	Teissier (1931)
Tenebrio molitor	26.5	−0.76	−2.87	−3.57	−2.52	I	Teissier (1931)
Gerris lacustris	20	−1.82	−3.10	−3.57	−2.52	I	Teissier (1931)
Notonecta glauca	20	−1.82	−2.56	−3.03	−1.98	I	Teissier (1931)
Notonecta glauca	23	−0.82	−2.43	−2.90	−1.85	I	Toth & Chew. (1972)
Crustacea							
Asellus aquaticus	15	−1.60	−3.77	−4.04	−2.99	C	Marcus *et al.* (1978)
Gammarus pulex	15	−1.60	−3.88	−4.15	−3.10	C	Willoughby & Sutcliffe (1976)
Archnida							
Tetranychus urticae	20	−4.70	−5.30	−5.77	−4.72	A	Mitchell (1973)

$T°$ C = Temperature at which measurements were made
§1 = Host is a mammal; temperature assumed to be 38°C
§2 = Measured out of doors; temperature assumed to be 10°C
§3 = Temperature variable but approximated to 10° C

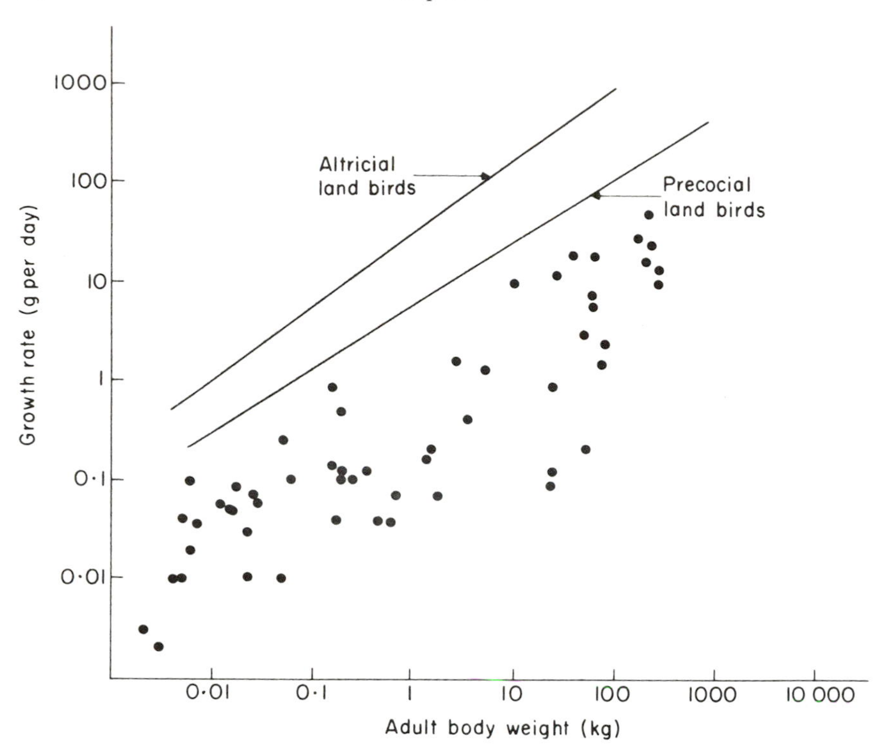

Fig. 9.5 The growth of poikilothermic vertebrates as a function of body weight compared to avian growth (after Case, 1978).

of the rates that have been used were means of more than five (usually 10) measurements. Exceptions were the slugs and the insects where the measurement for each species represents a single individual only. Apart from the parasite, all rates derived from the original work (LGR_T) were corrected to 10° C (LGR_{10}) since this is reasonably typical of conditions under which invertebrate metabolism operates in the field in temperate regions. Rates were also corrected to 38° C (LGR_{38}) for purposes of comparison with the data for mammalian homeotherms. These corrections were calculated using Krogh's method (Krogh, 1914).

The LGR_{10} data are plotted as points in Fig. 9.7. The results from Case (1978) are summarized in the same figure as regression lines. The following points of comparison should be noted:

1 most of the invertebrate data scatter around the lines for the poikilothermic vertebrates;

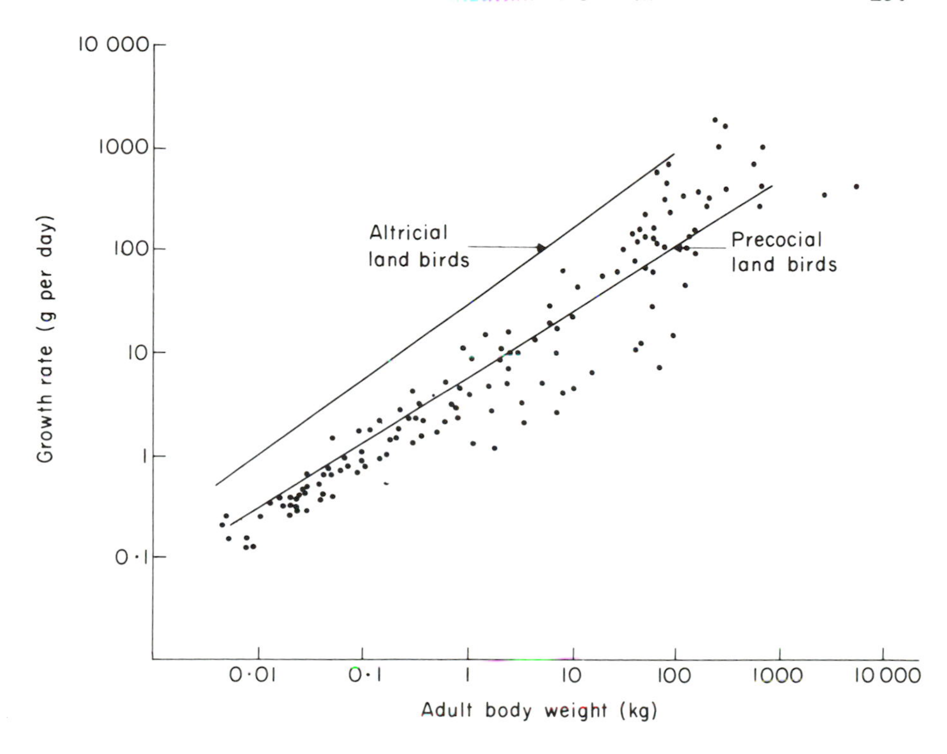

Fig. 9.6 Mammalian growth as a function of body weight and compared with avian growth (after Case, 1978).

2 an exception is the endoparasite *Hymenolepis diminuta* (point P) which at mammalian body temperatures has a growth rate better than the other invertebrates and even better than that predicted for animals of this size by the mammal line. The reason for its divergence from the other invertebrate data is almost certainly due to the difference in temperatures at which the rates were measured. The divergence from the mammalian expectation can be attributed to two other non-exclusive factors: (i) *H. diminuta* may derive metabolic benefits from its homeothermic host without having to pay the costs of endothermy; (ii) being bathed in a non-limiting source of food, this animal does not have to invest energy in food-getting. Clearly, the inefficiency associated with anaerobic metabolism (section 9.2.2) does not adversely affect the growth rate of this species;

3 another exception is *Octopus cyanea*. This grows as rapidly as mammals of equivalent size. A possible explanation is that it is a 'sit-and-wait'

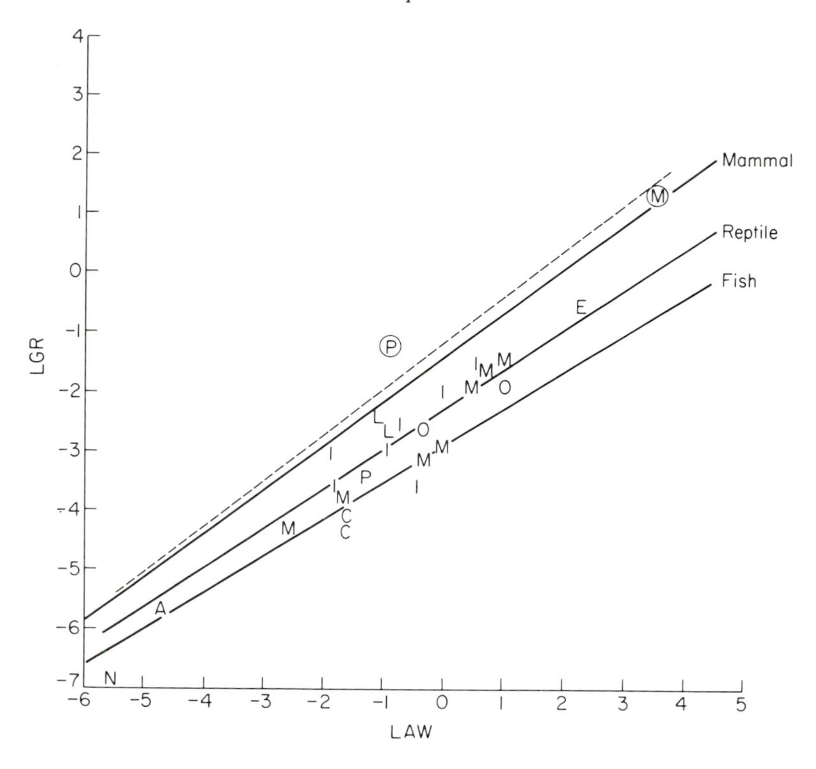

Fig. 9.7 Invertebrate growth as a function of body weight compared with the vertebrate data from Case (1978). Each letter represents a species (for key, see Table 9.1). The broken line represents the regression for the invertebrate data corrected to 38° C.

predator which, in consequence, does not expend a great deal of energy in finding and capturing food.

Also included in Fig. 9.7 is the regression line for the LGR_{38} data. Unfortunately, the slope of this line differs significantly from that of the mammal line so that no statistically valid, general statement can be made about the rates of growth of invertebrates at this temperature as related to those predicted by the data for mammals. Clearly, though, the invertebrates' rates are as good as, if not better, than those of the mammals over the upper part of the size range.

A number of interesting conclusions follow: first, it seems likely that invertebrate growth rates are constrained under normal conditions by temperature; second, the fact that some invertebrates have better growth rates than those predicted by the mammal data suggests that mammals carry extra metabolic costs, perhaps associated with endothermy. Hence the mammals, and also birds, are able to escape the constraints of temperature but only at the expense of reduced conversion efficiencies. Nevertheless, and as already noted, such a strategy brings both metabolic and fitness benefits because the rates of metabolic processes in these animals can be maintained at high levels which are independent, at least in part, of external conditions. Finally, the proportionately lower growth rates of smaller invertebrates may be due to the increased metabolic costs per unit weight of tissue associated with small animals (Chapter 2).

9.3.3 Plants

Grime & Hunt (1975) measured the growth rates of 132 species of terrestrial plants under standard, favourable laboratory conditions and expressed them as the maximum potential growth rate R_{MAX}, where

$$R = \frac{1}{W} \times \frac{dw}{dt}$$

$$= \frac{d(\log_e W)}{dt}$$

where W is fresh weight in grams.

A computer was used to calculate R at short intervals during the entire growth period of each plant. The highest value of R obtained in this way was taken to be R_{MAX}. This index is, therefore, the maximum size increment (in fresh weight) per unit time per unit fresh weight of plant; very similar, therefore, to the index used by Case for vertebrates. The use of R_{MAX} as an index for comparing growth rates of plant species has been criticized by Harper (1977) who points out that differences within species may often be greater than differences between species so that biological properties gauged from study of a single plant, or a few samples, may be unrealistic. This criticism may also be levelled at the approaches described above, although there will be less inter-individual variation among invertebrate and, in particular, vertebrate species. Nevertheless, in all instances clear patterns of a gross nature emerge, and these can be accepted as fair generalizations about growth rates in species groupings which represent particular habitats or taxonomic affinities.

Distributions of R_{MAX} from Grime & Hunt's analysis are summarized in Fig. 9.8. The overall distribution has a single 'hump' with the right-hand tail

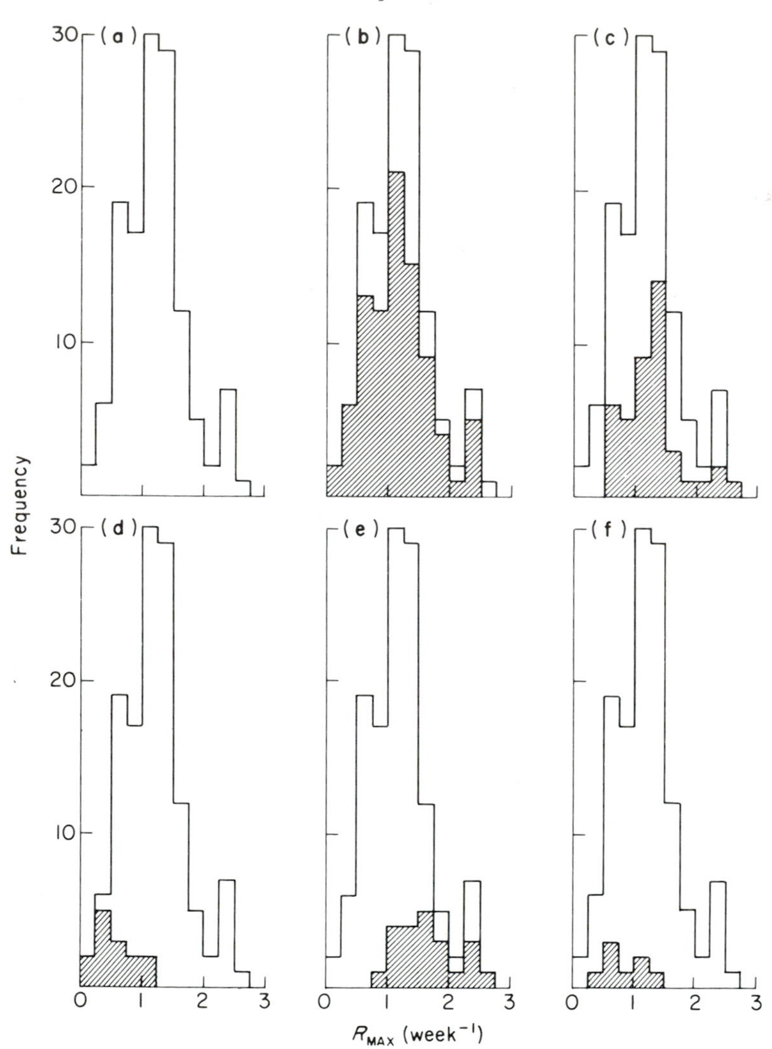

Fig. 9.8 The frequency distribution of R_{max} in: (a) the whole sample; (b) dicotyledons; (c) monocotyledons; (d) woody species; (e) annuals; (f) legumes. For ease of comparison the distribution pattern in (a) is repeated each time. (After Grime & Hunt, 1975.)

slightly more extended than the left. There were few differences between monocotyledonous and dicotyledonous plants, both groups showing distributions with single humps and wide ranges. Lowest values of R_{MAX} were found in woody species, while the highest values occurred in the annuals.

9.4 ECOLOGICAL CORRELATES OF GROWTH RATE

While much of this chapter is concerned with the physiological mechanisms by which more rapid growth is realized in certain organisms, we may equally well ask whether there is selective value in the different rates of growth that characterize organisms in different ecological circumstances.

9.4.1 HIGH MAXIMUM GROWTH RATES

The possession of a high maximum growth rate should allow a more rapid completion of the life-cycle. In the case of birds and mammals it has been postulated that in those species whose young are habitually exposed to high levels of predation there will be an advantage accruing to individuals who complete the vulnerable juvenile part of their life as quickly as possible (Williams, 1966a; Lack, 1968; Case, 1978). Reliable evidence on mortality rates is notoriously hard to obtain, but much of the information available generally supports the postulated relationship between growth and predation rates (Lack, 1968; Cody, 1973; Case, 1978; but cf. Ricklefs, 1969). It has also been argued (Case, 1978) that an explanation for the fact that young altricial mammals grow more slowly than altricial birds is that the former nest almost exclusively below ground in burrows, in caves or in tree holes, and are likely to be less vulnerable to predation than the birds which are not pre-adapted for a burrowing existence. (Two further partial explanations for this difference may be, first, that the birds go through a more prolonged ectothermic period of development, during which time more assimilated energy may be channelled into growth; and second, that both parents usually take part in feeding the birds whereas in most cases only female mammals nourish their young.)

Another circumstance where rapid growth might be predicted to confer a particular selective advantage would be in an environment where duration of the breeding season is very short. Case (1978) was unable to detect a general latitudinal pattern in maximum growth rate. However, in those high-latitude species whose breeding activity is especially restricted, notably high growth rates were evident. This was clear in the case of the arctic ground squirrel, *Spermophilus undulatus*, which needs to produce and rear young before it hibernates (Mayer and Roche, 1954), and for seals, such as *Halichoerus grypus* and *Histriophoea fasciata*, which breed on ice floes and must rear their young before the ice melts (Tikhomirov, 1971). Finally, perhaps the most fully discussed general case where rapid completion of the life-cycle is at a premium relates to disturbed habitats (early successional stages, *r*-selecting conditions, etc.). Plants which have a high maximum growth rate and come into this category are referred to by Grime (1979) as *ruderals*. These comprise annuals and short-lived perennials which rapidly exploit disturbed, but productive, environments, growing fast and allocating a large proportion of photosynthate directly into seeds. Within a relatively short time other species will come to dominate the area, unless a fresh disturbance occurs. 'Ruderal' life-histories also occur in certain invertebrates and vertebrates whose populations expand and contract rapidly in response to temporary abundance in food supply (Southwood, 1977). In addition, some fungi, particularly species in the Mucorales which are ephemeral colonists of organic substrates, grow rapidly until the supply of soluble carbohydrates declines, after which they sporulate profusely.

Grime (1979) identifies another category of plants with high maximum growth rates as *competitors*, capable of occupying extensive areas of productive, relatively undisturbed habitat and apparently excluding the majority of others by virtue of a superior competitive ability both above (light capture) and below ground (nutrient and water capture). Such species, including *Epilobium hirsutum* and *Chamaenerion angustifolium*, are often characterized by their tall stature and capacity for rapid lateral spread of shoots and roots. Harper (1977) has pointed out that in crowded communities such as meadows and forests 'the fit individual is one that pre-empts to itself a disproportionate share of the resources.' (These ideas have often been discussed under the heading of *K* selection. See Chapters 1 and 11.) Grime (1979) suggests that 'competitive' characteristics are also evident in fungi such as *Serpula lachrymans* and *Armillaria mellea* which are involved in long-term infections of timber and produce a mycelium which may extend rapidly by the production of rhizomorphs.

9.4.2 Low maximum growth rates

Low maximum growth rates in plants may be associated with predictably low resource availability (Grime & Hunt, 1975) and such plants have been classified by Grime (1979) as *stress-tolerators*. Such species are successful in environments subject to stress, particularly involving nutrient deficiencies and severe shading, first because they have only modest demands and are less likely to exhaust their local resources, second because their low rates of incorporation of photosynthate and mineral nutrients into structure may allow a build-up of reserves in the plant, and third because they are better fitted to survive periods of protracted or extreme stress when little or no growth is possible. Typically, these species demonstrate low respiration rates and extended life of individual leaves and roots. They are not common in productive environments or in disturbed situations because they are poorly adapted for rapid growth responses to environmental variations or fast seedling development. (Conversely, Grime's *competitors* are likely to possess characteristics, such as inflexibly high respiration rates, which will be disadvantageous in environmental extremes.) Grime (1979) points out that 'stress-tolerant' fungi include the slow-growing Basidiomycetes which often take part in the terminal stages of fungal succession on decaying matter when the most readily metabolizable constituents have been exhausted.

Case (1978) has argued that slow maximum growth rates will also confer an advantage on vertebrates in conditions of predictably poor resource availability. This is exactly analogous to Grime's contention regarding plants. Some evidence in support of the hypothesis derives from data on sea-birds (Lack, 1968; Nelson, 1969; Cody, 1973). For example, birds that feed offshore exploit food which is both far from nesting sites and difficult to capture. Thus the offshore-feeding Procellariformes, Pelecaniformes and Sphenisciformes produce altricial young which grow more slowly than the more precocial young of the onshore-feeding Laridae, Sternidae and Stercorariidae. Other possible parallels with stress-tolerant plants occur in the long life-histories and delayed and intermittent reproduction in reptiles such as the Aldabran giant tortoise (*Geochelone gigantea*).

9.5 IS GROWTH RATE MAXIMIZED OR OPTIMIZED?

The differences in growth rate between woody and non-woody plants are mainly due to the unavoidable extra costs of wood production (Grime & Hunt, 1975). The contrast in growth rate of ectothermic and endothermic animals can be explained by temperature constraints on the metabolism of the former. The lowest avian growth rates occur in precocial and self-feeding young. In all these cases it could be argued that the organisms grow as fast as they are physiologically capable of doing. Ricklefs (1968) has postulated this specifically for birds.

Another view is that growth rates are *optimized* and not *maximized* (Case, 1978). Perhaps the best evidence for this alternative hypothesis consists of the apparent 'fine-tuning' of growth rate to the organisms' ecological circumstances, discussed in the previous section. Particularly strong support comes from those animals and plants which demonstrate low maximum growth rates. The argument can also be made on *a priori* grounds. In animals with parental care, for example, the energy demands of the young will increase with growth rate. Hence, if there is a ceiling on the rate at which parents can gather food or supply milk then the intensity of parental care per offspring per unit time must decline as growth rate increases; in other words there are costs associated with rapid growth (Case, 1978). A more general argument might be that in organisms whose rate of growth is under genetic control the fittest individual will not necessarily be the one whose maximum growth rate is closely geared to conditions of abundant resources. It may be more advantageous to possess a low growth rate appropriate to average or poor conditions, since such an organism may be at a competitive advantage over others with inflexibly high demands, and the lower risk of mortality in poor conditions may more than compensate for higher fecundity in good ones.

There are at least two other possible costs associated with high maximum growth rates. The first derives from the argument, used in support of the maximization hypothesis, that rapid development reduces the time spent by organisms in the vulnerable small stages of the life-cycle. Note that the reverse would apply if small-sized animals were less vulnerable than large. Predators do sometimes choose bigger food items (e.g. see Chapter 4) and high predation rates are known to select for smaller size in the freshwater amphipod *Hyalella azteca* (Strong, 1972). Second, given that the resources which can be dealt with by a consumer are determined to some extent by its size, then it may be necessary to synchronize size changes in the consumer to seasonal fluctuations in the size–frequency distribution of food.

It is difficult, on the basis of experimental evidence, to distinguish between the optimization and maximization hypotheses. In vertebrates, at least, both optimum and maximum growth rates can be expected to be fixed genetically, if for different reasons, so that augmenting the supply of resources to young animals should not affect their growth rates–which seems to be generally the case for birds (Case, 1978) and mammals (Minot, 1908). However, as far as the maximization hypothesis is concerned, the fixedness derives from developmental and physiological constraints whereas the optimization hypothesis suggests that rates are fixed below those determined physiologically by *active control*. Hence, the very occurrence of growth-controlling factors (growth hormones) has been taken as lending some support to the optimization argument. Case (1978) suggests that it should be possible on the basis of the optimization, but not the maximization, hypothesis to enhance growth rates by artificially raising the concentration of growth hormones. According to Case, nearly all vertebrates tested display an increased growth rate following this kind of manipulation.

There are also physiological manifestations of active control. First, growth does not respond in a direct and proportional fashion to reductions in food supply (Calow, 1976). This outcome depends on making metabolic economies during periods of complete or partial starvation, a phenomenon which has been most thoroughly studied in the rat by Westerterp (1977). Second, growth rate sometimes spurts, in a compensating fashion, when food is replenished after a period of partial or complete starvation. This is common in mammals (Tanner, 1963) but has also been recorded in invertebrates (Calow, *in press*). Growth spurts are difficult to reconcile with the maximization hypothesis which suggests that growth rates are *always* maximized.

Growth rates of plants are certainly under less rigid control than those of most animals, and supplements of nutrients or light usually enhance growth. However, the low maximum growth rates of some plants (e.g. stress-tolerators of Grime, 1979) are of particular interest in the optimization/maximization debate. Even when supplied with high light levels and an abundance of nutrients, their growth remains remarkably slow (Grime & Hunt, 1975). It seems that the growth rates of these plants are limited to a level which is appropriate to the stressed environments they inhabit and are dominant in.

9.6 GROWTH CONTROL

The maximization and optimization distinction in growth strategies corresponds approximately to the distinction between passive and active control of metabolism. Maximization is likely to lead to systems which are passively controlled and optimization to systems which are actively controlled. Models of metabolism involving active or passive control have been discussed in detail elsewhere (Calow, 1976, 1977a) and only a brief account will be given here.

The most straightforward model of growth simply considers the organism to be an energy store (G) into which flows organized energy from food (A) and out of which flows disorganized heat (H) thus:

$$\mathrm{d}G/\mathrm{d}t = A - H$$

This is the basis for the *energy budget equation*. Now consider a slightly more complex model in which the output parameter H is some function of G (i.e. respiratory rate is dependent on body size—Chapter 2) then:

$$H = f(G)$$

and

$$\mathrm{d}G/\mathrm{d}t = A - f(G)$$

by rewriting the integral of this equation in block-diagram form we find that it contains a crucial, regulatory mechanism—a feedback loop (Fig. 9.9). The circle with a sum sign is a comparator which sums $-H$ and A to give the net energy input N. Integrating, or summing N with respect to time$\int(\ )\mathrm{d}t$ and including the initial size at birth a gives G and this multiplied by f gives H. The latter feeds back to the comparator and the whole process continues iteratively throughout life. There is no physical entity which corresponds to the feedback loop and it is often referred to as *fictitious feedback*. However, it does introduce a limited degree of passive regulation; for example, a reduction in A will cause a reduction in G which will cause a reduction in H and thus feed back as a reduced negative term on to the comparator. G and $\mathrm{d}G/\mathrm{d}t$ are thereby buffered to some extent against fluctuations in energy supply. This kind of model represents a linear version of the famous Bertalanffy growth equation (Calow, 1976).

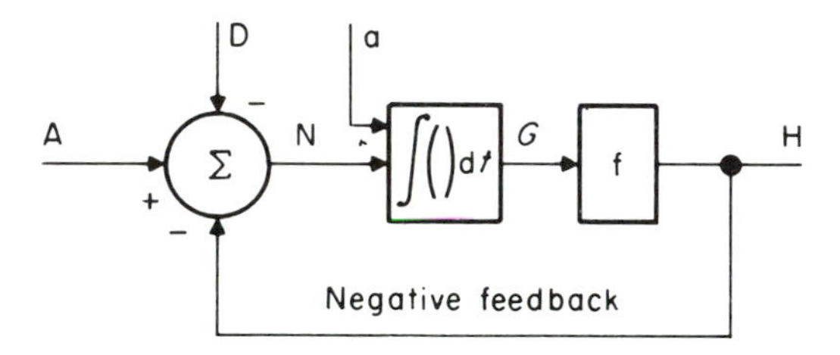

Fig. 9.9 An energetic model of growth with fictitious negative feedback. Here growth (G) i obtained by integrating ($\int(\ )\mathrm{d}t$) net energy input. The latter is obtained as the difference between energy input (A) and respiratory output (H). Since the output term is proportional to G then a feedback loop is introduced. There is no real feedback substance involved here. This is why it is referred to as a fictitious feedback system. D represents disturbances, a initial size.

A model with active control is illustrated in Fig. 9.10. This differs from the passive system in having an explicit growth program—in other words, growth is not simply determined by the parameter values of the physiological subsystems, as would be expected from the maximization hypothesis. The control and feedback signals are now *real* and correspond to hormonal factors. There is also a physically definable comparator which may correspond to an endocrine organ. Parameter E in the feedback channel is inserted to indicate that since this depends on a real link it will not normally be 100% efficient in transmitting information ($E < 1$). Once again this is only an extremely simplified model of metabolism but it has been exploited by Hubbell (1971) to model the bioenergetics of invertebrates to various degrees of realism.

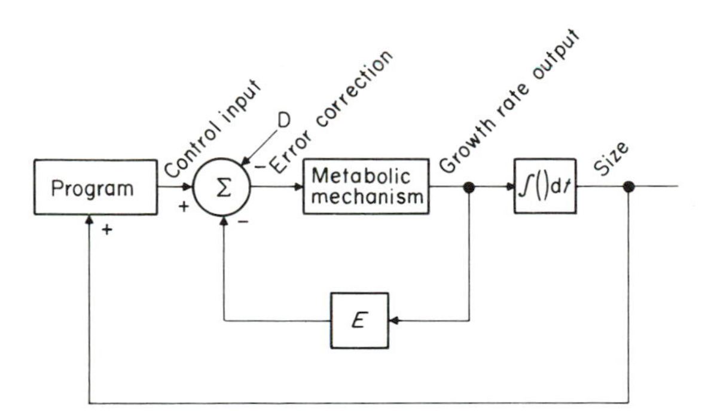

Fig. 9.10 An energetic model of growth with real negative feedback. Here a program specifies a 'desired' growth rate. The actual growth rate is compared with the desired simply by subtracting one from the other and this is achieved by a real negative feedback loop. The difference is proportional to the misalignment—the proportionality (E) depending on how faithfully the feedback loop can represent the misalignment. Under these circumstances the metabolic mechanism will be subjected to adjustment until the real growth rate approximates to the 'desired'.

It is as difficult to ascertain whether metabolism is actively or passively controlled as it is to determine whether growth rate is optimized or maximized. Ultimately, a solution will require careful study of the control machinery, if it is present.

9.7 FORM

Developmental processes of differentiation, pattern formation and morphogenesis continue to present some of the most profound problems in biology with respect to the mechanisms of control behind them and the selective forces which have produced them. Morphogenetic changes are effected, at least in part, by relative growth; that is, certain components increase in size at higher or lower rates than others; and so the problem of morphogenesis is relevant here.

It is convenient to begin with the concept of allometry. For a wide variety of organisms, Huxley (1932) discovered that the specific growth rate of an organ (y) stood in a constant ratio to the specific growth rate of another organ or the whole organism (x) thus:

$$\frac{\mathrm{d}y}{\mathrm{d}t} \times \frac{1}{y} : \frac{\mathrm{d}x}{\mathrm{d}t} \times \frac{1}{x} = \alpha$$

Integrating gives:

$$y = \mathrm{b}x^{\alpha}$$

or

$$\log y = \log \mathrm{b} + a \log x$$

α is the coefficient of allometry and is derived from the slope of the straight line obtained by plotting log y against log x. When $\alpha > 1$, y grows more rapidly than x (positive allometry); when $\alpha < 1$, the reverse is true (negative allometry); when $\alpha = 1$, x and y are said to be related isometrically. Not all relationships between organs and organisms are of this kind—curves may be obtained from logarithmic plots—but the majority conform.

That the specific growth rates of parts and wholes should be related in this way is not obvious. Indeed, because of the tremendous complexity of growth processes it might rather have been expected that parts should start and cease growing in an irregular fashion. The principles of allometry can, in fact, be explained from two points of view, both of which are appropriate to our approach here.

9.7.1 Allometric growth and resource allocation

The physiological basis of allometry can be clarified if the basic equation is rewritten:

$$\frac{dy}{dt} = \alpha \frac{dx}{dt} \times \frac{y}{x}$$

That is, part y receives from the increase of the total system a share which is proportional to its ratio to the total (y/x). Alpha can therefore be understood as a distribution coefficient indicating the capacity of y to take a share of the total increase. Allometric growth in this sense is represented as an expression of the competition between parts for the finite resources available to the whole organism. That the proliferation of tissues is sensitive to nutrient availability is suggested by the fact that mitosis itself is very sensitive to energy supply (Bullough, 1952).

9.7.2 Allometric growth and function

From the point of view of maintaining the proper functioning of the whole system any change in its overall size makes changes in its shape and proportions necessary. This is as true for machines as it is for animals and plants. These scaling effects (Alexander, 1971) thereby set tolerance limits within which any requirements for biological function must usually operate. McMeekan (1940) stated that those parts of the body essential to life processes and body function are relatively well developed at birth and make smaller increments in post-natal life than the body as a whole, whereas those organs primarily concerned with the movement or storage of resources appear ill-developed at birth but grow more during post-natal life. The organs that function and undergo most of their chemical development during foetal life, and whose continued function during post-natal life is vital to the existence of the organism (for example, liver, kidneys and heart), are the ones which change least in chemical composition as a result of undernutrition (Widdowson, Dickerson & McCance, 1960). Skeletal muscle, on the other hand, undergoes substantial changes in chemical composition after birth and varies most in chemical composition as a result of underfeeding. Although all these changes must usually take place within the constraints set by the scaling effects, in some organisms the building and operation of certain organs may move beyond these limits (e.g. reproductive organs, section 10.2.2).

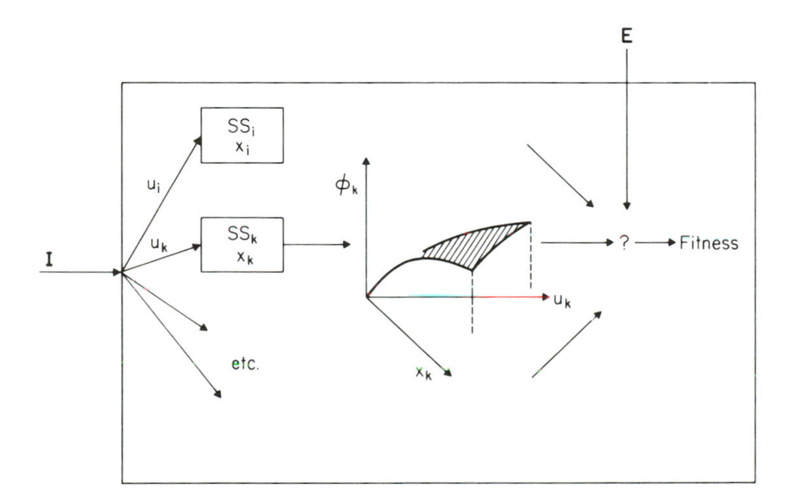

Fig. 9.11 A more sophisticated model of resource partitioning in organisms. I is a vector, representing the whole gamut of resources which are important to the organism (the large square box). The u's are vectors showing how I is partitioned amongst a number of physiological subsystems (SS_i, SS_k, etc.). The state of each subsystem (x—another vector) depends upon u and in turn influences the physiological performance (Φ) of the subsystem. This kind of dependency is represented hypothetically for SS_k in the graph. Subsystems interact and determine how the organism interacts with its environment (E). These interactions determine the survival and reproductive success of the organism and hence its fitness. However, this part of the model is very 'fuzzy' and is represented by a question mark. If we knew about all the functional dependencies and interaction terms we could rigorously and precisely examine the effects of a particular physiological strategy on fitness.

9.7.3 Evolutionary loss of useless features

In the present context, phenomena such as the loss of eyes and pigmentation in cave animals and the 'simplification' of parasites are of interest because many of the lost characters have no obvious disadvantage for the bearer. Several authors have proposed, therefore, that the loss or reduction of useless characters serves to conserve resources. In other words, there is no advantage in allocating resources to a defunct organ while there may be positive advantages in using the resources elsewhere (Barr, 1968; Heuts, 1953). Thus, Zamenhof & Eichorn (1967) have demonstrated the competitive superiority of mutant microbes that lack the capacity to make tyrosine over wild types when both are cultured in media containing the compound.

9.7.4 Control of form

In summary, the allocation of resources can be considered to be affected by both scaling and functional requirements. Interestingly, the actual method of control remains unclear after more than fifty years of research effort. It might either occur on the basis of actual competition for resources between parts with those functioning at the highest rates taking the most (Goss, 1964), or on the basis of hormonal control programmed to allocate according to functional demand (Bullough, 1967). Furthermore, these two alternatives are not exclusive (Calow, 1978b). Although this question remains unresolved it is, nevertheless, possible to envisage what kinds of models will be helpful in its solution. A sketch of one, formulated by Sibly (pers. comm.), is illustrated in Fig. 9.11.

Chapter 10

Resource Utilization and Reproduction

PETER CALOW

10.1 INTRODUCTION

Like the 'ideal gas', the 'ideal neo-Darwinian organism' does not exist. If it did, it would invest all the energy at its disposal in the formation and support of offspring and would partition this energy into gametes of the smallest possible size. In the real world, however, constraints operate against such extreme strategies. The benefits of maximizing reproductive output must balance the costs of concomitant reductions in parental survivorship and long-term fecundity and the low individual fitness of midget offspring. Since Lamont Coles' pioneering work much has been written about the relative demographic benefits of particular reproductive strategies (Cole, 1954; Stearns, 1976, 1977; May, 1977; Horn, 1979) but costs have been treated more superficially and more theoretically (e.g. see Gadgil & Bossert, 1970). Here I place less emphasis on the demographic arguments (but see Chapters 11 and 12) and more on the physiological basis of the costs and constraints associated with energy allocation in reproducing organisms. What are the consequences of investing energy in reproduction rather than in the maintenance and growth of the soma? What constraints are imposed on the miniaturization of gametes? How does investment in the care of offspring influence the well-being of parent and progeny? What are the relative metabolic and evolutionary merits of asexual, sexual and hermaphrodite reproductive strategies?

One initial note of caution is necessary with this kind of approach. Physiological processes are not the only phenomena which act as a causal link between a reproductive strategy and its consequences for the survivorship and fecundity of the system. For example, because of their behaviour or appearance reproductively active individuals may be more prone to agents of mortality like accident and predation than non-reproductive counterparts (Tinkle, 1969). Hence the physiological account that follows should not be considered as complete in itself. A sister chapter remains to be written on the behavioural basis and consequences of reproductive phenomena.

10.2 RESOURCE INVESTMENT IN GAMETES

10.2.1 Reproduction versus growth

Growth and reproduction are fuelled by the energy remaining (E_W) after the metabolic 'levy' has been taken from the input to the organism. The way E_W varies throughout life, therefore, depends on the way income and costs themselves vary and much evidence suggests that both tend to increase, but at a decreasing rate, with size (W); i.e. they can be modelled by the allometric equation, $Y = k\,W^c$ where Y = income or expenditure, k and c are constants and usually $0 < c < 1$. For example, in a wide range of animals, plants and micro-organisms, oxygen inspired, carbon dioxide expired and heat loss are proportional to weight raised to the power of between 0.66 and 0.88 (Chapter 2). The relationship between energy intake and body weight is not as well documented but in several animals is functionally related to body surface (see review in Calow, 1975) and this, in turn, is directly proportional to the two-thirds power of body weight ($W^{0.67}$). Similarly, in plants the rate of photosynthesis is functionally related to leaf surface which is often related, in turn, to the square root of the weight of the plant; (i.e. $W^{0.5}$; Evans 1972).

If, in general $c_{\text{output}} > c_{\text{input}}$ then the curves of respiratory output and energy input against W must intersect at a point (W_{max}) which represents the maximum metabolic limit to body size. This is most clearly visualized by plotting the logarithm of the input and output terms against the logarithm of size since, upon logarithmic transformation, equations of the type $Y = k\,W^c$ are converted to 'straight lines'; i.e. $\log Y = \log k + c \log W$. Hence with $k_{\text{input}} > k_{\text{output}}$ (i.e. input begins at a higher level than output) and the slope (c) for input less than that for output, we have two intersecting trajectories as shown by solid lines in Fig. 10.1 (a). The difference between input and output at any W is E_W. Taking antilogarithms of inputs and outputs for given values of W and subtracting one from the other yields a curve for E_W against W of the form shown in Fig. 10.1(b).

Using this kind of approach Sebens (1979) has considered at what value of W, the energy available for production, E_W, should switch from use in the growth of somatic tissue to use in building the gonads and gametes. Clearly organisms should not grow to W_{max} because although they would become very large they would produce no offspring and this strategy has zero neo-Darwinian fitness (Chapter 1). Hence, W_{opt}, the best size for switching, must be

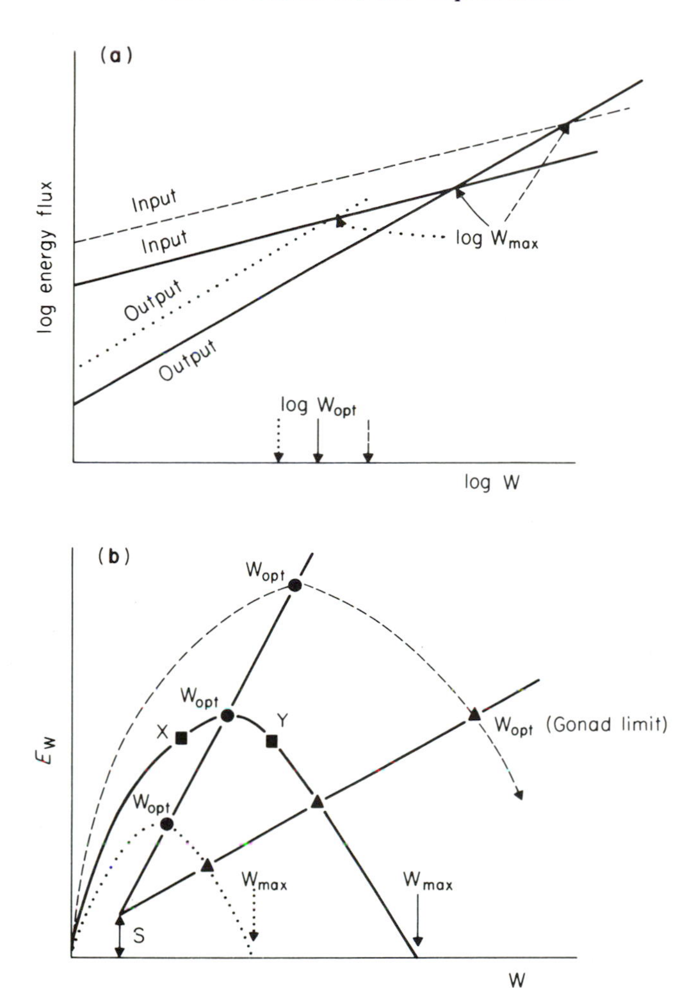

Fig. 10.1 (a) Relationship between the size of organisms (log mass, W) and their rates of input and output of energy (expressed logarithmically). Consider first the solid lines. Their point of intersection defines the maximum size achievable by the organism. If resource availability is improved the input line is raised (— —) and W_{max} is increased whereas the reverse occurs if costs are increased (. . . . .). (b) Relationship of the difference between input and output (E_W) and body mass. The peaks (●) define the proposed optimal size for switching E_W from use in growth to use in reproduction. The sloping line to these peaks represent what happens if E_W is switched gradually from growth to reproduction. Points denoted by ▲ represent the switching points that might be imposed when body size puts a constraint on gonad size. X, Y and S are discussed in the text. The argument is from Sebens (1979).

less than W_{max}. Seben's solution to the problem of when to switch is based on a metabolic argument; namely that switching should occur at that size which maximizes the energy available for reproduction. W_{opt} therefore occurs where the difference between intake and cost is maximum (Fig. 10.1(b)). An earlier or later switch, for example at points X and Y in Fig. 10.1(b), would make less energy available for reproduction.

The model can be used to gain some insight into the effect of resource availability on W_{opt}. If resource availability influences only the elevation of the input line (i.e. k_{input}), then an increase in resources would be likely to cause an increase in the elevation, as illustrated by the broken line in Figs. 10.1(a) and 10.1(b). In consequence, W_{opt} like W_{max} moves to the right and this suggests that the optimal size for switching E_W from growth to reproduction will be bigger in richer habitats. Conversely a reduction in resources or an increase in costs (e.g. perhaps brought about by increased searching in nutritionally dilute conditions) causes W_{max} and W_{opt} to move to the left (dotted lines in Figs. 10.1(a) and (b)). Hence animals in trophically poor circumstances should begin to breed at a smaller size than those in trophically rich circumstances (Sebens, 1979). This is true of a variety of marine invertebrates (Sebens, 1979, and others in press) and also of some fishes (Weatherley, 1972).

No doubt these models define rather fundamental constraints on reproductive strategies, but there are a number of complications.

1 *Demographic factors can frequently be expected to intervene in the metabolic model.* Whenever the chances of parents surviving to W_{opt} are small the switch to reproduction should occur at a smaller size and earlier age. That is to say, W_{opt} is equivalent to the optimal body size for reproduction only when mortality is independent of body size. E_W must be weighted by the probability of survival (P_W) to estimate its relative ecological fitness (F_W; Lynch, 1980), viz.:

$$F_W = P_W \times E_W$$

(Lynch distinguishes between ecological and genetic fitness since an organism with a high F_W only has a high genetic fitness if it is using its energy supplies to support reproduction and not growth). Figure 10.2 plots an estimate of E_W (i.e. total energy used in growth and reproduction) and an estimate of the proportion of E_W used in reproduction against the body length of freshwater cladocerans (data from Lynch, 1980). *Daphnia schodleri* and *D. pulex* have larger adults and live in habitats where there are invertebrate but not vertebrate predators. Since the invertebrate

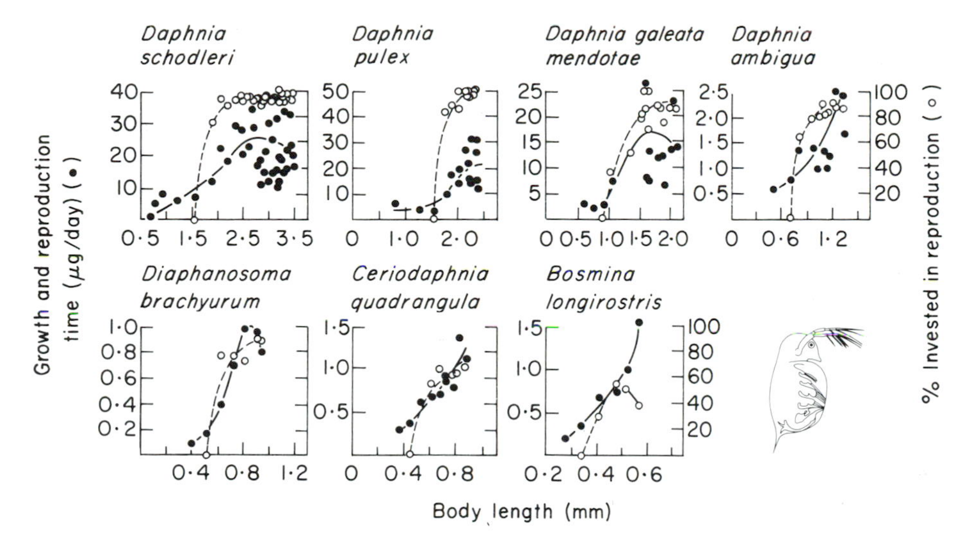

Fig. 10.2 Size-specific production (growth + reproduction)—as solid lines—and % energy invested in reproduction—as dashed lines—for seven species of planktonic cladocerans. (After Lynch, 1980.)

predators of these species feed preferentially from the small and young cladocerans, P_W reduces with body length. Here F_W is maximized at or after E_W is maximized and we see that reproduction is initiated and maximized (i.e. in proportional terms) when E_W is maximized in these species. The other species have smaller adults and live in habitats with both vertebrate and invertebrate predators. Invertebrates feed throughout the size range of these cladocerans but vertebrates concentrate on adults. Hence P_W will decrease as length increases such that F_W will be maximized before E_W and in general in these species we do find that reproduction is initiated and maximized before peak E_W values are achieved. *Daphnia galeata mendotae* and *Diaphanosoma brachyurum* fit least well into these patterns.

Another demographic complication arises from the fact that in growing populations, more may be gained from early breeding than from delaying until W_{opt} is reached because here populations of genes grow by compound interest and those coding for early breeding will collect interest more frequently, and hence increase more rapidly (i.e. will be fitter), than others (Lewontin, 1965). Genes delaying reproduction could only keep pace with those coding for early reproduction if they were also associated with an exponential increase in reproductive output with age. Furthermore, since they would be unlikely to sustain exponential increase in the size of the bearer for long periods of time, reproductive output would have to increase disproportionately, perhaps exponentially, with size, and this is rare (Calow, 1977a; Kaplan & Salthe, 1979). On the other hand, in a non-expanding population, where established parents have better survival chances than offspring, breeding may be delayed beyond W_{opt} if the act of reproduction puts the parent at risk. This will be considered in greater detail below (section 10.2.2).

2 *The switch from growth to reproduction may occur gradually and not sharply.* In plants, where the diversion of resources from soma to reproduction causes a reduction in the growth of photosynthetic tissue (cf. Bazzaz *et al.*, 1979) and hence in resource input per unit weight of tissue, the optimal theoretical solution does seem to be a sharp switch (Cohen, 1971) and this is often, though not always, observed in plants in nature (Cohen, 1971; Amir, 1979). Alternatively, in heterotrophs the diversion of resources from soma to reproductive processes need not have such profound implications for nutrient input as in autotrophy and gradual shifts in energy allocation can and do occur (e.g. Fig. 10.2). This kind of strategy is represented as a straight line in Fig. 10.1(b). Clearly the beginning of reproduction (at S in Fig. 10.1(b)) will inevitably cause a reduction in growth rate and thus in the size that is reached in future breeding seasons. Since reproductive output is often intonately related to the body size of the parent the 'choice' is between the production of fewer offspring early or more offspring later. Lawlor (1976) has studied these options in the terrestrial isopod, *Armadillidium vulgare*, and his results are considered in Chapter 12.

3 *In the case where organisms produce a single gonad per season its size depends not only upon E_W, but also on the body space available to accommodate it.* It is feasible that E_W accumulates so rapidly that the available body space is filled before W_{opt} is reached. Under these circumstances further growth would be the only possible way of creating more space. Size might, therefore, be pushed beyond W_{opt} to maximize gonad production (Fig. 10.1(b)). Growth, in this situation, should proceed until E_W and the maximum energy which could be allocated to the gonad (based on morphological constraints) are equal, and then terminate (Sebens, 1979). Such trade-offs may be going on in lizards where body shape (which in many cases is related to predator escape and foraging strategies) plays a more decisive role in determining clutch size

than other demographic and possibly metabolic factors (Vitt and Congdon, 1978; Chapter 12).

4 *Finally, the metabolic characteristics of organisms do not always follow a smooth course of change but may alter radically and abruptly at the onset of reproduction.* For example, the efficiency with which food is converted to gametes is often much higher than the efficiency with which it is converted to somatic tissue by a similar-sized organism. This is illustrated in Fig. 10.3(a) & (b) for a short-lived, annual triclad, *Dendrocoelum lacteum* and a longer-lived, perennial triclad, *Polycelis tenuis*. In these species, growth and reproduction are virtually exclusive so that there is a sharp switch from one to the other (see above). The conversion efficiency of food to gametes is between 3–5 times greater than the growth efficiency over the last half of the growth cycle. There are several possible reasons for this:

a undifferentiated gametic tissue may be produced more easily and more quickly than differentiated somatic tissue so that more production can occur per unit basal metabolism;

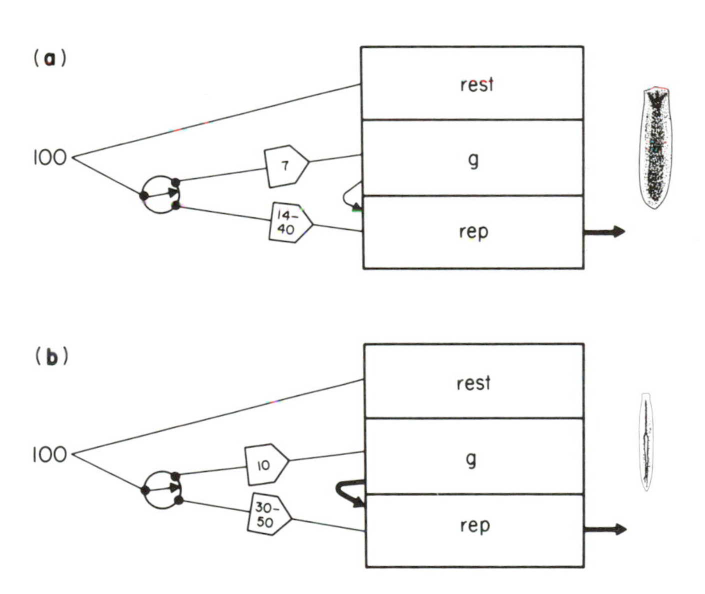

Fig. 10.3 Switching energy from growth to reproduction in (a) *Polycelis tenuis* (an iteroparous triclad) and (b) *Dendrocoelum lacteum* (a semelparous triclad). Energy input is represented as 100 units and the figures in the arrowed boxes are conversion efficiencies. g = growth; rep = reproduction; rest = the rest of metabolism—e.g. respiratory heat loss. (After Calow, Beveridge & Sibly, 1979.)

b less energy may be used in the maintenance and repair of the soma during reproduction;
c there may be an increased input of food without a proportional increase in basal metabolism.

For triclads, a and b seem the most likely candidates since food input does not appreciably increase during reproduction in these animals (Calow, Beveridge & Sibly, 1979), but strategy c is adopted by many animals (e.g. Calow, 1979; see also section 10.4).

10.2.2 Gonadal versus somatic activity

It is reasonable to expect that reproductive output per unit food input (Rep/I), as discussed in the last section (page 251), should have been maximized by natural selection. However, conversion efficiencies do show some variability between species and, in general, as with the triclads (Fig. 10.3), short-lived, semelparous organisms (die after first breeding season) have higher Rep/I values than long-lived, iteroparous organisms (survive through several breeding seasons): a review is supplied in Calow (1979). The main reason for this is that the excessive diversion of resources to reproduction at the expense of the metabolism of the parent (i.e. Rep/I increased by method 4(b) above) will exert a cost in terms of their survival chances (Calow, 1979). Vital somatic systems, like muscles, may be starved of substrate and even 'parasitized' to support the demands of the gonads. When reproduction is initiated in the semelparous *Octopus vulgaris*, for example, protein synthesis in the muscles is suppressed, the amino-acid pool of the muscles is increased and there is a rapid growth of ovaries with a concomitant loss of weight elsewhere (O'Dor & Wells, 1978). The general repair processes of organisms, discussed in Chapter 7, may be denied resources by reproductive activity so that the accumulation of molecular damage and hence the rate of senescence is accelerated (Calow, 1978a). The gonads, by increasing in size, may mechanically impede normal activity and may cause physiological stress (Orton, 1929) and the need to collect more food to produce gametes or to support developing offspring might have the same effect (section 10.4). Finally, the reproductive expenditure is likely to upset the general homeostasis of parents and render them more prone to disturbance from any environmental perturbation (Calow, 1973). There may also be behavioural consequences of reproduction which increase parental mortality and these are discussed in Tinkle (1969) and Shire (1980).

If there were no risks associated with reproduction then 'big bang, iteroparous reproduction' would be the adaptive norm. Since there is a cost, the evolutionary choice is between big-bang, high-effort reproduction with short, parental life, and low-effort reproduction which can be repeated over a longer period. Which strategy is favoured by natural selection depends on the environment and particularly on its effects on age-specific mortality. This has been considered rigorously by Murphy (1968), Gadgil & Bossert (1970), Charnov & Krebs (1973) and more recently by Goodman (1979) and Law (1979a). Stated simply, whenever as a result of *extrinsic* mortality factors small offspring have a much lower chance of survival than larger parents the latter should not put themselves at risk to *intrinsic* mortality factors for the sake of offspring and thus iteroparity is favoured (Wittenberger, 1979). This is likely to occur in situations where competition is intense since smaller offspring are usually less good competitors than large parents (i.e. K conditions of selection; Chapters 1, 11 and 12) but may also occur when young are more susceptible to the vagaries of weather and other physical, density-independent factors (i.e. r conditions of selection; Chapters 1, 11 and 12). Alternatively semelparity will be favoured whenever the survival chances of parents and offspring are similar, as is likely with calamitous density-independent effects, like spates in rivers or with floods on land, or in K conditions of selection where offspring are better at withstanding resource limitations than parents (as they are in some triclads; Calow & Woollhead, 1977). Hence, semelparous and iteroparous reproduction cannot be assigned strictly to r and K selection, as they sometimes have been (Pianka, 1970), but depend more particularly on the differential susceptibility of parents and offspring to extrinsic mortality factors (Stearns, 1976; Goodman, 1979).

'Big-bang iteroparity' *is* observed in a few exceptional groups; for example in many endoparasites (Calow, 1979). The solution to this apparent paradox is that high gamete output and indeed high Rep/I values do not always incur a cost if they are supported by a proportional increase in resource input. This is easily achieved in endoparasites which are often bathed in a superabundant food supply. Hence increased absorption of nutrients can sustain high levels of reproduction by insulating the soma from inanition. A cost is only incurred when the difference between the resource input to the organism and the reproductive output does not meet the requirements of parental metabolism. This can be expressed precisely by the following cost index:

$$C = 1 - \left(\frac{\text{input} - \text{reproductive output}}{\text{metabolic demands of soma}}\right)$$

When $C > 0$ a cost is incurred and this metabolic condition is referred to as reproductive 'recklessness' in that the parent is being 'reckless' with its own soma for the sake of reproduction. When $C < 0$ a cost is not incurred and this condition is referred to as reproductive 'restraint'. Endoparasitic platyhelminth tapeworms have high levels of egg production and high Rep/I values but their C values tend to zero (Calow, 1979). Semelparous triclads (free-living platyhelminths), though having similar Rep/I values to the tapeworm, have much higher C values (Fig. 10.4). Similar differences between Rep/I and C values, as found in endoparasites, may explain the poor correlation noted between Rep/I (or simply egg output) and shortened lifespan in other animals; for example, in some gastropods (Calow, 1978c), bivalves (Thomson, 1979; Houkioja & Hokala, 1978) and beetles (van Dijk 1979). To test this interpretation more research is required on the energetics of reproductive as compared with non-reproductive adults in these key, problem species, so that C values can be computed and relative positions of species on the 'recklessness–restraint spectrum' (Fig. 10.4) can be defined.

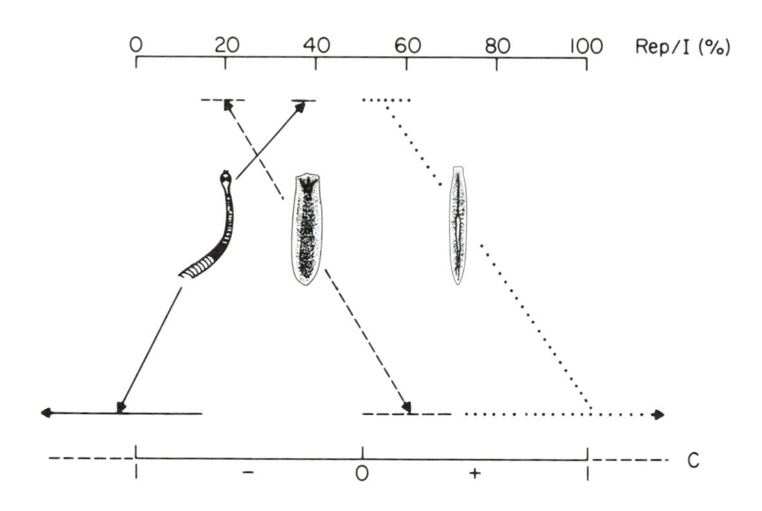

Fig. 10.4 Comparison of two indices of reproductive effort computed for three species of Platyhelminthes. (Data from Calow, 1979a.)

Finally, it is important to note that a given species may not always deploy the same reproductive strategy but may be influenced by variations in personal circumstances and conditions and by the way these differentially influence the survival chances of parent and offspring (Ballinger, 1977; Calow, 1979; Goodman, 1979). Thus as trophic conditions deteriorate the triclad, *D.*

lacteum, becomes more reckless and *P. tenuis* more restrained. The reason for this is that the hatchlings of *D. lacteum* are better adapted for dealing with a nutritive crisis than those of *P. tenuis* (Calow & Woollhead, 1977). Similarly, the marine gastropod *Nassarius pauperatus* produces more eggs and more egg capsules (within which they are packaged) as food supply becomes reduced (Mckillup & Butler, 1979). This is because *N. pauperatus* has a planktonic larval stage which can escape from the parents' habitat to adjacent areas differing significantly in food availability. Desert rodents, on the other hand, appear to be more reckless under good than under poor feeding conditions (Nichols *et al.*, 1976) presumably because the survival of offspring is particularly favoured under good conditions which, in the desert, may be rare. In iteroparous, usually restrained, reproducers, the effort put into reproduction may increase in a systematic fashion with the age of parents as their chances of surviving to, and making a reproductive contribution in, (i.e. residual reproductive value) the next breeding season become reduced (Pianka & Parker, 1975a). Alternatively, when adults are unable to 'predict' the quality of the environment for their juveniles they may vary their reproductive output in a random fashion from season to season (Hirshfield & Tinkle, 1975). For example, this happens in *Mytilus edulis* which produces planktonic juveniles (Thomson, 1979). Some iteroparous fishes, amphibians and reptiles regularly skip breeding seasons, reproducing only in alternate years. This seems to be correlated with reproductive patterns which involve accessory activity, like migrating to a breeding site, entailing fixed metabolic or survival costs that are independent of the direct effort put into reproduction. Hence saving on the fixed accessory costs in one year may sufficiently enhance reproduction in the following year to be selected as a regular pattern (Bull & Shine, 1979).

10.3 GAMETE SIZE VERSUS NUMBERS

In many organisms the resources made available for reproduction are partitioned into a number of smaller propagules; usually haploid gametes (cf. section 10.5). Since there is a finite amount of energy available for this process some kind of trade-off must occur between the sizes and numbers of gametes that are produced. The prime function of these gametes is not to carry material and energy but to transmit information (in particular that needed to build and control a new individual) and since information can be transmitted economically in something as small as a nucleus (ca. $10^{-5} - 10^{-7}$ g), gametes can, in principle, be very small and hence very numerous.

Suppose the total amount of resources made available for gamete production is M and that all gametes are of the same size (D), then the number of gametes produced per parent is M/D. Fitness might be expected to increase as D reduces, becoming maximum when D tends to nuclear size (say d). However, a complicating constraint is that the survival of the zygote will be posively related to its initial size. This is because embryological systems are often isolated from an external food source and as well as requiring information to control the developmental sequence, also require energy reserves to power it. Hence survivorship of the zygote (denoted as S with appropriate subscript to signify size) will be functionally related (Φ) to the size of the zygote, e.g.:

$$S_{2d} = \Phi(d + d)$$

The fitness of the d-producing system (W_d) is therefore given by:

$$W_d = \frac{M}{d} \times S_{2d}$$

The first question is: can the d system persist against a mutant with some larger gamete size (Δ) or, in other words, is it an ESS (Chapter 1; Maynard Smith & Price, 1973)? Now:

$$W_\Delta = \frac{M}{\Delta} \times S_{2\Delta}$$

For d to be an ESS, $W_d > W_\Delta$ and since it is assumed that M is the same between systems S_{2d}/d must be greater than $S_{2\Delta}/\Delta$. Whether or not this inequality holds depends crucially on the relationship between survivorship and gamete size and this has been discussed in detail by Parker, Baker & Smith (1972) and Bell (1978).

If, as is illustrated in Fig. 10.5(a), survivorship is a continuously reducing function of size, then S_{2d}/d will always be greater than any $S_{2\Delta}/\Delta$ and microgamete formation will be favoured ($W_d > W_\Delta$). Maynard Smith (1978b) believes that this is the primitive condition for eukaryotes; i.e. the earliest sexual eukaryotes were isogamous and produced gametes as small as was compatible with effective information transmission and this is found to occur in some colonial Protista. With increasing adult size, however, the chance of a zygote surviving to be an adult when formed from two microgametes is likely to be very small. In the extreme case survivorship might increase continuously with gamete size as shown in Fig. 10.5(b), when macrogamete formation would be favoured; i.e. any $S_{2\Delta}/\Delta > S_{2d}/d$. However, the more likely, intermediate condition, where the survivorship curve pulls away

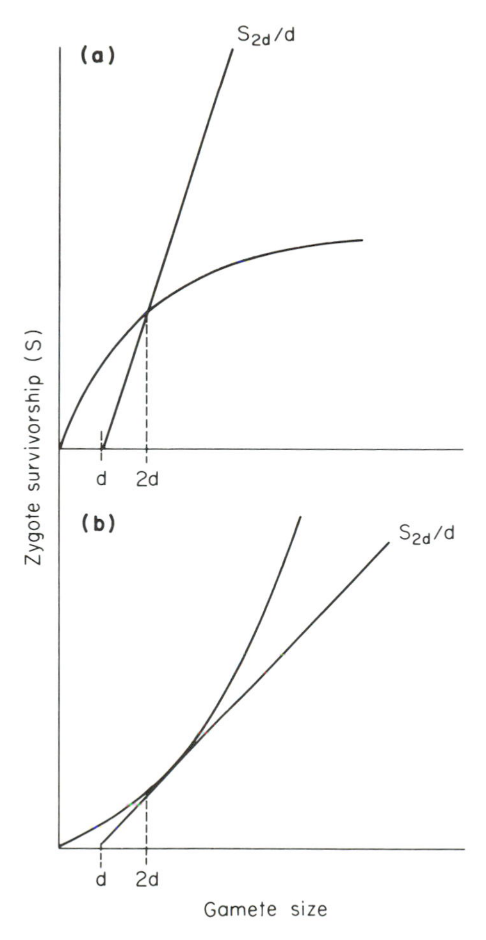

Fig. 10.5 Possible extreme relationships between the size and survivorship of gametes. (a) Favours microgametes because any increase in gamete size above d (curve) results in lower zygote survivorship per unit investment than the d system (straight line), (b) favours microgametes for opposite reasons.

exponentially from S_{2d}/d at small zygote sizes (because small additions of cytoplasm bring great advantages) but turns back at large zygote sizes (because as size increases beyond a particular point it brings less and less benefits), must favour zygotes of intermediate size. Intuitively, it seems reasonable that these might be formed out of a combination of small and large gametes and rigorous proof that anisogamy is the most likely outcome in these intermediate, disruptive conditions is forthcoming from simulation (Parker, Baker & Smith, 1972), analytical (Bell, 1978) and geometrical (Maynard Smith, 1978b) analysis. Hence, theoretical reasons have been found for the almost universal distribution of anisogamy throughout the Metazoa. These ideas are given some support by the Volvocidae (Knowlton, 1974). Here small colonies are isogamous, medium-sized ones are slightly anisogamous and the largest colonies are completely anisogamous.

Once the distinction has occurred between microgametes (male) and macrogametes (female) rather different selection pressures must have operated on each to determine their precise sizes. As far as the microgamete is concerned, selection should favour the maximum number of fusions. Therefore, what is important is the propagation of information and not the provisioning of the zygote. The only constraint is the provision of some energy for survival before fusion and usually some resources have to be allocated to the apparatus needed to find and fuse with the egg. In animals, spermatozoa tend to a nuclear size though some of the cytoplasm usually becomes modified into a flagellum (sometimes several) for locomotion and an acrosome for penetration. The same kind of microgametes occur in ferns but in the gymnosperms and angiosperms the male gametes are contained within pollen grains. Here resources are used in the formation of an outer protective coat and an inner gametophyte which forms the pollen tube (necessary to effect fertilization) as well as for the gametes themselves. There are usually two gametes per pollen grain and these have little cytoplasm, and tend to a nuclear size. Indeed early cytologists were of the opinion that these cells were naked nuclei, and not until 1965 was limited cytoplasm found using electron-microscope techniques.

For the macrogametes, selection should favour that size which will yield the maximum number of surviving progeny. Hence provisioning is important and there is likely to be a real compromise between the size and number of gametes produced by a parent. Egg and seed size shows much interspecific variation—ranging over approximately 10 orders of magnitude for extant plants (Harper, Lovell & Moore, 1970) and animals (Fig. 10.6). In general, however, the within-species variation for egg and seed size is much less than the variation in numbers produced per organism and this is particularly the case for angiosperms (e.g. Harper, Lovell & Moore, 1970). The implication to be drawn from these observations is that gamete size has been more subject to strict selection than the number of gametes produced per parent.

It is almost tautologous to say that large gamete size is advantageous when circumstances favour more complete development. Conversely, smaller macrogametes should be produced where extended development is not at a premium and possibly where dispersal is. Also, small gametes, with few

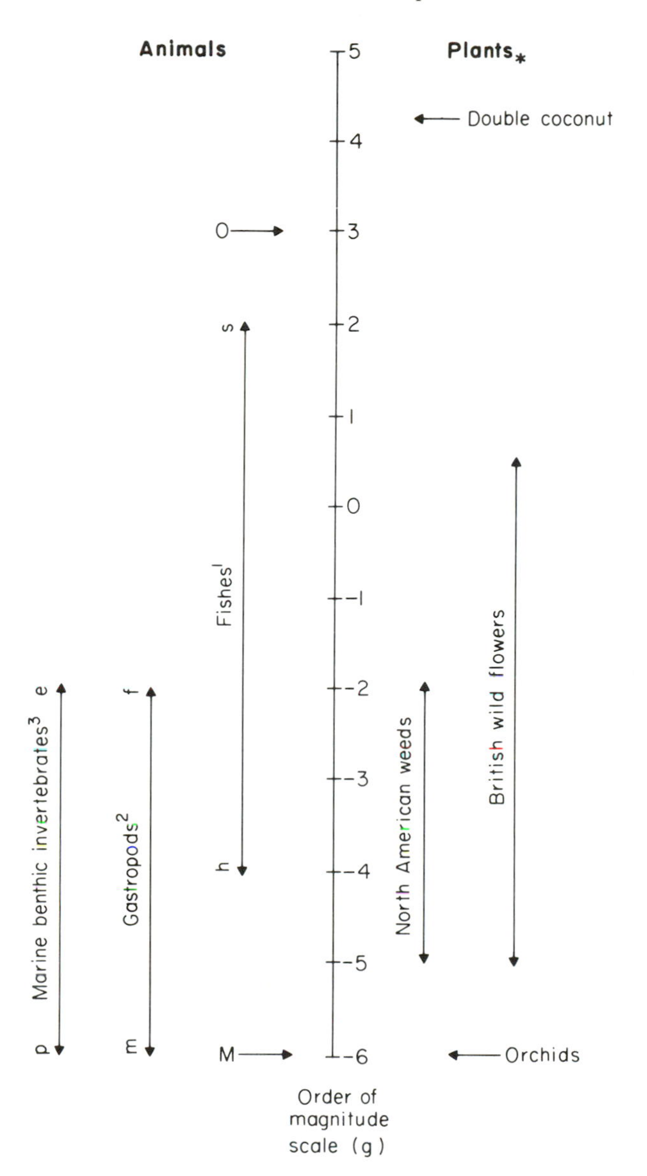

Fig. 10.6 Range of egg/seed sizes in a variety of Metazoa. *—summary of data from Harper, Lovell & Moore (1970); 1—from Nikolskii (1969); 2—from Calow (1978b); 3—from Thorson (1950); M = mammalian egg (alecithal); O = ostrich egg; p = planktrotrophic; e = lecithotrophic (yolk-feeding); m = marine species; f = freshwater species; h = herring; s = shark.

reserves (alecithal eggs), are likely to be found in those animals which show a measure of viviparity. In the latter case, however, resources that might have been used to form more gametes are used instead via the placenta to promote development. Hence the argument is the same; invest in development or concentrate on producing large numbers of gametes.

It is possible to define general criteria which favour this or that size of eggs and seeds but it is difficult to be very precise (see attempts of Smith & Fretwell, 1974, and Pianka & Parker, 1975a) and for every rule exceptions seem to abound. Food shortage during the breeding season is likely to favour the provisioning of eggs to give embryos a good start. Cold-water animals generally lay fewer and yolkier eggs than warm-water relations; for example this occurs in frogs in general (Moore, 1942), *Rana pipiens* in particular (Moore, 1949) and in barnacles (Barnes & Barnes, 1964). Competition may also favour large zygotes so that offspring are in a better position to compete for limited resources. Thus seeds in climax communities are often much larger than those produced by plants typical of earlier seral stages (Harper, Lovell & Moore, 1970) and competition for hosts seems to promote the production of larger eggs in commensal and parasitic copepods (Gotto, 1962). Alternatively the requirements for dispersal in marine habitats where competition for space is intense, favours a reduction in the size of zygotes in animals with dispersal-phase larvae. Similarly, seed size seems to be reduced in woodland trees and shrubs as an aid to dispersal. Predation also affects gamete size. Where predation is intense and the size of large eggs and/or young is in itself not sufficient as a deterrent against being eaten, small eggs and seeds are favoured. Among marine invertebrates, those with the longest planktonic phase and hence most predation-prone planktonic larvae often produce the largest number and smallest eggs (Thorson, 1950). Central American legumes that are most attacked by beetles produce smaller seeds in larger numbers than those which are less exploited (Janzen, 1969). On the other hand marine turtles alleviate predation on eggs by burying them in the sand, and since larger hatchlings can crawl more rapidly back to the sea, predation pressure here has resulted in the selection of large not small eggs (Emlen, 1973).

10.4 PARENTAL CARE

This section is concerned with parent–offspring relationships in animals; i.e. with the metabolic aspects of care offered by parents after the reproductive propagules are released into the outside world. Parental care prior to that, as in the nourishment of embryos in viviparous animals and the provisioning of seeds with endosperm, was the topic of the last section.

The effort involved in provisioning and guarding offspring is largely metabolic but these activities may also increase the parental risks of mortality by accident and predation. The gains from parental care are derived from the enhanced survival of offspring. Maynard Smith (1977) has examined how these opposing forces might determine which system of parental care evolves and the main components of his argument are:

1 the probability of survival of the offspring with zero, one or two parents;
2 the effect of egg production on the ability of the female to care for her offspring;
3 the ability of the male to find and mate with other females.

If the probability of survival of the lone offspring is not very much less than that if either one or two parents are involved in caring for the young, parental desertion of offspring is favoured. In many invertebrates which produce well-provisioned and protected eggs both parents commonly do desert. The duck strategy (only female cares for young) is expected to evolve if males have a good chance of finding other mates and if females are not incapacitated by the production of eggs so they are capable of caring for young. Also where fertilization is internal the female is committed to care for the young while she carries the zygote around inside her and so the male may have an earlier opportunity to desert. The female is here said to be caught in a 'cruel bind' (Dawkins & Carlisle, 1976). The opposite, i.e. stickleback strategy (only male cares for young), occurs where the male has a poor chance of mating again and/or where the female is incapacitated by egg production—as is likely in many fishes which may produce hundreds and thousands of eggs in one breeding episode. There is no 'cruel bind' on the female with external fertilization and if she can 'arrange' to lay her eggs before the male is ready to fertilize them, she might impose a 'cruel bind' on him (Dawkins & Carlisle, 1976). If, finally, the chance of survival of the offspring is very much greater in two- than in one-parent 'families' then care by both parents is expected, as occurs for example in several species of primates.

In mammals the early stage of parental care is dominated by lactation. Surprisingly, males never lactate (Maynard Smith, 1978b), perhaps because internal fertilization leads to difficulty in establishing paternity, making the metabolic effort of lactation of questionable fitness value to them, or because (particularly in monogamous species where paternity assurance is greater) they invest in other things which enhance inclusive fitness more than would an alternative investment in lactation, e.g. territorial defence, food gathering for the family, infant-carrying and so on (Daly, 1979). The effort invested by females in lactation is considerable, and may be much greater than the combined metabolic effort they invest in producing gametes and in nourishing embryos in the uterus. For example, food intake sometimes increases by 100% in lactating as opposed to non-lactating females and the proportion of this intake which ultimately appears in the milk may be more than 60% (Baldwin, 1968). The amount of energy required to support the metabolism of females during lactation is often more than twice the amount needed by non-lactating females (Stenseth *et al.*, 1980). Millar (1977) refers to this ratio as reproductive effort and has found that it exhibits much interspecific variation; small, short-lived mammals put more effort into reproduction in this sense than larger, longer-lived species (Daly, 1979).

If parental care in mammals is dominated by lactation, in birds it is dominated by the provisioning of young in the nest. Many birds can produce more eggs than they do. There is, for example, a yellow-shafted flicker which subjected to continuous removal of eggs from its nest, was induced to lay 71 eggs in 73 days (cited in Emlen, 1973). Lack (1954) suggested that clutch size was adapted to the maximum number of offspring that parents could provision and he explained commonly encountered tropico-polar clines in clutch size in this way. Larger clutches occur at lower latitudes, he supposed, because there were more daylight hours for collecting food. However, some parent birds do not forage for as long as they might and in some species the females alone rear clutches as large as those found in other species with 'two-parent families' (Ricklefs, 1977). At least some birds could (and in manipulated circumstances do) rear more fledglings than observed and a slight modification of Lack's hypothesis, to accommodate this observation, is that parents do not rear the maximum number possible but the optimum number consonant with their own survival (Williams, 1966b; Charnov & Krebs, 1973). The main assumption, here, is that provisioning young causes the parents to incur metabolic costs and that, as suggested above (section 10.2.3), this adversely influences parent survivorship.

Using a technique involving doubly labelled water (D_2O^{18}) some actual estimates have been obtained of the costs of parental care in free-living birds (for the rationale of the technique see Lifson & McClintoch, 1966). In brooding parents, the ratio of average daily metabolic rate to resting metabolic rate was 2.7–3.0 for mocking-birds and 2.9–3.4 for purple martins (Utter, 1971) and 2.22–5.27 for house martins (Hails & Bryant 1979). In non-brooding birds this ratio is usually between 1.6 and 2.6 (Utter, 1971). Hails &

Bryant (1979) found that male house martins expend between 0.41 and 0.46 cm^3 CO_2 $g^{-1}h^{-1}$ (CO_2 release being a measure of energy expenditure) during each foraging trip in the brooding period. Here the total cost of brooding increased with the number of offspring in the brood and the total brood weight. For females, Hails & Bryant found that total metabolic rate also correlated with brood size but, for reasons not fully understood, not with brood mass. Predation might be another cost of brooding in that more numerous visits to a nest might draw more attention from predators (Skutch, 1949).

There is some indirect evidence supporting the assumption that the increased foraging costs associated with larger broods takes a toll on parent survivorship (e.g. Kluijver, 1970; Bryant, 1979). For example, the loss of weight regularly observed in brooding parents has been taken as a sign of loss of condition and as an indication of a greater probability of death (Kluijver, 1952; Newton, 1966). Two cases of the experimental manipulation of brood size also lend support to the hypothesis that brooding exerts metabolic and mortality costs. Weight loss amongst snow buntings that were required to tend to artificially enlarged broods was faster than for normal broods (Hussell, 1972) and the return rate of pied flycatchers to nest boxes following a season in which brood size had been increased artificially was less than in control, non-manipulated individuals (Askenmo, 1979).

10.5 NON-SEXUAL REPRODUCTION

A number of reproductive processes can take place without sex. Parthenogenesis is the formation of a new individual from an unfertilized egg. Asexual reproduction involves the formation of new individuals from fragments or outgrowths of the parent and, though particularly common in plants, also occurs in animals (Fig. 10.7). Often, but not always, non-sexual reproduction depends exclusively on mitotic cell multiplication and does not involve a meiotic phase (cf. automictic forms of parthenogenesis; Maynard Smith, 1978).

It is frequently argued that parthenogenetic females enjoy a twofold fitness advantage over sexual equivalents because they waste no resources in producing sons. Other things being equal the descendants of the non-sexual process should multiply more rapidly than those of the sexual process. Other things, of course, are rarely equal. For example, through heterozygosis, meiosis and the mingling of genetic material at syngamy, sex generates diversity and the offspring from a single parent may vary appreciably amongst

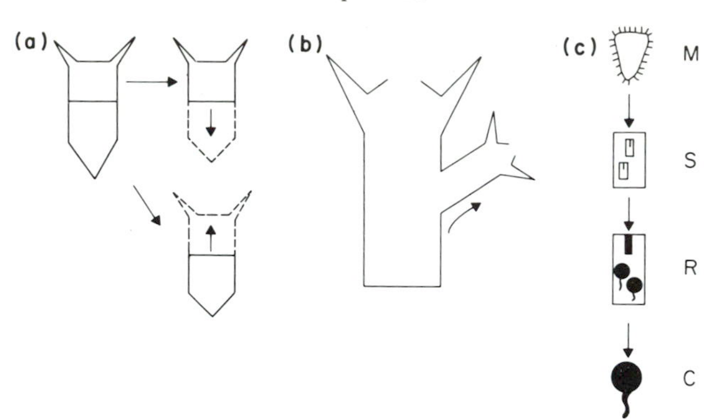

Fig. 10.7 Examples of asexual reproduction in animals. (a) Binary fission of triclads; (b) budding in *Hydra*; (c) asexual multiplication in the life-cycle of a trematode parasite. M = miracidium; S = sporocyst; R = redia; C = cercaria. (After Calow, 1978b.)

themselves. Parthenogenetic progeny, on the other hand, when they are produced by a mitotic copying mechanism from a single parent should, unless there have been mutational changes (which are rare), be identical. Hence by producing several different kinds of offspring, the sexual system has several different kinds of chance of successfully coming to terms with challenges posed by the environment. No matter how numerous they are, progeny of a parthenogenetic parent have just one chance in this existential game. The sexual system, though less productive, may therefore be better able to keep pace with the inevitable changes that take place in the world at large. The crucial question is whether the advantages due to this genesis of variety can pay for the 50 % cost of sex? The question is deceptively simple, for the answer has proved particularly elusive. There is, of course, no difficulty in seeing the long-term advantage of sex, once it has evolved, but the advantage must be apparent on a generation-by-generation basis for selection is not clairvoyant and only operates on the variation which turns up generation by generation. It has been suggested that most habitats are not sufficiently variable in physical and chemical characteristics to make sex worthwhile (Maynard Smith, 1078b), though biological variability in highly diverse communities may be sufficient to favour sexuality once it has originated (Glesener & Tilman, 1978). For example, the continuous generation of variety by sex in predators, parasites or pathogens is likely to necessitate a compensatory generation of variety in hosts and prey to keep pace, and this in turn is likely to promote the generation of variety in the attackers and so on.

This debate continues (for an excellent review see Maynard Smith, 1978b), but in it surprisingly little attention has been directed towards asexual reproduction. For example, what are the relative merits of asexual propagation, parthenogenesis and sexual egg production? One of the major problems with these kinds of question is the dissimilarities between sexually and asexually produced offspring. When is the asexual increase in somatic tissue just growth and when reproduction? What, in these circumstances, is meant by an individual? This question is provocatively considered by Janzen (1977a). Recently we have compared sexual reproduction with fission in freshwater flatworms (Calow, Beveridge & Sibly, 1979). Here, reproduction may occur by a binary fission which yields separated head and tail halves that go on to regenerate the missing part (Fig. 10.7a). The sexual strategy leads to egg cocoons containing several to many developing embryos and here hermaphroditism is the rule but self-fertilization is rare and parthenogenesis is absent. Hatchlings and regenerants are similar in form if not in size (see below). The degree of sexuality and asexuality may vary between species, between populations of the same species occupying different habitats, and between individuals at different stages in the same life-cycle.

In the fissiparous species, food energy cannot be converted to reproductive propagules directly but is first converted to somatic tissue before the 'progeny' are separated by binary fission. This means that fission depends exclusively on a growth process and cannot make use of the same highly efficient conversion process as gamete production (Fig. 10.8). In principle, therefore, the gamete-

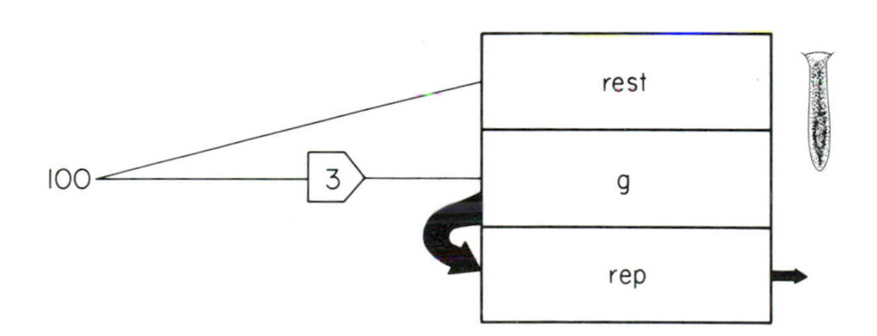

Fig. 10.8 The diversion of energy from soma to 'offspring' produced by fission—see Fig. 10.7(a). Compare this figure with Fig. 10.2. (After Calow, Beveridge & Sibly, 1979.)

producing strategy can render more energy to reproduction than the fissiparous strategy. Alternatively, because of investment in sperm and other costs discussed below, gamete production, and particularly sexual gamete production, is less efficient in transforming reproductive energy into progeny.

The costs of producing progeny from gametes as opposed to asexual propagules are illustrated in Table 10.1. This compares the conversion process of an equivalent amount of reproductive energy to offspring in an exclusively fissiparous flatworm, *Polycelis felina* (real data), and an equivalent gamete-producing triclad (hypothetical example designed to represent *P. felina* as if it were a gamete producer). Both parthenogenetic and sexual egg production are considered. Gamete production carries definite extra costs in building the apparatus necessary to form the gametes and to effect fertilization, but as yet it is not possible to be precise about the amounts involved. The cost for the sexual form might be as much as twice that of the parthenogenetic form, however, because in triclads the male part of the hermaphrodite system is as extensive as the female (Hyman, 1951). Actual data on *P. felina* suggest that of all the reproductive energy invested in the tail fragment, only about 5 % is used up in the initial non-feeding, developmental stage when the tail is reorganized into a complete worm. By contrast as much as 40 % of the reproductive energy may be used up in the more extensive embryonic development that goes on within the egg capsule (Calow, 1977c). In addition the sexual system carries

Table 10.1 Comparison of efficiencies of converting reproductive energy to offspring and then to reproductive adults for different modes of reproduction of *P. felina* at 10°C. All energy data are in joules. (Data from Calow, Beveridge & Sibly, 1979).

	Fission	Parth.*	Sex*
1 Energy available	3.5	3.5	3.5
2 Cost of reproductive apparatus	0	X	2X
3 Cost of sperm†	0	0	0.5 (3.5)
4 Energy available for development	3.5	3.5	1.75
5 Cost of development†	3.5 (0.95)	3.5 (0.6)	1.75 (0.6)
6 ∴ Energy/offspring	3.33	2.1	1.05
7 Final size to be reached by adult	14	14	14
8 ∴ Approximate J that must be ingested by offspring of size given in 6 to reach size given in 7‡	300	396	430
9 Minimum generation time (days)‡	75	85	90

* Data imaginary.

† See text for further explanation.

‡ 8 and 9 are calculated on the basis of an assumed best conversion efficiency.

X An unknown amount of energy—see text.

extra costs in sperm production and the search for mates. The sperm cost is put at 50% of the reproductive energy because, despite hermaphroditism, the relative size of gonads in triclads seem to suggest an equal distribution of energy between ovaries and testes. However, it is not known to what extent energy lost in sperm can be made good in hermaphrodites by absorption of unused sperm derived from a mate and we have no estimates of the cost of searching for mates.

As a result of the definite extra costs incurred by gamete production, the final size of offspring is likely to be smaller for an equivalent amount of reproductive energy in the gamete producers than in the fissiparous triclads. This assumes that both types produce equal quantities of progeny per unit reproductive energy but the difference is accentuated if, as happens, the asexual form produces fewer 'offspring'. In fact multiple fission is rare in the triclads and binary fission predominates whereas sexual triclads usually produce many offspring per reproductive bout. Hence the developmental time of offspring from fission or hatching to their own reproduction, and the energy required to convert them to adults are likely to be greater for those produced from gametes than for those produced from fission. For *P. felina*, for example, which has a final size of 14 J, the products of fission take about 25% less energy and 17% less time to develop than might the products of gamete production.

The trade-off between the two modes of reproduction is therefore between the ability of gamete-producing triclads to exploit good conditions of food supply to make energy available for reproduction rapidly, and the ability of fissiparous flatworms to convert this reproductive energy efficiently into large, reproductively competent adults. In consequence the success of egg production depends on the free availability of food. In conditions of poor food supply the effect of the inefficient conversion of reproductive energy to offspring is likely to become prohibitive. Consequently, as trophic conditions deteriorate gamete production is likely to be favoured less strongly and fission favoured more strongly.

Reynoldson (1961) suggested that trophic conditions are worse for triclads in streams (lotic habitats) than lakes (lentic habitats) and noted that fission is most common in stream-dwelling as opposed to lake-dwelling triclads. It would be a mistake to suggest, however, that the littoral regions of lakes are consistently richer habitats than the headwaters of streams. There is good evidence from both physical measurements and biological observations that conditions in lakes range from very poor in autumn and winter to very rich in spring and summer. Conditions for at least half of the year are unsuitable for growth and may even cause degrowth in triclads. So over a large part of the year tissue production cannot occur. In spring and summer rapid growth and production is possible in both adults and hatchlings, and this favours gamete production because the latter can exploit the pulse of nutrient enrichment to effect the rapid production of offspring. Why gamete production should occur sexually rather than non-sexually refers back to the question raised at the beginning of this section and remains something of a mystery.

10.6 HERMAPHRODITISM

Superficially, hermaphrodites would appear to have the best of all worlds; sex without separate males. However, externally fertilizing hermaphrodite animals and self-incompatible hermaphrodite or monoecious plants still require to invest a considerable amount of reproductive energy (Maynard Smith, 1978b, would argue 50 %) in the male system. This is unnecessary in parthenogenetic equivalents which therefore still enjoy an advantage (perhaps even twofold).

With internal fertilization, however, the position is less clear for there is now the possibility of investing a larger share of the reproductive energy in egg production (Altenberg, 1934). Furthermore, there is also the possibility of sperm absorption after transference. Hence if both partners donate the same quantity of sperm, each may retrieve by absorption only a little less than it loses in sperm formation. This is certainly a real possibility in triclads where sperm resorption has been observed histologically and where ducts even occur between the reproductive and gut systems whereby excess sperm may be transferred for digestion, absorption and recycling (Hyman, 1951).

If hermaphroditism with internal fertilization derives benefits approximating to those of parthenogenesis whilst also allowing the genesis of variety through meiosis and syngamy, why does it not replace the gonochoristic or two-sex condition? To do this and to persist against gonochoristic mutants, hermaphrodites must as a minimum produce as many gametes and hence fertilized zygotes from the reproductive energy as the separate males and females (Charnov, Maynard Smith & Bull, 1976). Heath (1977) suggested that this cannot generally be possible because each hermaphrodite must build and maintain two sets of reproductive apparatus whereas each gonochorist carries and therefore requires to support only one set of gonads. Hence, each hermaphrodite has less reproductive energy to spend on gamete formation and hence, in principle, will produce fewer gemetes. In hermaphrodite flowers, of course, where a single floral apparatus contains both male and female reproductive parts, a saving is made in the building and maintenance of this accessory apparatus. Significantly this kind of arrangement *is* the most common reproductive strategy of all in the flowering plants, whereas gonochorism is most common in animals and, here, organ-sharing in hermaphrodites is less extensive.

The extra costs for hermaphrodite animals may be worth bearing if, because of lower population density and sluggishness, mating contacts are rare. This is because hermaphroditism improves by twofold the chance of a successful contact since all contacts with mature individuals may lead to successful cross-fertilization.

In principle the cost of maintaining two-organ systems can be offset if one is reduced relative to the other so that more than 50 % of the reproductive energy is then available for the non-reduced system. This reduction will usually involve the male parts but sometimes the female parts may be affected. For example, if constraints like brood-pouch size limit egg production, hermaphrodites can channel more resources into sperm production (Heath, 1979). Such unequal partitioning of reproductive energy is not possible in externally fertilizing organisms. In these cases the advantage goes to those who maximize the number of ova available for fertilization and the number of sperm available to effect these fertilizations and this requires equal investment in the male and female systems (Maynard Smith, 1978b). For the same reasons savings may not be possible at high population density, because here mating contacts are frequent and selection will favour the ability to take each opportunity for fertilization and hence will maximize sperm production. In these circumstances, hermaphroditism would incur the full metabolic cost and would be disadvantaged with respect to gonochorism. Alternatively, in sluggish forms or those at low population density, when mating contacts are infrequent, the advantage of hermaphroditism can be reinforced by an unequal partition of the reproductive energy.

Molluscs provide some documentation of these principles (Calow, 1978b). Most are sluggish and throughout the phylum hermaphroditism is a primary and dominant adaptation. In hermaphrodite molluscs, however, both male and female systems share common ducts and, to some extent, a common gonad, the 'hermaphrodite gland', so that the metabolic costs of hermaphroditism should be minimized. Interestingly, gonochorism *is* widespread in this phylum in the fast-moving and visually acute cephalapods. On the other hand, many opisthobranchs are rapid movers and yet all are hermaphrodite. Similarly, some of the sluggish and immobile bivalves are gonochoristic. Perhaps in these instances population density is a more important factor than activity *per se*. For example, low population densities and poor acuity may mean that mating contacts are rare in opisthobranchs despite their high mobility. Alternatively, high population densities of even sluggish or immobile organisms will mean that mating contacts, either between externally liberated gametes or individual adults, are sufficient to favour gonochorism.

Of course, factors other than metabolic ones may act against the benefits of hermaphroditism and these include the advantages that might arise from division of labour between the sexes (see section 10.4), and disadvantages arising from dangers of self-fertilization with associated reduction in the viability of progeny (Maynard Smith, 1978b).

Part 4

Integrated Studies of Bioenergetic Strategies

Introduction

The adaptational features of one physiological system depend upon others—the organism is a tightly integrated and precisely controlled system that has evolved as a whole. The three chapters in this section will help to draw together the principles enumerated in preceding sections by considering the range of allocation options as they operate in selected species. In Chapter 11 we are offered a review of the role and success of optimization models in understanding plant evolution. Two contrasting optimization problems are considered—optimal leaf shape and size in a given light environment and, on a more complex level, optimal life-history strategies. Chapter 12 presents an overview of resource acquisition and allocation in individual animals. By means of a series of specific studies the problems involved in making various trade-offs are elucidated, and directions for future research are suggested. In the final chapter (Chapter 13) the objective is to consider, using a cost/benefit approach, the extent to which the individual animals in social systems make optimal use of the resources in a particular environment.

Chapter 11

Energy, Information and Plant Evolution

OTTO T. SOLBRIG

> 'Science is the attempt to make the chaotic diversity of our sense-experience correspond to a logically uniform system of thought'.
>
> A. EINSTEIN

11.1 INTRODUCTION

A zygote is a promise of more zygotes; it is a set of instructions for the formation of offspring; in other words, a zygote possesses a genetic code for the formation of a machinery we call a plant, that can bloom and form eggs and pollen grains, that upon uniting with other pollen grains and eggs, or with each other, will give rise to other zygotes. But a zygote possesses not one single genetic code, but a great variety of codes. It is the environment that the zygote encounters that determines which of the many possible alternative codes is used for the production of the phenotype. A plant can produce a phenotype capable of producing new zygotes only in certain environments. We will say that a zygote (and by implication the individual, population or species) is *adapted* to an environment if it can reproduce in that environment. The environment is the sum of the environmental phenomena which have or may have an operational relation with any organism. Following Mason and Langenheim (1957) we will call this the operational environment.

In a population of zygotes that are formed at a particular place and time, some zygotes produce phenotypes that are non-adapted to the environment and that leave no offspring. Among those that leave offspring, some leave more descendants than others. Consequently the frequency of some genotypes increases with time and that of others decreases. Furthermore, because of genetic recombination, at each generation new genotypes are formed by new combinations and permutations of parental genes. The logical consequences of these processes, i.e. mutation, recombination and selection at the gene level, have been worked out in detail (Wright, 1968–78), and will be referred to as the *genotypic form* of the theory of natural selection. The genotypic form of the theory is complementary to the *demographic* or *phenotypic* form of the theory of natural selection, which is concerned with changes in absolute and relative numbers of individual phenotypes in populations over time. The phenotypic form of the theory is not yet as well developed as the genotypic form. Geneticists have been concerned primarily with the development and testing of the genotypic form of the theory; ecologists have worked within the framework of the phenotypic form.

The genotypic form of the theory of evolution by natural selection is based on the thesis that given two reproductive phenotypes, the genes of the one that produces the largest number of offspring surviving to reproductive age will inevitably increase in frequency in the population. Since those genes are the basic components of the genetic codes carried by phenotypes, it follows that the property of a genotype of producing a phenotype that leaves a large number of surviving offspring will tend to be maximized by the very nature of the process. This is what we call the genetic definition of *fitness*.

A plant can be viewed as a system for gathering, transforming, and transmitting energy and information (Fig. 11.1). The process of photo-

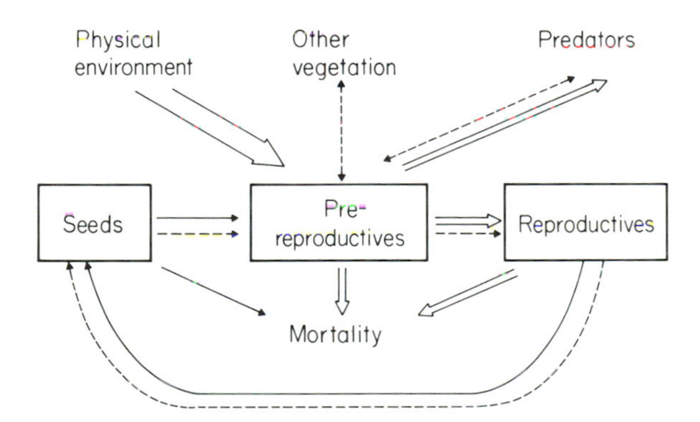

Fig. 11.1 Energy and materials (solid arrows) and information flow (dashed arrows) through the life cycle of a plant.

synthesis (Chapter 3) gathers energy, which is apportioned to the various parts of the plant in relation to the relative strengths of sinks and sources (Wareing & Patrick, 1975), while the processes that lead to the production of mature seeds control the transmission of energy, materials and information to the next generation. However, it is the realized genetic code, through a system of very complex and poorly understood processes involving environmentally modifiable chemical signals (i.e. plant hormones and vitamins) that ultimately determines actual photosynthetic rates, details of carbon apportionment and reproductive events. Information to build this machinery is inherited in the form of chromosomal DNA. As this code is translated into metabolic machinery, the speed and mode of translation, and to a certain extent the kind of enzymes produced (i.e. the primary gene products), are modulated by other genes, by other primary products already in the cell (Paigen, 1979), by metabolic products and, directly or indirectly, by environmental factors such as water, light and temperature (Hochachka & Somero, 1977). These interactions are extremely complex.

Recent discoveries in molecular biology (Gilbert, 1977; Tonogawa *et al.*, 1978; Crick, 1979; Wang *et al.*, 1979) show that in eukaryotes the arrangement of genes (= cistrons) along the chromosome is not analogous to that of beads on a string as classically described in genetics textbooks. Neither are mutations truly random as usually assumed by evolutionists, but certain classes of mutation are much more frequent than others (Chinnici, 1971; Kidwell, 1972). For these and other reasons there is an increasing feeling that only a part of the genetic material codes for cytoplasmic enzymes, while the rest of the DNA is directly or indirectly involved in regulatory functions (Paigen, 1979). Layzer (1980) has provided theoretical evolutionary arguments for the existence of such regulatory mechanisms, as otherwise it is difficult to explain observed evolutionary rates and existing organic diversity. Nevertheless, natural selection is expected to operate on both regulatory and structural genes. It is interesting to note that one of Darwin's original critics (Jenkin, 1867) foresaw the problem of accounting for all evolutionary change on the basis of small random changes. It still has to be seen whether these new discoveries are going to invalidate the elaborate mathematical models of the genotypic form of the theory of evolution, although it is already evident that the present form of the theory will have to be modified to accommodate the new evidence (Karlin & Nevo, 1976; Roughgarden, 1979).

Natural selection, i.e. differential mortality and differential reproduction, affects phenotypes and only indirectly genotypes. Because nearly identical phenotypes may represent different genotypes, there is little detailed evidence on how genes are affected by differential mortality and reproduction of phenotypes. In the absence of that information, the assumption is made that the genes themselves have characteristic fitnesses. This has allowed a more precise and comprehensive development of the genotypic form of the theory. It is even claimed (Orgel & Crick, 1980) that 'selfish' genes create phenotypes only as vessels to insure their survival. But under the genotypic theory there is no way of predicting phenotypic behaviour. We now need predictive and testable hypotheses at the phenotypic level as well as a way to map precisely events at the genotypic level into events at the phenotypic level. This is especially important, since it is abundantly clear that a phenotype is not the result of the simple addition of individual gene expressions.

A comprehensive phenotypic theory of evolution by natural selection has to take into account all the components of the life cycle of the organism. An understanding of the action of natural selection will not be possible unless a dynamic demographic and quantitative view is taken. Furthermore the physical and biological environment must also be taken into account as it provides the constraints within which the organism exists.

The ways by which a large number of offspring are produced and survive to reproductive age are varied, and knowledge of them is important for understanding evolution in plants. Natural selection in its phenotypic form can be visualized as a process that maximizes or optimizes the production of reproducing offspring. Optimization is a mathematical technique that economists and engineers use to resolve problems of allocation of scarce resources so that maximal utility is obtained. For example, an engineer may be asked to design a permanent bridge, made of steel or stone, that is convenient for users and that is aesthetically pleasing. This creates a host of subsidiary problems, i.e. how permanent does the bridge have to be, what is aesthetically pleasing, how much inconvenience can motorists tolerate, and what is a reasonable cost? In formal terms, what the engineer is faced with is to find a function, called a cost functional, which assigns to each factor involved in building a structure a cost or value. The problem of finding an optimal solution is to find the solution that corresponds to the minimum cost within the constraints (load, permanency, aesthetics) that are given. Needed also are detailed descriptions of the soil, the strengths of materials, the weather and other pertinent characteristics—called the *state space*—and a knowledge of all alternative building techniques, or *strategies*. In translating optimization techniques into biology there are serious conceptual and methodological difficulties that have been enumerated by Oster & Wilson (1978), and Calow & Townsend (Chapter 1). These problems stem from the fact that it is usually not possible to produce a complete and rigorous description of the relevant state space nor is it possible to specify *a priori* a complete set of strategies or to enumerate all relevant constraints. Consequently, although modelling how organisms maximize the production of offspring within certain available energy and materials resources is not different in principle from the problem of designing a bridge, our knowledge of the relevant parameters is very limited. Therefore, to develop a general phenotypic theory of evolution by natural selection, educated guesses regarding the right population descriptors and operating constraints have to be made, before the set of possible strategies (within the criterion that offspring production is being optimized) can be specified and tested.

There are a variety of ways by which an organism can increase reproductive output, and probably every species has its own particular mechanism. These processes can be arbitrarily classified into two categories:

1 processes that lead to an increase in the reproductive capacity of the plant, i.e. that increase the number of seeds produced;
2 processes that lead to a higher probability of survival of the plant, so that it can reach reproductive age and stay reproductively active for a longer period of time.

We can represent this by

$$w = m \times l$$

where w = fitness (i.e. number of reproducing offspring); m = number of offspring per unit time in years; and l = lifespan of the individual in years. The first group of processes are those that increase m, and the second are those that increase l.

Since the production of offspring is an energy- and materials-consuming process, whenever energy and/or materials are in limited supply increased reproduction will affect survivorship of the parent in a negative way, and vice versa. We consequently expect adjustments that maximize w, the product of fecundity and lifespan.

Nevertheless, fitness cannot be equated solely with total number of offspring as there are non-energetic considerations that have to be taken into account. The offspring are normally not identical genetically to the parent plant. Consequently, the extent to which a given number of offspring contributes to the multiplication of the genes of the parent plant depends on the degree of their relationship, which in turn is a function of the breeding system and the genetic structure of the population.

Another non-energetic trade-off relates to the number and size of seeds (Chapter 10). A large number of small seeds may not be more demanding energetically than a small number of large seeds but one or the other may be evolutionarily more advantageous. We refer to these non-energetic aspects of plants as *control mechanisms*. Other examples of control mechanisms are processes that sense photoperiod, be it to time germination in seeds, or to time flowering in adults, or developmental processes, as for example those that control the number and size of seeds in different environments.

In order to complete its life cycle a seed must germinate and produce a phenotype that can:

1 compete successfully with other vegetation, both congeners and non-congeners;
2 provide adequate defences against herbivores (Chapter 6);
3 reproduce.

These processes—competition, defence, and reproduction—require energy and materials, which are also needed for growth and maintenance. They also demand information, to interpret environmental signals correctly and anticipate environmental changes (i.e. defences must be provided before herbivores injure the plant; growth rate and plant morphology have to be adjusted to fit prevailing light and water regimes) and control mechanisms to respond to those signals. The energetic demand of the machinery that provides information and control (i.e. energy content of DNA, hormones, etc.) is probably so low as to be insignificant. The value of information and control cannot be assessed in joules. It is therefore impossible to develop comprehensive models of evolution based solely on energetic considerations.

11.2 THE ROLE OF ENERGETIC MODELS IN EVOLUTIONARY THEORY

The thesis of this book is that organisms can be viewed as systems for the capture, transformation and transmission of energy, and that survivorship of a lineage depends on the effectiveness with which organisms process energy. That a certain degree of efficiency is necessary for the continued survival of individuals and, that—all other things being equal—the organisms that are most efficient in handling energy (and materials) will leave more offspring, is undeniable. But all things are never equal and changes involving sacrifice of efficiency take place in the course of evolution. Given that there is no way of establishing precisely what organisms are optimizing ('fitness' in the phenotypic form of the theory of natural selection being impossible to define rigorously given the trade-offs between number and an indefinable 'quality' of offspring that take place), and that furthermore, there is no way of rigorously defining the state space or the constraints that operate on an organism, it follows that it cannot be established precisely whether organisms are built, or function, following criteria of optimality. Does this mean that optimality models have no place in biology?

In my opinion optimality has an important role to play as a technique to produce predictive and testable theory, even though organisms may not function in an optimal way. The principal role of theory in population biology is to suggest systematic tests and observations. The traditional research approach in the field of plant population biology comes from taxonomy and involves comprehensive, in-depth studies of individual species followed by *ad hoc* explanations arrived at by a process of induction: Model building starting with basic principles with subsequent testing of deduced consequences ('predictions') allows the researcher to disprove hypotheses, which is not the case within the taxonomic tradition. Nevertheless, optimality models have to be applied judiciously, and the phylogenetic history of species must also be considered.

Although the theory of evolution deals with the history of organisms, paradoxically, population biologists have not always given sufficient attention to the constraints that past history places on the evolution of a lineage. So, for example, capturing and transforming solar energy severely constrains the evolutionary possibilities of a plant. The systematic investigation of the morphology and physiology of the process of energy capture, allocation and transmission in species of different phyla and trophic levels is a necessary prerequisite to comprehend the constraints that form and function place on evolution. In this more limited context, optimization models based solely on energetic considerations, or solely on information or control criteria have a role. In the following pages we shall discuss such models, to illustrate the points just made.

Two different optimization problems will be considered. First, models regarding optimal harvest of solar energy will be presented. This is an apparently simple problem: what is the optimal leaf size and shape for a given light environment? The second problem deals with a consideration of life-history strategy models. Given the conflict between growth and reproduction, under what circumstances should reproduction be favoured at the expense of lower competitive ability and shorter lifespan, the so-called r strategy, and when should the opposite, K strategy be favoured? In each of these cases the existing theory or theories are reviewed briefly, explicit or implicit assumptions made are discussed, and data and observations to test the models presented. The underlying general question that is being asked is whether this functional ecological approach has aided materially in understanding plant evolution. I shall close with a discussion of the value and limitations of optimization models in physiological ecology.

11.3 LEAF SIZE AND SHAPE

Plants growing in different environments have leaves of characteristic sizes and shapes. So, for example, the leaves of trees growing in the temperate zone are normally 5–10 cm in length by 3–5 cm in width, entire and with toothed or serrated margins; evergreen plants from warm semidesert regions have smaller leaves 1–3 cm long and about 1 cm wide with entire margins; sub-canopy tropical trees have very large leaves with entire margins and pointed apices; many trees in tropical and subtropical savannas have compound leaves, and so on. Within a vegetation type, leaf sizes are not uniform. Species with larger and smaller leaves than the norm are always present. Furthermore, even within a tree, leaves that are exposed to the sun tend to be smaller than those in the shade, there are even changes in leaf size with time.

Leaves are the main biochemical factories of the plant. They intercept light and use that energy to fix carbon dioxide (CO_2) and synthesize five and six carbon sugars (for details of this process see Chapter 3). In addition, organic acids, amino acids and a number of specialized small molecules (vitamins, plant hormones, carotenoids, anthocyanins, etc.) are synthesized in leaves. These products are exported to growing points, such as shoots, roots, young leaves, storage organs and reproductive organs. A mature leaf imports vast quantities of water and dissolved mineral ions, especially nitrogen (N). To function effectively as a chemical laboratory, the leaf needs:

1 adequate light;
2 ample supply of 'raw' materials, primarily CO_2 and N;
3 plenty of water;
4 appropriate temperatures.

Securing these conditions presents difficult problems to the plant. Photosynthesis is inherently inefficient in its use of light. Only 1–3% of incoming light, and light of only certain wavelengths, can be utilized. This means that a large portion of incident light must be dealt with effectively and much of the absorbed light energy has to be dissipated. Carbon dioxide is obtained by passive diffusion from the atmosphere. Since CO_2 concentration is very low in the atmosphere, for every molecule of CO_2 that is fixed anywhere from 300 to 1000 molecules of H_2O vapour are lost. This water has to be replenished from the roots, through the xylem. The roots are also the organ through which minerals and N_2 are supplied to the leaf. The adaptive problem with which the plant is faced is how to maintain an adequate supply of water, N and CO_2 and an adequate temperature to maximize photosynthate production. The independent environmental variables are light, CO_2, soil-water potential, air humidity, wind, temperature and soil-mineral availability. The principal plant dependent variables are:

1 the biochemical and physiological characteristics of the photosynthetic apparatus, including the protein (carboxylase) to chlorophyll ratio; the ratio of P770 to P750; the carbon-fixing system (whether C_3, C_4, or CAM), and the shapes of the light and temperature curves (Tenhunen *et al.*, 1976; Chapter 3);
2 the number and characteristics of the stomata;
3 the ability of leaf cells to withstand water deficits;
4 the extent of the root system;
5 the form and size of the leaves.

The large number of environmental and plant variables emphasizes the difficulties involved in modelling the system.

The question that arises is whether the characteristic leaf shapes and sizes of plants associated with specific habitats, presumed to be of adaptive significance since they are associated with the process of energy capture, can be predicted *a priori*. Several authors have addressed this problem (Rashke, 1956; Gates, 1962; Taylor, 1975; Givnish & Vermeij, 1976; Orians & Solbrig, 1977; Givnish, 1979; Mooney & Gulmon, 1979). I shall now briefly review the problem and the difficulties inherent in developing an optimality model that predicts leaf size and shape.

The basic elements necessary to understand leaf size and shape from a purely energetic point of view are presented in Fig. 11.2. According to the surface characteristics and other morphological and anatomical properties of the leaf, a certain proportion of incoming radiation (usually 50% or more) is absorbed (Birkebak & Birkebak, 1964) the remainder being either transmitted or reflected. As light energy is absorbed, the leaf tends to heat up. This increases substomatal water vapour pressure, which in turn increases the rate of transpiration. Under many environmental conditions (e.g. in forest understory), the rate of transpiration may be high enough to dissipate most of the heat from the absorbed light energy and keep the leaf close to ambient temperature or even below it. The rate of transpiration is controlled in part by the boundary-layer resistance which is proportional to leaf size and wind speed, but primarily by the stomatal resistance, the value of which depends on the number and shape of stomata and whether they are open, or partially or completely closed. As water is lost through transpiration, leaf water potential decreases. This water must be replaced with water extracted from the soil by

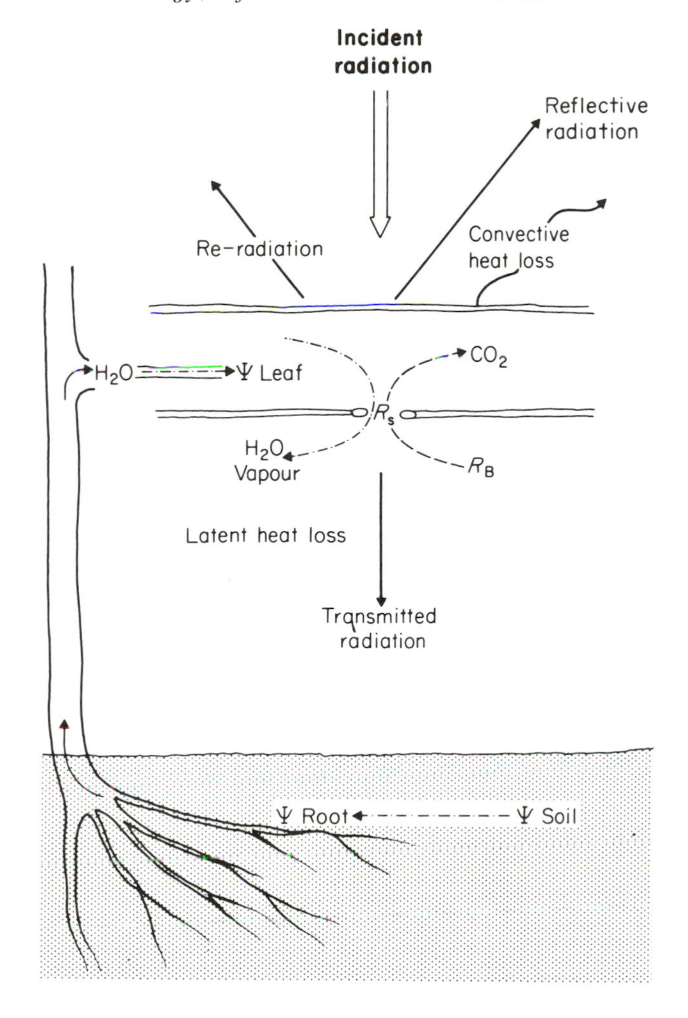

Fig. 11.2 Diagrammatic representation of radiation (solid arrows), carbon dioxide (dashed arrows) and water flow (dot-dash arrows) in a plant leaf and surrounding environment. Ψ = water potential; R_s = stomatal resistance; R_B = boundary layer resistance

the roots and transported to the leaf via the xylem. The ability of the plant to extract water from the soil depends on soil-water potential, the extent of the root system and the resistance to flow provided by the xylem, which depends on length and diameter of the conducting elements. Leaf-water potential will always be lower than soil-water potential in any actively transpiring leaf. Reduction in leaf-water potential reduces photosynthetic rates.

As leaf temperature increases, the rates of chemical reactions in the leaf also increase, especially respiration and photosynthesis, up to thermal optima. Increased rates of carboxylation increase the demand for CO_2. The diffusion rate of CO_2 into the leaf depends on the values of the boundary-layer resistance, the stomatal resistance and the mesophyll resistance, as the concentration of CO_2 in air is more or less fixed, and the internal concentration approaches zero in an actively photosynthesizing leaf.

High photosynthetic rates will be sustained only when light flux density is saturating the photoreceptors, temperature is close to optimal, leaf-water potential is close to zero and the rate of carbon dioxide diffusion is close to maximal. Light and carbon dioxide fluxes, and leaf temperature and water potential are mutually interdependent and their interactions are complex. Therefore, there is probably more than one set of leaf characteristics that provides adequate supplies of light, water, and carbon dioxide to photosynthesizing leaves. To find out what those alternative strategies are, and to learn which of them may be favoured by natural selection in a particular environment, the relevant state space, constraints and cost functionals of the system must be identified and described.

The first decision facing the modeller is to decide what factor natural selection is maximizing. Fitness, that is, differential mortality and differential reproduction, is not a practical function, as it is not possible, with present knowledge, to translate leaf size and shape into fitness. Instead the function that has been used in most cases is the production of photosynthate per unit time, under the assumption that individuals that produce more photosynthate will have more energy and materials available for predator defence, growth and/or reproduction, and thereby should realize a fitness gain. The use of photosynthate production per unit time has the further advantage that it is fairly well understood how photosynthesis is affected by temperature, heat, light and water. I will assume here that net photosynthate production per unit time is what is being maximized. Next, the constraints of the system must be determined.

The first constraint to consider is the effect of environmental parameters on leaf temperature. As a leaf is exposed to solar radiation its temperature is determined by its ability to dissipate the energy it absorbs. Heat is lost from a leaf by transpiration, convection and radiation. Of these, transpiration and convection are the dominant processes at moderate temperatures. The relationship between radiation and leaf size can best be understood in terms of the energy budget of a leaf, that can be written as:

$$R_n - LE - C \pm J = 0 \tag{11.1}$$

where R_n = absorbed radiation (in $Wm^{-2}K^{-1}$); L = latent heat of vaporization (2450 J g^{-1}); E = transpiration rate; C = convective heat flux; J = rate of change of stored heat. The amount used for metabolic functions is negligibly small and can be ignored.

The storage term J (Monteith, 1973) represents the change in heat content of a leaf. It can be determined from the Equation 11.2:

$$J = d(kT)/dt \tag{11.2}$$

where k = heat capacity of the leaf and T = absolute temperature. Unless the stomata are closed and there is no wind, J will tend to have a small value. Nevertheless, in hot, dry environments, such as deserts, where stomata are often closed, J may be important.

Convective heat loss can be represented by the Equation 11.3 (Gates, 1962):

$$C = K(T_a - T_l)\left(\frac{u}{d}\right)^{1/2} \tag{11.3}$$

where u = wind speed; d = characteristic leaf dimension (usually 0.7 × maximum dimension in direction of flow) and K = constant that depends on the leaf size. Convective heat loss is proportional to the difference between the temperature of the leaf (T_l) and that of ambient air (T_a) and to wind velocity, and inversely proportional to leaf size (d). Convective heat losses will be more effective with strong winds, narrow leaves and large temperature differences between air and leaf.

Although the leaves of some plants, most notably desert plants, may spend a significant part of the day with their stomata closed, most species rely on transpiration rather than convection to dissipate most of the heat they absorb. Transpiration consumes relatively large quantities of water that must be constantly extracted from the soil. The rate of water vapour loss through transpiration is given by the Equation 11.4:

$$\Delta E = \frac{P_{vs} - P_{va}}{r_b + r_s} \tag{11.4}$$

where P_{vs} (in g cm^{-3}) = water vapour concentration inside the leaf; P_{va} = water vapour concentration in air; r_b = boundary layer resistance; r_s = stomatal resistance. As leaf temperature increases, the water vapour pressure in the intercellular spaces rises, increasing the water vapour differential with the atmosphere, and the water loss through transpiration. Since in an irradiated leaf, temperature increases with leaf size faster than the increase in the boundary-layer resistance (which also increases with leaf size), under the same radiation environment large leaves tend to lose more water per gram than small leaves.

With increased leaf temperature transpiration is augmented, leading to a reduction in leaf-water potential. If the leaf-water potential reaches a very low level (about -1.5 MPa (-15 bars) for most species) it can have a lethal effect on the photosynthetic apparatus. Long before that point is reached, transpiration is reduced due to the closing of stomata. However, the closing of stomata will also affect CO_2 diffusion as follows:

$$\Delta CO_2 = \frac{CO_{2\,air} - CO_{2\,Chl}}{r_b + r_s + r_m} \tag{11.5}$$

where r_m = mesophyll resistance. The closing of the stomata reduces photosynthesis by reducing CO_2 diffusion. Furthermore, lowering of the leaf-water potential has also a direct and negative effect on chloroplast performance (Boyer & Bowen, 1970). That is, water loss and CO_2 uptake are closely coupled because both pathways are linked through the stomata.

Carbon dioxide diffusion into the leaf is not affected by leaf temperature, as the CO_2 concentration inside the leaf is lower than that of air. As the rate of CO_2 consumption increases as a result of higher carboxylation rates resulting from increased temperatures, the maximal rate of CO_2 diffusion is reached when the stomata are wide open.

In summary: leaves must absorb light energy for use in photosynthesis. However, of the absorbed energy only a small amount is utilized. The remainder has to be dissipated. In many environments, the primary loss of heat is through evaporation and convection, the amounts lost by re-radiation are small in comparison. Evaporation consumes large quantities of water that must be replaced. Photosynthetic rates increase with temperature like any other chemical reaction. Because the increase in leaf temperature with increased leaf size is steeper than the increase in the boundary-layer resistance, photosynthesis and transpiration increase at different rates with increased leaf size. As leaf temperature increases, eventually the photosynthetic rate will level off, as CO_2 diffusion reaches its maximum rate, but not transpiration rate. Beyond that point as leaf temperature increases, transpiration and respiration increase without any photosynthetic gain. Beyond a certain point photosynthetic rates actually go down. Higher transpiration rates demand a greater supply of water to the leaf. Otherwise, leaf-water potential drops, which has a depressing effect on photosynthesis. To maintain this greater supply of water, energy has to be devoted to the production of more roots and conducting tissues, thereby reducing the photosynthetic energy and materials that are available for reproduction, defence or further leaf production.

Gates (1968), Gates *et al.* (1968) and Taylor & Sexton (1972), have tested the precise effects of incident radiation (and other microclimatic factors) on the performance (temperature, photosynthetic rate and water use) of leaves of different dimensions. Using a model based strictly on the energy-balance equation as presented by Gates (1968), Taylor (1975) studied how leaf temperature, transpiration rate (Fig. 11.3(a)), net photosynthesis (Fig. 11.3(b))

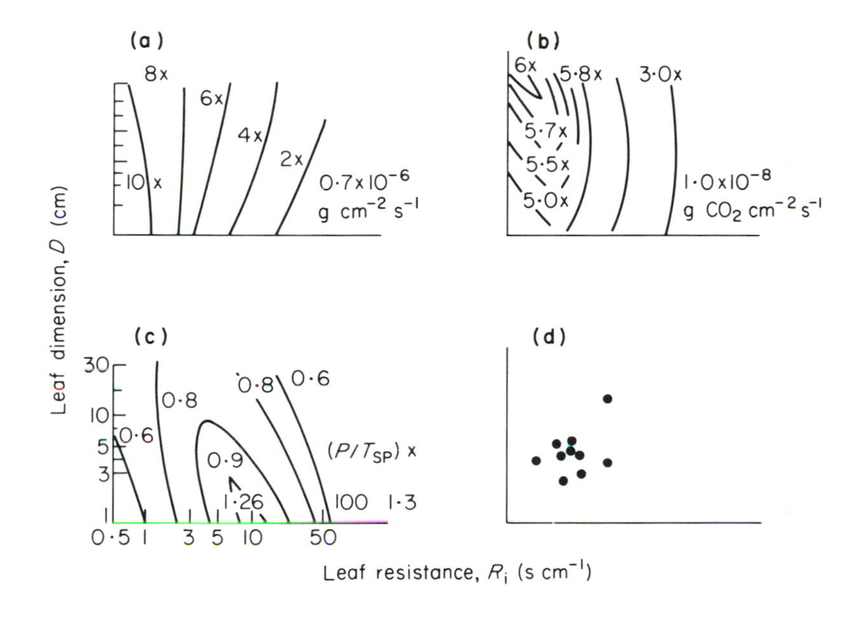

Fig. 11.3 (a) Isolines representing transpirational rate; (b) isolines representing net photosynthesis; (c) isolines representing water-use efficiency expressed as the net weight of carbon dioxide fixed per weight of water expended times 10^2, all as influenced by leaf dimension and resistance to transpiration, for a warm day, moist air, moderate wind, bright sun and moderate to high radiation absorbed by leaves in the conditions of the Michigan Biological Station.
(d) Actual data for a number of major species of the region (dots). (From Taylor, 1975.)

and water use (Fig. 11.3(c)) are affected by the characteristic leaf dimension (the characteristic leaf dimension is a function of the maximal dimension of a leaf in the direction of wind flow) and leaf resistance. In his calculations, Taylor used the environmental conditions present at the University of Michigan's Biological Station situated in northern Michigan. The question asked was whether species would maximize photosynthetic rate or minimize water use.

The leaf characteristics (Fig. 11.3(d)) and resistances of the dominant species of the study area exhibited characteristics intermediate between those that gave maximal photosynthetic rates or most efficient water use. Clearly, net photosynthetic rate is not being maximized in this area. Taylor's study shows that there is no severe environmental limitation to leaf form in the studied environment. Since leaf dimensions and resistances tended to converge (Fig. 11.3) towards a central value of approximately 5 cm effective leaf size and resistance of around 2 $s\,cm^{-1}$ something other than maximal photosynthetic rate or water use efficiency *per se* appears to be favoured.

Parkhurst & Loucks (1972) proposed a model to predict optimal leaf size in relation to environment, based on the assumption that natural selection maximizes the ratio of photosynthesis to transpiration (P/E)—the familiar water-use efficiency ratio (Slatyer, 1967). The model is based on seven independent variables (a coefficient of convection, air temperature, relative humidity, absorbed radiation, stomatal resistance, mesophyll resistance and position of stomata—whether hypo- or amphi-stomatous. Their model makes three general predictions (Fig. 11.4):

1 When the radiation flux density absorbed by the leaf is high, smaller leaves have a higher water-use efficiency and should be selected;
2 When the radiation flux density is low, and air temperature is low, the tendency towards small leaves should prevail;

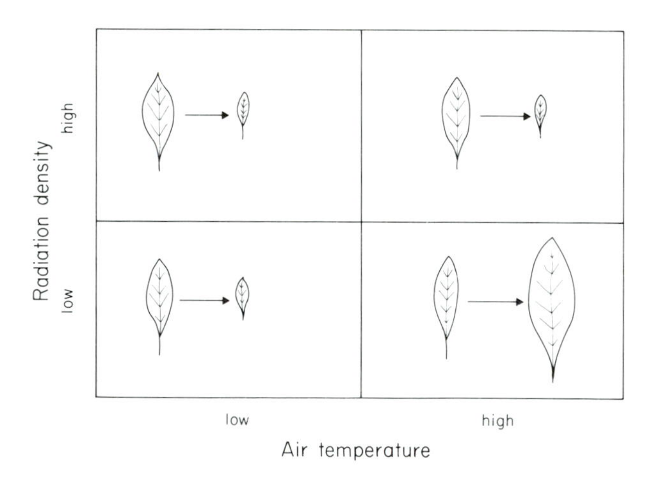

Fig. 11.4 Diagrammatic summary of the expected trends in leaf size as predicted by the model of Parkhurst & Loucks (1972).

3 When radiation flux is low and the temperature is high, the trend is reversed and larger-sized leaves should be favoured.

The predictions of Parkhurst & Loucks' model are borne out insofar as the 'tendencies' they predict correspond with reality, since the larger plant leaves are found in the understory of tropical and temperate forests, where air temperature is high and radiation flux density low. However these are very general predictions, and furthermore, the 'optimal' leaf according to the mathematics of their model should be either infinitely large or infinitesimally small, according to conditions.

Another way of looking at this problem is to consider that photosynthesis is a process that provides 'income' (in photosynthetic 'carbon' units) but that has also 'costs' (in carbon units), these being the costs of making leaves, and roots and conducting tissue to maintain an adequate transpirational stream (Givnish & Vermeij, 1976; Orians & Solbrig, 1977; Givnish, 1979). The process can be represented by the following equation (Givnish & Vermeij, 1976):

$$N = (P - R_1 - bE)A \qquad (11.6)$$

where N = net carbon assimilation; P = net photosynthate production per unit area, per unit time; R_1 = night respiration (per unit area and unit time); E = rate of transpiration (per unit area and unit time); b = proportionality constant that measures the metabolic cost of arranging for the supply of a unit flow rate of water; and A = total leaf area. The problem now reduces to finding the point at which the derivatives of N with respect to P, R_1 and E vanish. At that point the leaf and root characteristics of the plant balance the income of photosynthesis with the costs of construction and maintenance, so as to maximize the profit to the plant in terms of carbon gain .

Using this approach Givnish & Vermeij (1976) made the following predictions.

1 In a sunny environment small leaves will be favoured over large leaves. The reason is that transpirational costs rise steeply with increased leaf size (Fig. 11.5) but the gain from photosynthesis, while increasing at low or moderate sizes, eventually levels off and then declines.
2 In a shady environment, the effects of leaf temperature and leaf resistance on photosynthesis are small. Transpiration is likely to cool leaves below ambient temperature, thereby reducing the photosynthetic gain with increasing leaf size. Consequently effective leaf size will again tend to be small.
3 In an environment intermediate in light flux density between sunny and

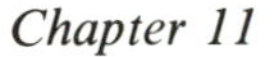

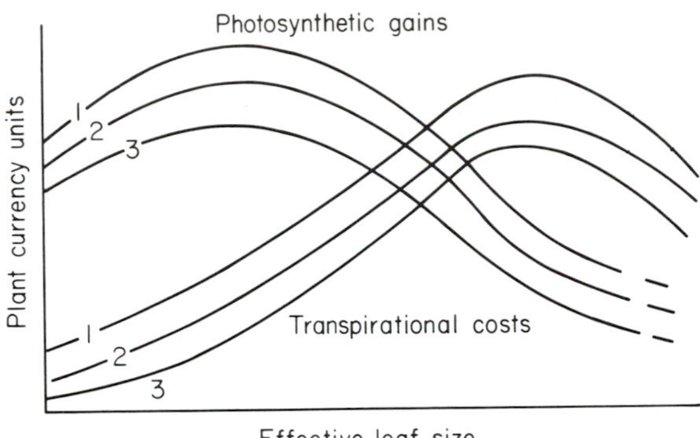

Fig. 11.5 Benefit and cost curves of photosynthesis and transpiration for leaves in a sunny environment as a function of the effective leaf size, according to Givnish (1979). The different curves indicate more sunny (1) or less warm, (3). The optimal size for a given environment is where benefit most exceeds cost (after Givnish, 1979).

shady, transpiration tends to be independent of leaf size (Fig. 11.6). Thus the effective leaf size should be close to the point where photosynthetic gain is maximized.

Givnish & Vermeij have checked their qualitative predictions against actual field data on vines growing in Venezuela and Costa Rica, and found the field data to be in good agreement with predictions.

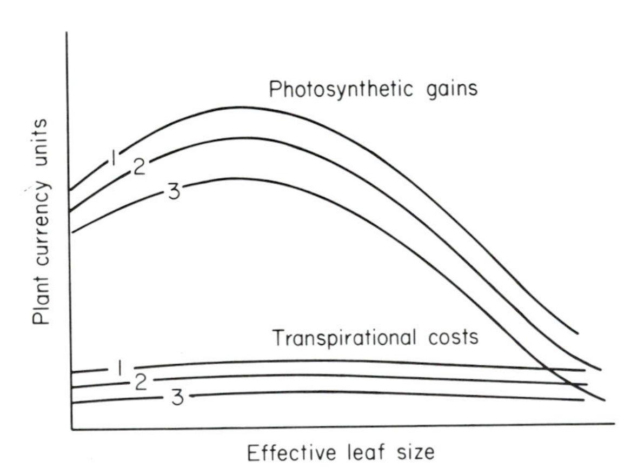

Fig. 11.6 Benefit and cost curves of photosynthesis and transpiration for leaves in an environment with open shade. Labels as in Fig. 11.5 (after Givnish & Vermeij, 1976).

Orians & Solbrig (1977) developed a model that predicts leaf sizes of plants of semidesert areas with different phenologies. Their graphical model is again based on the tight linking of carbon uptake and water loss through stomata. They show that leaves of plants exposed to extensive periods of drought, will not only be small, as predicted by all three of the models so far discussed, but that their construction costs will be higher than those of leaves of plants that do not have to be subjected to such drought periods. This is related to the fact that the ability to withstand high negative leaf-water potentials is apparently enhanced by small cells and thick cell walls, as well as by biochemical changes (Stocker, 1968; Boyer, 1976). These changes reduce the photosynthetic rates per unit area even when water is available. There is, therefore, an inverse relation between the ability to photosynthesize at low leaf-water potentials and the maximal rate of photosynthesis when soil-water potential is close to saturation (Fig. 11.7). Thus plants with mesic leaves require less time before the costs of

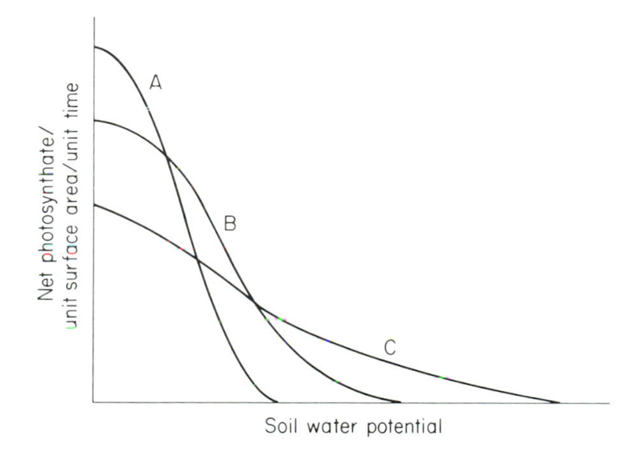

Fig. 11.7 Predicted net photosynthate/unit area/unit time, as a function of soil-water potential for three different kinds of levels: A, mesic; B, intermediate; C, xeric. Each of these leaf types has a zone where it outperforms the others (after Orians & Solbrig, 1977).

production are amortized than plants with xerophytic leaves (Fig. 11.8). They predict that desert ephemerals, which are functional only during the rainy season, should have less specialized leaves, capable of higher absolute photosynthetic rates than leaves of perennial plants, which will be longer lived and possess a lower maximal photosynthetic rate; predictions that are borne out by the available data (Solbrig & Orians, 1978).

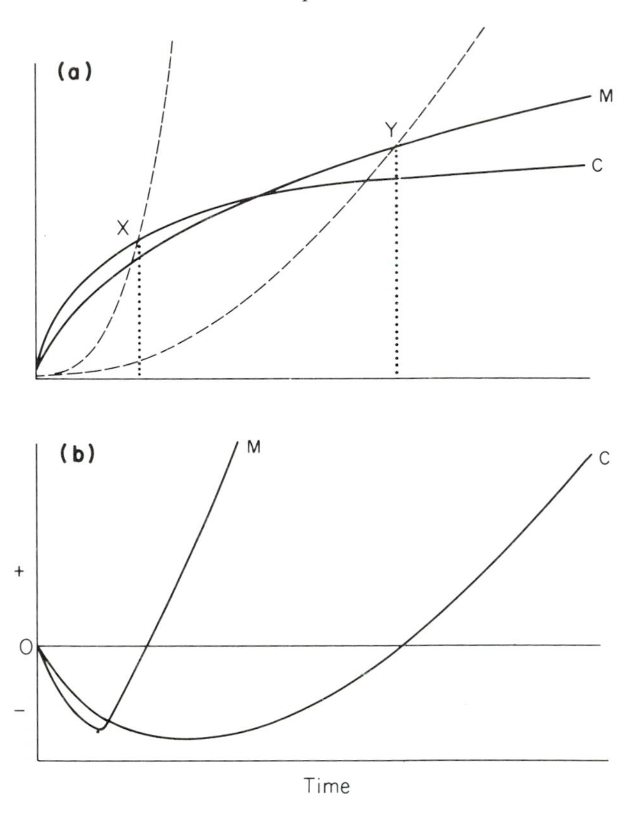

Fig. 11.8 (a) Hypothetical cost-income curves for a mesophytic (M) and xerophytic leaf (C) showing break-even points at X for the mesophytic and at Y for the xerophytic leaf. Cost is defined in terms of energy expenditure for building and maintaining leaves, and income—is the energy derived from photosynthesis. (b) profit (income minus cost) curves for mesophytic (M) and xerophytic (C), which have the cost–income properties shown in the upper graph. Mesophytic leaves show a profit earlier in time (after Orians & Solbrig, 1977).

This brief analysis shows that indeed some realistic qualitative predictions have been made, tested and verified regarding the size and shape of plant leaves. Not mentioned here are the role of nutrients on leaf size and shape, considered by Mooney & Gulmon (1979) and Givnish (1979), different photosynthetic system (Teeri, 1979) or the role of predators. Nevertheless, for the precise quantitative modelling of leaf size and shape these factors will have to be considered.

11.4 LIFE-HISTORY

In the preceding pages it was shown that when reasonable guesses can be made regarding the state space and operating constraints of a plant, realistic qualitative predictions can be proposed and tested. I now consider a more general problem, namely a plant's life-history characteristics.

The life-history of an organism is an epiphenomenon which may be roughly equated with its 'occupation' or 'trade'. For many species there is enough factual knowledge to define reasonably unambiguously its life-history. The information needed are data on the size of the soil-seed pool; germination rates of seeds; survivorship of juveniles and adults; fecundity as a function of age or size; and reproductive effort. The life-history of plants varies with environment. So, for example, species growing in transient environments, such as highly disturbed ones (ploughed fields, lawns, riverbanks), or those that are available for brief periods (vernal pools, rainy periods in deserts), have short lifespans, high reproductive effort, large soil-seed pools and small seeds. On the other hand, species growing in areas with predictable resource levels are long-lived, show high seedling mortality, low reproductive effort, small soil-seed pools and large seeds. Very often they reproduce vegetatively. However, not all species in the same environment have identical life-histories, which sets up interesting problems regarding the exact environmental factors that regulate and select a species' life-history. The question I wish to address here is, whether, given a set of environmental parameters, the optimal life-history of an organism can be predicted *a priori*. The subject has been discussed at length since first broached by Cole (1954) and MacArthur (1962) (MacArthur & Wilson, 1967; Gadgil & Solbrig, 1972; Stearns, 1976, 1977; Grime, 1977, 1979; Hickman, 1979).

One of the problems with which life-history theory is concerned is that of the trade-off between investment of energy and materials in reproduction and vegetative growth (Chapters 9 and 10). Plant species differ in the proportion of energy and materials they allocate to reproduction (White & Harper, 1972; Abrahamson & Gadgil, 1973; Solbrig & Simpson, 1974, 1977; Hickman, 1975, 1977, 1979). The question therefore is not whether such trade-offs occur, but whether they can be predicted *a priori*. This reduces to identifying the relevant selective factors (the state space) and the significant constraints, and enunciating all possible viable alternatives (strategies).

The proportion of energy that species allocate to reproductive functions is correlated with the successional stage of the vegetation (Newell & Tramer, 1979) (Table 11.1). A most interesting theoretical notion that explains in part

Table 11.1 Number of seeds per plant per year for herbaceous species growing in a recently ploughed (1 year) field, a weedy field (10 year) and a forest (after Newell & Tramer, 1978).

Species	Number of plants	Mean number of seeds/plant	Range in seeds/plant
1-year field			
Ambrosia artemisiifolia	12	1190	110–2690
Chenopodium album	57	4820	90–50100
10-year field			
Solidago altissima			
S. cancadensis	11	3070	960–5330
Forest			
Dentaria laciniata	9	24	9–42
Cardamine bulbosa	10	33	5–90
Sanicula gregaria	27	26	3–50
Prenanthes alba	7	118	95–170

this correlation is the theory of *r* and *K* selection (MacArthur, 1962; Cody, 1966; MacArthur & Wilson, 1967; Hairston, Tinkle & Wilbur, 1970; Pianka, 1970; Roughgarden, 1971; Gadgil & Solbrig, 1972). The central idea of *r* and *K* selection is that populations living in environments imposing high density-independent mortality (such as in a ploughed field) will be selectively favoured to allocate a greater proportion of resources to reproductive activities at the cost of their capabilities to propagate under crowded conditions. These are the *r*-strategists. Conversely, populations living in environments imposing high density-dependent regulation (such as later successional stages) will be selectively favoured to allocate a greater proportion of resources to non-reproductive activities, at the cost of their capabilities to propagate under conditions of high density-independent mortality (Fig. 11.9).

Solbrig (1971; Solbrig & Simpson, 1974, 1977) studied three populations of dandelions that grew along a gradient of disturbance. It was found that seed number per gram of plant biomass, as well as the proportion of biomass in reproductive tissues (reproductive effort), was correlated with the level of disturbance. It was shown that the proportion of energy allocated to reproduction is genetically determined. Furthermore, it was shown that dandelion plants that allocate the smallest proportion of energy to reproductive activities are the best competitors, and that this higher competitive capacity is related to a higher growth rate made possible as a result of the trade-off between investment of energy and materials in reproductive versus

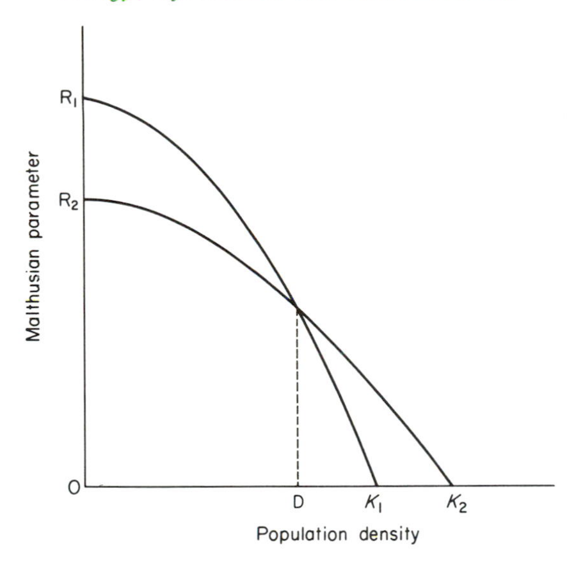

Fig. 11.9 Malthusian parameter (or population growth rate) as a function of population density for two biotypes. R = maximum growth rate of the population; K = carrying capacity; D = density at which the Malthusian parameters of the two biotypes are equal in their value (after Gadgil & Solbrig, 1972).

vegetative structures. The basic assumption of the theory of *r* and *K* selection (which is testable) is that a species will increase its fitness by enlarging its seed production as well as by increasing its lifespan. However, in most situations the plant can only allocate energy and materials to seed production at the expense of vegetative growth, and vice versa. Consequently there must be an optimal solution that maximizes overall seed production over the life of the individual. This optimal solution depends primarily on the sources of mortality. If a species grows in an environment where lifespan is restricted to a fixed period by outside forces (such as dry or cold seasons, or catastrophic events such as ploughing of fields, or floods) then any investment in vegetative growth at the expense of reproduction beyond a certain point will not increase lifespan. But if the species grows in an environment where such mortality factors do not exist, investment in vegetative growth at the expense of seed production will be favoured as long as the gain in reproductive life (as a proportion of total reproductive lifespan) is greater than the loss in seed production (as a proportion of total seed production).

Two principal criticisms have been levelled at the theory of *r* and *K* selection. The first type of criticism (Hairston, Tinkle & Wilbur, 1970; Wilbur, Tinkle &

Collins, 1974; Wilbur, 1976), maintains that the trade-off between reproductive and vegetative growth is an oversimplification, as is the distinction between density-independent and density-dependent mortality. These authors assert that r and K selection is a misnomer, that what is being selected is the best life-history to maximize r. Clearly, *r and K strategy* is a misnomer. The carrying capacity K is a population parameter and consequently individuals cannot be under direct selection pressure to increase K. Unfortunately, the term has entered the literature, including elementary textbooks, and is probably there to stay. That reproduction or vegetative growth is not the only trade-off in energy allocation is also self-evident. Size as well as number of seeds is an important consideration, as is the way seeds are dispersed, including investment in dispersal structures (awns, pappusses, hairs, etc.), seed coats, and so on. Other possible trade-offs are between investment in reproductive structures and defences against herbivores. Gadgil & Solbrig (1972) pointed this out when they maintained that r and K selection makes sense only in the context of very closely related taxa, preferably individuals within a population or populations within a species. Clearly, differences between species or between higher taxa are likely to involve many adaptive changes.

A more cogent criticism has been presented by Stearns (1976). The theory of r and K selection assumes a deterministic environment. If a fluctuating environment is assumed, with populations near equilibrium, and if fluctuations in the environment result in highly variable juvenile mortality, then populations should evolve a smaller reproductive effort and greater longevity (Schaffer, 1974; Schaffer & Rosenzweig, 1977), that is, they will show the same syndrome of characters as a K-selected taxon. Table 11.2 presents a summary of assumptions and predictions made by deterministic (r and K theory) and stochastic theories of life-history. Since most environments in which plants grow are stochastic, the criticism is valid.

For a number of years I, and collaborators, have been conducting studies on the demography of populations of several species of violets. These studies include analysis of seedling and adult mortality patterns, seed production and reproductive effort. It was shown (Solbrig, Newell & Kincaid, 1980; Newell, Solbrig & Kincaid, 1981; Cook, 1979) that juvenile mortality is high but varies greatly with environmental conditions from year to year and from site to site, while adult mortality is much more regular and predictable. Of the five species investigated all show a low reproductive effort (5% or less of biomass is invested in reproduction) and adults have long lifespans (over 10 years). The life-history of these violets fits equally well the predictions of a deterministic model of a K species and of the stochastic model for a low-reproductive-effort

Table 11.2 Assumptions and predictions made by the deterministic (r and K selection) and stochastic models of life-history evolution (after Stearns, 1977) modified.

	Assumption	Prediction
	Deterministic model	
a	Exponential population growth Repeated colonizations or fluctuations in population density due to density independent mortality (r selection)	Earlier maturity Larger reproductive effort Shorter life Poor competitor
b	Logistic population growth Population near equlibrium density Density-dependent mortality (K selection)	Later maturity Smaller reproductive effort Longer life Good, efficient competitor
	Stochastic model	
	Environment fluctuates, population near equilibrium	
a	Juvenile mortality or birth rate fluctuates, adult mortality does not	Later maturity Smaller reproductive effort
b	Adult mortality fluctuates, juvenile mortality or birth rate does not	Earlier maturity Larger reproductive effort

species. The major difference in life-history between the species studied is in the mode of reproduction, one group of species (*Viola sororia*, *V. fimbriatula*) reproducing almost exclusively by seed, while the other group (*V. blanda*, *V. pallens*, *V. lanceolata*) produces new plants through asexual means, by stolon production. Neither theory addresses itself to the exact mode of reproduction.

If asexual reproduction is not considered a form of reproduction but a form of growth as some authors maintain (Harper, 1977; but see Chapter 10), then the set of violet species that reproduces only by seed has a greater reproductive effort and should be considered an r-strategist in comparison with the stolon-producing species (Solbrig, *in press*). Since the non-stolon-producing species are more ubiquitous and are colonizers of mildly disturbed habitats, they fit the predictions of r and K selection theory better than those of stochastic models, as we have not found any significant differences in patterns of adult or seedling mortality (Solbrig, *in press*).

Theory is not yet at the stage where it can predict the life-history of an organism on the basis of independent environmental variables. As Stearns (1977) states 'neither the deterministic nor the stochastic models are empirically sufficient.' However, the existence of the theory, forces on the student of nature a set of criteria that guide him in the collection and testing of data and increase the level of rigour of the research. The value of the added rigour gained must not be overlooked.

11.5 SUMMARY AND DISCUSSION

In the introduction to this chapter, the need to develop a comprehensive, phenotypic theory of natural selection was stressed. Organisms are systems for the capture, transformation and transmission of energy, information and materials. Consequently such a theory must be based on the mechanisms of processing energy and materials. Because the theory of natural selection as presented by Darwin is an optimality theory, a comprehensive phenotypic theory should take that form. However, simple energy-maximization models are incomplete because plants also possess control functions and they also process information, mechanisms whose value cannot be measured by the energy and materials they consume. The precise physico-chemical processes of a plant are so complex and our knowledge of them so incomplete that the formulation of a precise predictive phenotypic theory of natural selection is still not possible.

However, enough progress is being made that provisional, qualitative models can now be proposed that predict specific plant characteristics. Models relating to two plant properties have been presented. First, a set of models to predict leaf size and shape of plants was introduced. The basic physics and chemistry of the processes of light energy absorption and dissipation, transpiration and photosynthesis are fairly well understood, probably better than any aspect of plant life. The better the functioning of a leaf is understood, the more complex it turns out to be, and the more it involves the total functioning of the plant and its coupling with the interfacial environment. While Schimper (1896) thought he could explain the characteristic leaf shapes of desert plants simply on the basis of the need to conserve water, we know today that leaf size and shape is dependent on many factors including radiation, temperature, water economy, CO_2 diffusion and nitrogen economy, and that the relationship between these factors is extremely complex. Some, such as the biochemistry of nitrogen, are still incompletely known.

When it comes to predicting even more complex phenomena, such as the proportion of energy allocated to reproduction, our knowledge of plant processes is so primitive that present models cannot even make a pretence of taking into consideration the underlying physico-chemistry of the functions involved. Nevertheless, the application of some econometric ideas permits the formulation of testable hypotheses.

In summary, the view that considers plants as systems for the capture, transformation and transmission of energy and materials promises to give us a better understanding of the forces that account for structures and processes of plants, for their variation within plant groups and for their distribution across environments. Clearly, we are just at the beginning, and by necessity present models are primitive and qualitative. They have already proven to be superior to the *ad hoc* explanations of the past, if for no other reason than that they make testable predictions. However, it must be remembered that when dealing with complex systems such as living organisms complete and precise descriptions of all aspects of the system are never possible, and statistical approximations must suffice. Science progresses by making increasingly more precise explanations of reality, and in the words of Einstein, attempting to make the chaotic diversity of our sense-experience correspond to a uniform system of thought.

Chapter 12

Resource Acquisition and Allocation among Animals

ERIC R. PIANKA

12.1 INTRODUCTORY CONSIDERATIONS

12.1.1 Resource budgets

Previous chapters have considered various aspects of the energetics of foraging as well as the budgeting of time, matter, and/or energy in digestive processes, predator avoidance, tissue maintenance and repair, storage, growth, and (last, but far from least) reproduction. These activities cannot be dealt with adequately in isolation since they usually make conflicting demands on an animal's finite resources and hence require trade-offs and compromises. Time spent foraging to gather materials and energy results in greater exposure to predators and thus requires increased expenditure of energy on predator escape behaviour. Similarly, energy and matter devoted to tissue repairs or growth are rendered unavailable for reproduction. Ultimately, of course, successful offspring are the only currency of natural selection so that any expenditures on non-reproductive (somatic) tissues and/or activities will enhance the animal's fitness only to the extent that they can in fact be translated into progeny at some future time. For example, investing in growth should be viewed in the context of how the resulting increase in size influences fecundity, perhaps indirectly by altering competitive ability and/or survivorship as well as directly by increasing the volume of the body cavity in which eggs and progeny develop. Constraints and interactions among the vital activities listed above are numerous and can sometimes be exceedingly subtle.

12.1.2 Units of measurement

Ideally, analyses of resource budgets would be couched in terms of relative lifetime reproductive success; however, fitness has proven to be exceedingly difficult or even impossible to measure (even if one were able to count progeny produced, each must be weighted by its own fitness!). Not all offspring are equivalent—those born late in the season may usually have more remote prospects of reaching adulthood than those produced earlier, even if they have similar genotypes. Comparable considerations apply to progeny of different sizes (for literature on this subject, see Brockelman, 1975; Pianka, 1976; Smith & Fretwell, 1974).

As pointed out in Chapter 1, energy is often used as a 'lowest common denominator' as a currency for such studies, almost by default. In vertebrates as disparate as red squirrels (Smith, 1968), cotton rats (Randolph (McClure) *et al.*, 1977), and stickleback fish (Wootton, 1977), females have been shown to be energy stressed during reproduction. Recognizing the failings and limitations of such an energetic approach, I nevertheless ultimately finish up overemphasizing energy myself. Only a few highlights in the extensive literature on animal bioenergetics can be considered here.

I shall proceed by outlining a series of specific studies, each of which illustrates a given trade-off that may not be particularly evident or discernable in the others. Since all animals face many of the same or at least very similar dilemmas, this disparate collection of seemingly unrelated cases may help in attaining an overview by providing some building blocks for the construction of a more general and complete unified framework. As you read through these various synopses, bear in mind that the individual animal must function smoothly as an integrated entity so that the entire phenotype is the true unit of natural selection.

12.2 CASE HISTORIES

Far from being static, energy budgets change both with immediate environmental conditions and with age. Indeed, altering the energy budget may often be the most effective way of coping with a dynamic environment, including unpredictable ones. During a mild winter on the British sea-coast, foraging rock pipits (*Anthus spinoletta*) spent about $6\frac{1}{2}$ h of the 9-h day feeding, but during a harsher winter, birds of the same species spent a full $8\frac{1}{4}$ h foraging (Gibb, 1956). In another study, Gibb (1960) estimated that English tits must find an insect on the average once every $2\frac{1}{2}$ s during daylight hours in order to balance their energy budgets during wintertime.

12.2.1 PEA APHIDS: AGE-SPECIFIC ENERGY BUDGETS

Even in a relatively constant environment, time and energy budgets change during ontogeny. Young (pre-reproductive) animals frequently allocate a greater fraction of the energy available to growth than do older animals, which typically show greatly slowed growth rates or even cease growing entirely in order to devote more to reproduction. One of the most detailed studies of such age-specific changes in energy budgeting is that of Randolph (McClure), Randolph & Barlow (1975) on pea aphids (*Acyrthosiphon pisum*) in the laboratory. These workers monitored daily energy flow to growth, reproduction, moulted exoskeletons, oxygen consumption (respiration) and honeydew production (Fig. 12.1) in a parthenogenetic strain of aphids. Total energy budgets of individual aphids varied with age by more than an order of

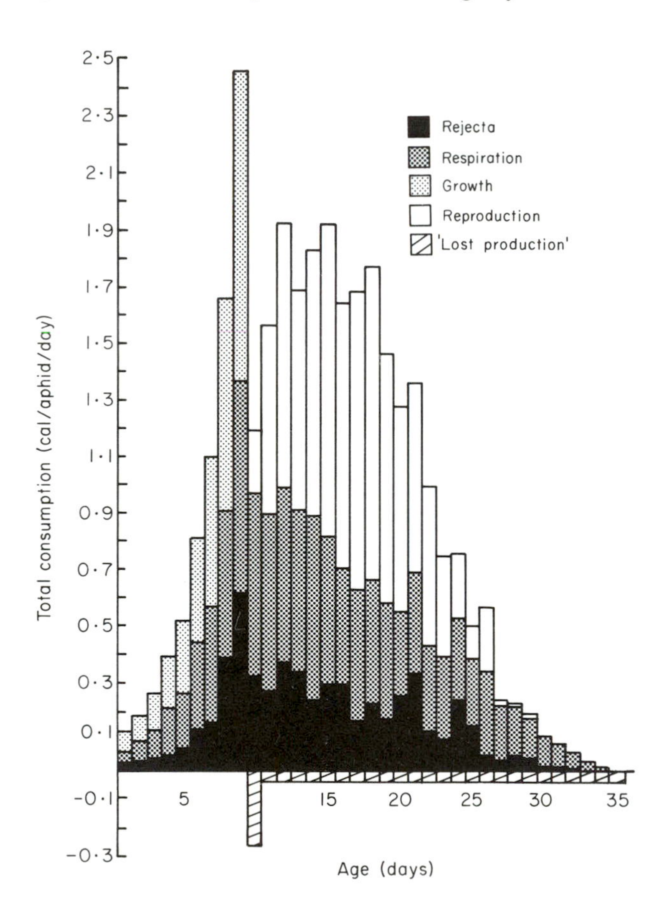

Fig. 12.1 Energy budgets of pea aphids at different ages. Total consumption is shown by the height of each bar, which is partitioned into five fractions: rejecta, respiration, growth, reproduction and 'lost production'. Note that all growth occurs prior to day 10, when reproduction starts (from Randoph (McClure) *et al.*, 1975).

magnitude: total consumption of energy increased approximately exponentially during the first 9 days of growth and then levelled off during the reproductive period and declined in old age (Fig. 12.1). The fraction of energy devoted to various activities changed greatly during the aphid's 35–39 day lifespan. All growth occurred prior to day 10, when reproduction began. Absolute amounts of energy expended on reproduction, growth, maintenance and rejecta all varied markedly with age, as did the efficiency of respiration, although efficiencies of assimilation and production did not.

As elegant as this study is, it unfortunately sheds little light on one of the central problems in the evolutionary approach to physiological ecology: *why* are resources (in this case, energy) partitioned exactly as they are?

12.2.2 Rotifers: costs of reproduction

In a similar laboratory study, age-specific survivorship and fecundity of the rotifer *Asplanchna brightwelli* were monitored by Snell & King (1977). These rotifers proved to be variable in lifespans and rates of reproduction: those that had a high fecundity tended not to live as long as those that reproduced at a lower rate (Fig. 12.2). Reproduction at any given age markedly decreased survival to subsequent age classes (Fig. 12.3). In these rotifers, reproduction is clearly deleterious to survival (and hence, future reproductive success).

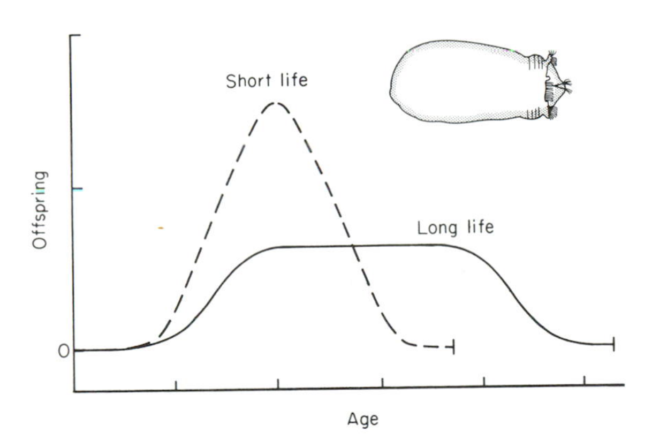

Fig. 12.2 Diagrammatic representation of the relationship between reproductive pattern and lifespan in laboratory populations of the rotifer *Asplanchna* (from Snell & King, 1977).

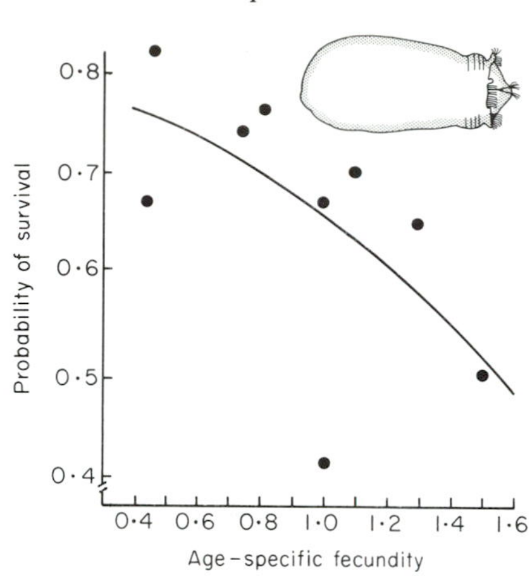

Fig. 12.3 Plot of probability of survival to the next age class versus fecundity at a given age (each data point represents ten rotifers) (from Snell & King, 1977).

12.2.3 Isopods: growth versus reproduction

Trade-offs between growth and reproduction in terrestrial isopods have been examined by Lawlor (1976). In these animals (*Armadillidium vulgare*), female fecundity increases with body size as it does in the majority of invertebrates and many other animals including most fish, amphibians and reptiles (perhaps birds and mammals are aberrant due to their determinant growth and/or parental care).

Female isopods that reproduce in the Spring do not grow as much as those that forgo this opportunity for reproduction. Further, the effect of such reduced growth on *future* reproductive success is considerably greater for small females than it is for larger ones. For example, a small 20-mg female not producing a Spring brood grows about 15 mg, hence increasing her body mass and resultant Summer fecundity by a full 75 %, whereas an equivalent growth increment of 15 mg (in lieu of Spring reproduction) would increase Summer fecundity of a large 100-mg female by only about 15 %. Hence small females have much more to gain from growth than do larger ones: such diminishing returns dictate that growth rates must decrease with size and age. Of course, growth and reproduction will usually vary inversely since costs in one constitute benefits in the other. In these isopods, reproductive females devote from 8 to 26 % more energy to growth plus reproduction than non-reproductive females expend on growth alone.

To maximize total number of progeny produced during their lifetimes, Lawlor (1976) argues that isopod females below a threshold spring weight (about 45–50 mg) should elect for spring growth rather than reproduction. Essentially, allocation of resources to growth during the springtime increases a small female's expectation of future offspring (her 'reproductive value') more than would the immediate expenditure of those same resources on a spring brood (current reproduction). By means of spring growth, these small females can bear more offspring in their single large summer brood than they could if they opted instead for reduced growth and both a spring and a summer brood (the sum of two such smaller broods is less than that of the single larger brood). However, above the size threshold, pay-offs for single-broodedness versus double-broodedness are reversed, with the low-growth two-brooded tactic producing more total offspring (Fig. 12.4). Hence, small females are single-brooded whereas larger (older) ones reproduce twice each year.

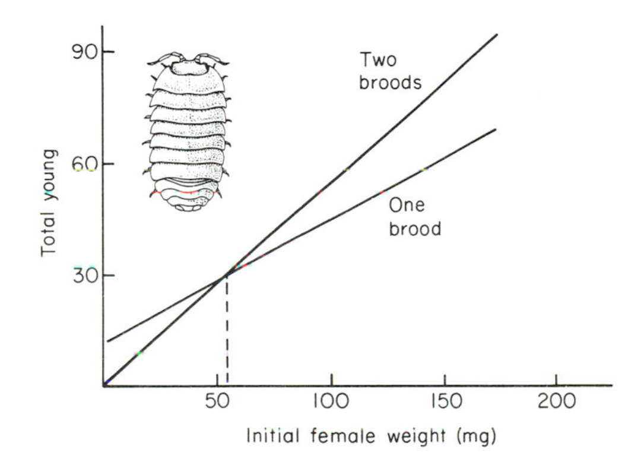

Fig. 12.4 Size–fecundity relationships for female isopods producing one and two broods. Below the threshold initial weight of about 50 mg, females leave more progeny by opting for growth and a single brood. Above this size threshold, females are double-brooded (from Lawlor, 1976).

12.2.4 Weasels: costs and profits of being long and thin

Next consider a somewhat more physiological example of resource allocation, involving thermal and energetic consequences of body shape in weasels (Brown & Lasiewski, 1972). Due to their long thin bodies and short fur, cold-stressed weasels must metabolize nearly twice as much energy to maintain body temperature than do more compactly shaped mammals of the same size and weight. Energetic costs of the weasel body form are therefore substantial and must, of course, be balanced by concomitant benefits. The major ecological advantage of the weasel body plan is that it allows these tiny predators to enter burrows of small mammals (their major prey), hence increasing access to food, facilitating foraging success and making an increased energy requirement possible. Factors involved in the evolution of the elongate body shape of weasels are summarized in Fig. 12.5. Still another ramification of the weasel body shape is sexual dimorphism in body size, enabling males and females to exploit prey of different sizes (other members of the family Mustelidae are not nearly so dimorphic in size).

12.2.5 Hummingbirds and bumblebees: time budgets and foraging energetics

Nectivorous animals are particularly well suited for certain observations on resource budgeting because time spent feeding can be readily converted directly into energetic gains. Pearson (1954) pioneered work on hummingbirds, keeping time budgets of individuals for a complete diurnal cycle of activity (Fig. 12.6). Subsequent studies have examined various aspects of the costs and profits of territoriality and foraging energetics (Wolf & Hainsworth, 1971; Wolf, Hainsworth & Stiles, 1972; Feinsinger, 1978). Analogous observations on the energetics of foraging in bumblebees (Heinrich, 1975, 1979) have underscored the intricate energetic interplay between plants and their pollinators.

12.2.6 Mammals: metabolic rates and reproductive tactics

In a recent review of mammalian energetics, McNab (1980) demonstrated that differences between species in basal metabolic rates varied in a systematic fashion with trophic habits: for a given body weight, insectivores have lower metabolic rates than frugivores or herbivores. Further, instrinsic rate of increase is correlated with rate of metabolism among mammal species. Finally, fluctuations in population size increase in amplitude with increasing rate of metabolism. McNab suggests that mammal species with larger energy budgets must have high metabolic rates. He implicates food type as the major causal factor limiting energy budgets and metabolic rates: in turn, these constrain reproductive tactics and thus determine maximal rates of increase, which dictate potential for population fluctuation. Under McNab's interpretation, interspecific variation in fecundity among mammals does not arise from

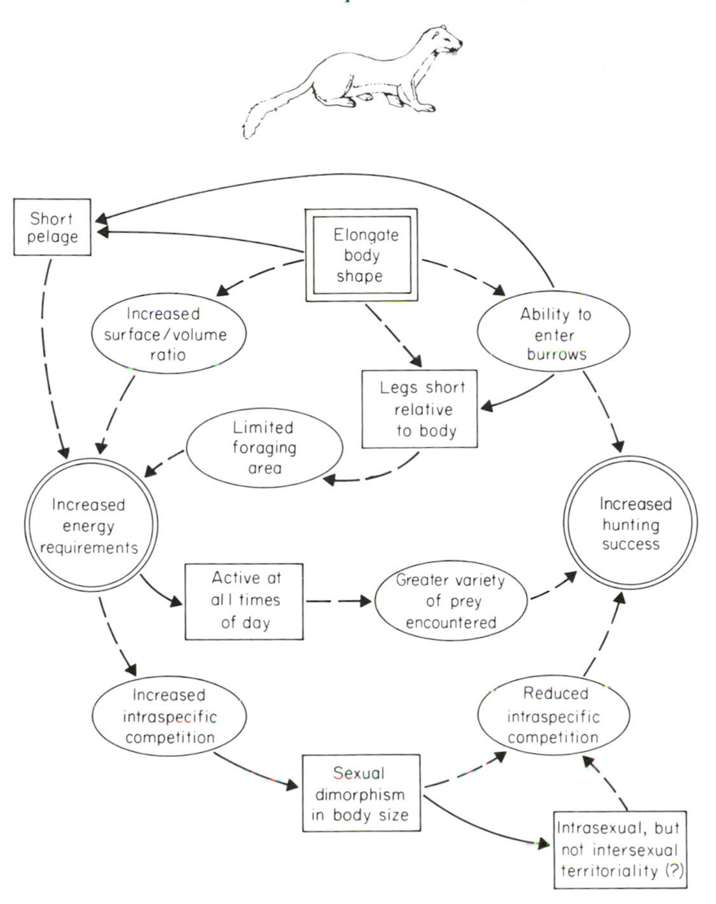

Fig. 12.5 Schematic representation of the factors involved in the evolution of elongate body shape in weasels. Primary consequences of evolving a long, thin body configuration are shown with circles; ellipses depict secondary consequences and rectangles show phenotypic characteristics affected by the evolution of the weasel body shape. Selective pressures portrayed by unbroken lines and causal sequences by dashed lines. As long as natural selection favours a more elongate shape, changes proceed as indicated by the arrows (from Brown & Lasiewski, 1972).

differences in energy allocation to reproduction but rather stems primarily from differences in diets, overall energy budgets and metabolic rates.

12.2.7 Desert lizards: consequences of foraging mode

Like many predatory animals, desert lizards separate into two natural groups:

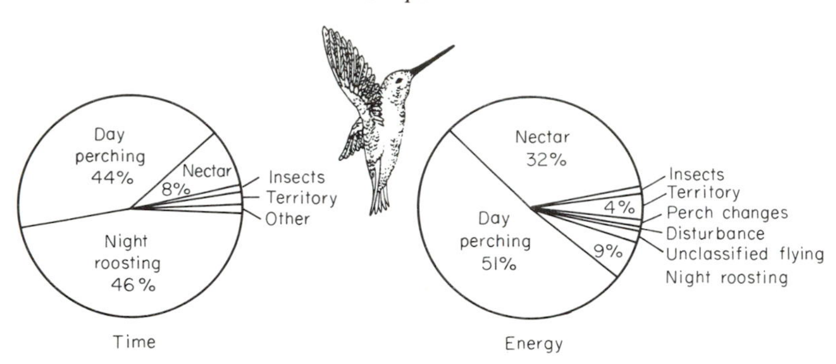

Fig. 12.6 Daily budgets of time (left) and energy (right) of a male hummingbird (from Pearson, 1954).

1 those that 'sit-and-wait' for their prey, capturing it by ambush;
2 those that hunt actively and 'forage widely' for their food (Pianka, 1966).

This dichotomy is fairly clean and few intermediates exist. In most lizard families, all members belong to either one category or the other; thus agamids and iguanids are invariably 'sit-and-wait' foragers, whereas skinks and teids tend to forage widely. However, lizards in the family Lacertidae exploit both foraging modes, even within the lacertid genus *Eremias*.

In the Kalahari semidesert of southern Africa (Pianka, Huey & Lawlor, 1979; Huey & Pianka, 1981), two species of lacertids, *Eremias lineo-ocellata* and *Meroles suborbitalis*, sit and wait for prey; several other syntopic species, including two other species of *Eremias*, forage widely for their food. Time budgets of these lacertids reflect their modes of foraging (Fig. 12.7). Foraging widely is energetically expensive and those species that engage in this mode of foraging appear to capture more food per unit time than 'sit-and-wait' species, judging from their stomach volumes. Indeed, Huey & Pianka (1980) estimate that overall energy budgets of widely-foraging species are from 1.3 to 2.1 times those of 'sit-and-wait' species. As might be expected, sedentary foragers tend to encounter and eat relatively mobile prey whereas more active widely-foraging predators consume less active prey. Compared with 'sit-and-wait' species, widely-foraging lacertid species eat more termites (sedentary, spatially and temporally unpredictable but clumped prey). One widely-foraging species, *Nucras tessellata*, specializes on scorpions (by day, these are non-mobile and exceedingly patchily distributed prey items).

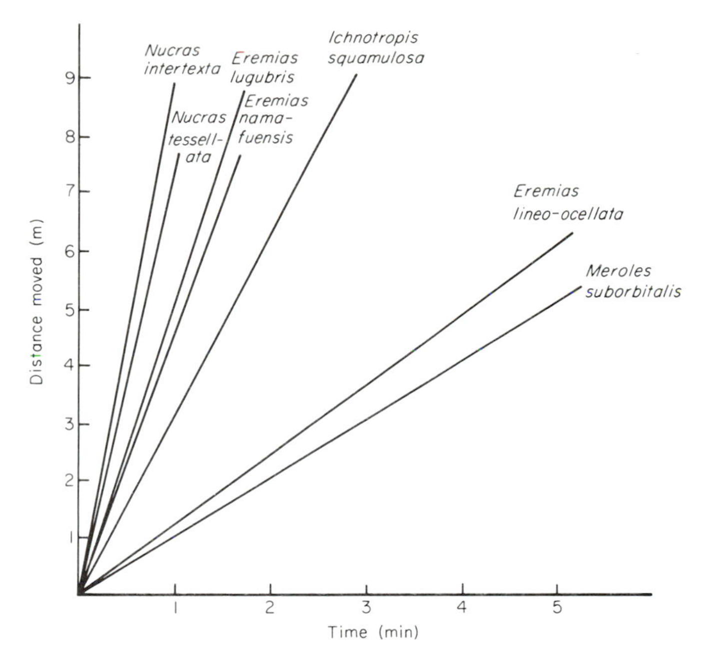

Fig. 12.7 Time budgets showing the average rates of movement of seven species of lacertid lizards in the Kalahari desert of southern Africa.

Another ramification of foraging mode in these lizards concerns exposure to their own predators. Widely-foraging species tend to be more visible and seem to suffer higher predation rates (frequencies of broken regenerated tails are high). However, widely-foraging species fall prey to lizard predator species that 'sit-and-wait', whereas 'sit-and-wait' lizard species tend to be eaten by predators that forage widely, so that 'crossovers' in foraging mode occur between trophic levels. Widely-foraging species tend to be more streamlined and to have longer tails than 'sit-and-wait' species.

Yet another spin-off of mode of foraging involves reproductive tactics. Clutch sizes of widely-foraging species tend to be smaller than those of sit-and-wait species, probably because the former simply cannot afford to weight themselves down with eggs as much as can the latter. Hence, foraging style (energy aquisition or input phenomena) constrains reproductive prospects (energy expenditure or output phenomena) in an important way. Some of the ecological correlates of foraging mode are summarized in Table 12.1. Similar patterns have been described in insectivorous birds (Eckhardt, 1979).

Table 12.1 Some general correlates of foraging mode (from Huey and Pianka, 1981).

	Sit-and-wait	Widely-foraging
Prey type	Eat active prey	Eat sedentary and unpredictable (but clumped or large) prey
Volume prey captured/day	Low	Generally high but low in certain species
Daily metabolic expense	Low	High
Types of predators	Vulnerable primarily to widely-foraging predators	Vulnerable to both sit-and-wait and to widely-foraging predators
Rate of encounters with predators	Probably low	Probably high
Morphology	Stocky (short tails)	Streamlined (generally long tails)
Probable physiological correlates	Limited endurance (anaerobic)	High endurance capacity (aerobic)
Relative clutch mass	High	Low
Sensory mode	Visual primarily	Visual or olfactory
Learning ability	Limited	Enhanced learning and memory, larger brains

12.2.8 Horned lizards: adaptive suites

In discussing the intricate physiological and ecological factors involved in water balance in desert-adapted kangaroo rats (genus *Dipodomys*), Bartholomew (1972) coined the term 'adaptive suite' to describe a constellation of integrated co-adapted phenotypic traits (several examples of such adaptive suites have already been considered above, e.g. see sections 12.2.4 and 12.2.7).

To illustrate, consider another, rather different, example. Desert horned lizards, *Phrynosoma platyrhinos*, are ant specialists, eating little else. Various features of the anatomy, behaviour, diet, temporal pattern of activity, thermoregulation and reproductive tactics can be profitably interrelated and interpreted to provide an integrated view of the ecology of this interesting animal (Pianka & Parker, 1975b; Fig. 12.8). Ants are small and contain much undigestible chitin, so that large numbers must be consumed. Hence an ant specialist must possess a large stomach for its body size (this horned lizard's stomach averages about 13% of the animal's overall mass, a substantially larger fraction than stomachs of other lizard species). This large gut requires a tank-like body form, reducing speed and decreasing the lizard's ability to escape from predators by movement. As a result, natural selection has favored a spiny body form and cryptic behaviour rather than a sleek body and rapid

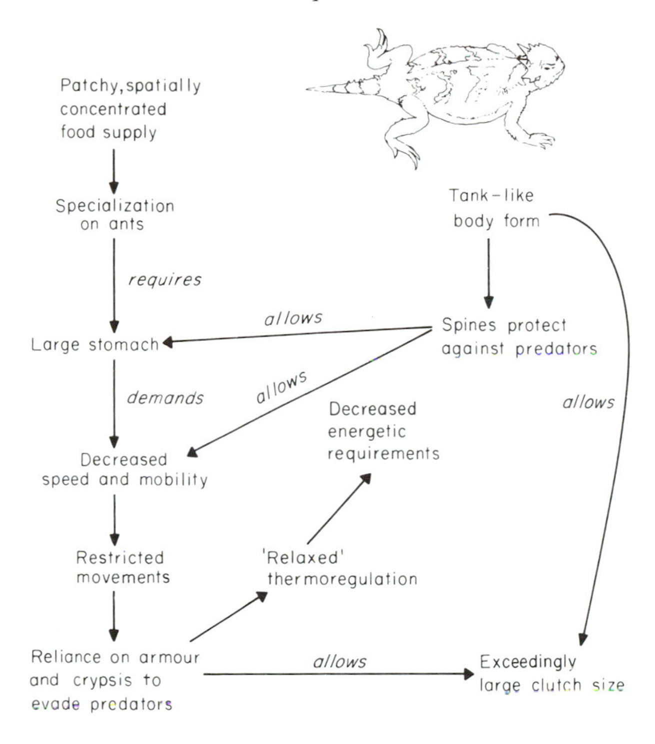

Fig. 12.8 Summary of the major factors influencing the ecology of the desert horned lizard, *Phrynosoma platyrhinos*. The complete constellation of co-adapted phenotypic traits represents this lizard's 'adaptive suite'.

movement to cover as in the majority of lizards. Long periods of exposure while foraging in the open presumably increase risks of predation. A reluctance to move, even when actually threatened by a potential predator, could well be advantageous: movement might attract the predator's attention and negate the advantage of concealing coloration and contour. Such decreased movement doubtless contributes to the observed high variability in body temperature of *Phrynosoma* (significantly greater than that of all other species of sympatric lizards). Wide fluctuations in horned lizard body temperatures under natural conditions presumably reflect both their long activity period and perhaps their reduced movements into or out of the sun and shade (most horned lizards are in the open sun when first encountered). A consequence is that more time is made available for activities such as feeding (foraging ant eaters must spend considerable time feeding). Food specialization on ants is economically feasible only because those insects usually occur in a clumped spatial distribution and hence constitute a concentrated food supply. To make use of this patchy and spatially concentrated, but at the same time not overly nutritious, food supply, *P. platyrhinos* has evolved a unique constellation of adaptations that include its large stomach, spiny body form, an expanded period of activity and 'relaxed' thermoregulation (eurythermy) (Fig. 12.8). The high reproductive investment of adult horned lizards is doubtlessly also a simple and direct consequence of their robust body form. Lizards that must be able to move rapidly to escape predators would hardly be expected to weigh themselves down with eggs to the same extent as animals like horned lizards that rely almost entirely upon spines and camouflage to avoid their enemies. The *Phrynosoma* adaptive suite is depicted in Fig. 12.8.

12.3 CONCLUSION AND PROSPECT

Understanding an animal's tactics of resource acquisition and allocation is clearly a massive challenge. Numerous dimensions demand simultaneous consideration. Input–output phenomena (foraging versus reproduction) are complexly intertwined and may severely constrain one another. Foods eaten and foraging mode may place fundamental limitations on metabolic rates as well as on overall resource budgets. This complexity is further confounded by temporal variation (including ontogenetic changes) in survivorship probabilities and conditions for acquiring resources (which are by no means independent themselves!). All the above factors interact in important, sometimes subtle, ways, both among themselves and with reproductive tactics. The latter in turn affect growth, foraging success, survivorship and prospects for future reproduction. Clearly it is premature to attempt to make generalizations about resource budgeting among animals. However, in lieu of such a synthesis, some useful directions for future research can perhaps be anticipated.

A fitting way to conclude this chapter would have been to compare the budgeting of energy (and/or other resource dimensions) among a variety of different animal species. Elements for such a summary are outlined in Table 12.2. Note that even two different studies on the same animal, the flour beetle *Tribolium castaneum*, gave markedly different estimates. Unfortunately, entries in this preliminary table are not strictly comparable for a number of reasons:

Table 12.2 Estimates of the fractions of total assimilated energy budgets allocated to three major activities for various animal species, with source references.

Animal species	Maintenance	Growth	Reproduction	Source
Pond snail (*Lymnaea stagnalis*)	64–67	22	11–14	Scheerboom (1978)
Pea aphid (*Acyrthosiphon pisum*)	42.0	23.7	34.3	Randolph (McClure) *et al.* (1975)
Rice weevil (*Sitophilus oryzae*)	73.9	11.8	14.3	Singh *et al.* (1976)
Granary weevil (*Sitophilus granarius*)	63.0	13.9	23.1	Campbell *et al.* (1976)
Flour beetle (*Tribolium castaneum*)	55.8	4.2	40.1	Klekowski *et al.* (1967)
Flour beetle (*Tribolium castaneum*)	88.3	11.7 (growth + reproduction)		Prus & Prus (1977)
Dobson fly (*Cordydalus cornutus*)	29.9	53.1	17.0	Brown & Fitzpatrick (1978)
Gila topminnow (*Poeciliopsis occidentalis*)	93.8 (89.9–95.4)	4.3 (3.9–7.0)	1.9 (0.7–4.6)	Constantz (1979)
Mountain salamander (*Desmognathus ochrophaeus*)	43.1	8.6	48.3	Fitzpatrick (1973)
Sagebrush swift (*Sceloporus graciosus*)	75.4	1.2	23.4	Tinkle and Hadley (1975)
Yarrow's spiny lizard (*Sceloporus jarrovi*)	76.8	11.3	11.9	Tinkle and Hadley (1975)
Side-blotched lizard (*Uta stansburiana*)	78.5	2.2	19.3	Tinkle and Hadley (1975)
Cotton rat (*Sigmodon hispidus*)	70.6–84.0	5.4–6.5	9.5–24.0	Randolph (McClure) *et al.* (pers. comm., 1977, 1980)

1 some studies were performed under laboratory conditions, with food supplied more or less *ad libitum*, whereas others were done in more natural food-limited situations;
2 some energy budgets are individual, others populational;
3 time-scales of different studies cannot be easily standardized either, since some are expressed as lifetime budgets whereas others are seasonal or annual.

Among these dozen or more studies, the fraction of energy allocated to reproduction varies by more than an order of magnitude (from about 2 to 48 %), as does that expended on growth (from 1.2 to 53 %). Expenditure on maintenance (respiration) is less variable, ranging only from about 30 to 94 %.

More informative comparisons could presumably be made if standardized techniques for studies of resource budgeting were adopted.

In addition to such comparative descriptive efforts, further work is needed on precisely why resources are allocated as they are. For example, studies like those on pea aphids and rotifers (sections 12.2.1 and 12.2.2) could lend themselves to fruitful experimental analyses. In a relatively constant laboratory environment, animals could be experimentally induced to devote *more* (and *less* in a companion experimental group) than the supposed 'optimal' amount (this would constitute the 'control') to various activities such as growth, maintenance and reproduction. (Some ingenuity and luck might be required to 'trick' many animals into adopting such suboptimal tactics!) Of course, experimental groups would be expected to suffer reduced reproductive success (fitness) compared to the control group. Yet another provocative line of investigation would be to monitor the performance of such experimentally treated animals under varying environmental conditions, as with altered levels of resource availability and/or intensity of predation, etc. These sorts of manipulations would certainly begin to elucidate ultimate factors underlying the causality of resource budgeting among animals. Indeed, such studies could easily become minor classics!

Chapter 13

The Economics of Social Organization

L. M. GOSLING AND M. PETRIE

13.1 INTRODUCTION

When starting to write this chapter, we asked the question: are social systems optimal for the environments in which they are found? The question soon revealed profound logical problems that stemmed from a fundamental incompatability between the level of organization that is assumed in the question and the level at which selection operates. This led us to redefine 'social organization' in a way that placed greater weight on the behaviour of the individual. In this way we hoped to provide a context for a meaningful analysis of the interaction between the social animal and its environment.

We have confined the chapter to the vertebrates partly because this reflects our own interests and partly because behaviour at the individual level has been considered in greatest detail within this class of animal.

13.2 SOCIAL ORGANIZATION AND THE INDIVIDUAL SOCIAL ANIMAL

Davies and Krebs (1978) define social organization by referring to 'the following characteristics of a species: breeding and feeding group size (e.g. solitary, flocking); spatial organization (e.g. territorial, overlapping ranges); and mating system (e.g. monogamy, polygyny).' Wilson (1975) reviews similar attempts to list properties of social systems and compiles his own 'ten qualities of sociality'. These include group size, cohesiveness, amount and pattern of connectedness, permeability, compartmentalization and others.

The problem associated with such definitions is that they frequently lead to questions about the evolution, for example, of particular group sizes. But such questions cannot be answered in a functional sense because they do not refer to the level at which natural selection operates. As Dawkins (1976) has remarked it is possible to discern group selectionist assumptions behind these treatments of social organization. In fact selection operates on individuals, but, since individuals have part of their genome in common with others, it is necessary to consider their inclusive fitness, as defined by Hamilton (1964).

Any definition of the behaviour of a social group, if it is to be useful in answering functional questions, must therefore be entirely based on the behaviour of the individual social animals that compose it. In other words the social system, if the concept is to be used at all, must be defined as the sum of the social behaviours of the individual animals, or, more correctly, as the sum of the individually determined behaviours. Features such as group size and permeability can then be clearly recognized as *consequences* of this summation and as features on which selection does not operate directly.

This is not to say that the analysis of pattern employing high levels of organization (the social qualities) cannot provide valuable insight. For example the comparative studies by Crook (1964, 1965, 1970), Clutton-Brock. & Harvey (1977) and Jarman (1974) on birds, primates and ungulates have been particularly useful in indicating the major environmental factors that could affect an individual's social behaviour. However Davies and Krebs (1978) have pointed out the difficulties of disentangling cause and effect in such analysis.

A related problem is that the 'qualities of sociality' vary in the ease and validity with which they can be transposed to answer functional questions. For example the frequency distribution of nearest-neighbour distance in a population has obvious relevance to the individual level of analysis, but properties such as 'permeability' and 'cohesiveness' (Wilson, 1975) are difficult to relate to the individual. Between these extremes of usefulness are a number of qualities that pose traps for the unwary. An example is 'group size' which may sometimes be of direct importance to the individual, for example in small primate groups with long-term inter-individual bonds, and sometimes be utterly misleading as a description of the environment of the individual. For example, a wildebeest (*Connochaetes taurinus*) in a group may be affected by such specific group properties as the number of animals that have passed over the patch of sward that it is feeding on, or by the fact that it is surrounded by animals, in its immediate vicinity, that might provide cover in the event of a predation attempt. But it is irrelevant, or of only the most obscure relevance, that the individual is in a group of two thousand animals.

We can now consider the behaviours that an individual must perform during its lifetime, adopting as a premise that it will attempt to maximize its inclusive fitness. These behaviours are directed towards one main goal: the animal must survive to achieve high reproductive success. Two main groups of behaviour can thus be distinguished.

One group falls under the general heading of survival, maintenance and achieving good physical condition:

1 avoiding injury and disease, for example, by grooming,
2 avoiding predation,
3 securing a food supply;
4 minimizing energy loss, for example, by resting.

The second group involves the need to reproduce:

1 finding and competing for a mate;
2 producing offspring of optimal number sex and quality;
3 rearing offspring.

Some of the activities listed are by definition social but all, under some circumstances, can be performed in a social context. In this chapter we consider the effect of energetic constraints on some of these behaviours by the individual social animal.

Space precludes an exhaustive treatment but we shall attempt to consider most of the important reproductive behaviours listed and we shall also consider anti-predator behaviour and securing a food supply.

13.3 THE ECONOMIC APPROACH

From what has gone before, and in keeping with the aims of this book, we can pose the central question: how does social behaviour affect the ergonomic efficiency of the individual and how far is ergonomic efficiency a corollary of inclusive fitness?

These questions can be answered using an economic approach which consists, most importantly, of considering what an animal can afford to do under particular circumstances in order to maximize a defined currency. This approach is derived from optimality theory which was first used in ecology by MacArthur & Pianka (1966). Our present objective is to consider the economics of resource utilization. An appropriate currency is thus energy and the appropriate maximizing assumption is ergonomic efficiency.

As reflected in the question above, it is often assumed that ergonomic efficiency is a corollary of inclusive fitness. This assumption has been the subject of extensive discussion (Lewontin, 1978; Maynard Smith, 1978a). Here we adopt the conventional position that this relationship is imperfect and that there are important determinants of inclusive fitness that are not affected by energetic considerations. Important examples are found in some aspects of anti-predator behaviour.

A limitation of a simple optimality approach is that there is often no single optimum strategy. Under some circumstances animals can adopt one of two, or more, alternative strategies which are equally successful. Mixed strategies have recently been investigated using the concept of evolutionary stability (Maynard Smith & Parker, 1976). When no one strategy can out-compete another and invade the population then the alternatives are evolutionary stable strategies, or ESSs. This approach has been most successful in considering the evolution of alternative mating strategies (Rubenstein, 1980).

13.4 AVOIDING PREDATION

In this section we ask to what extent social anti-predator behaviours are influenced by energetic considerations. We do not intend to review exhaustively the social anti-predator mechanisms that vertebrates employ but instead to point out where energetic considerations may be involved.

We would summarize the main social behaviours, currently thought to be important (e.g. Bertram, 1978) as follows:

1 using other animals as cover (Hamilton, 1971);
2 using other animals as an extended information system to detect predators (Pulliam, 1973);
3 group defence (including mobbing).

Bertram (1978) considered only the anti-predator advantages of group living. However, spacing is probably also important as a device to reduce the chance of predation and so we can add:

4 avoiding other animals, i.e. spacing out.

What are the energetic costs and benefits of these behaviours?

13.4.1 JOINING A GROUP, MAINLY IN OPEN HABITATS

First, we shall consider the anti-predator significance, to the individual, of joining a group. It has been predicted on theoretical grounds, and empirically confirmed, that an animal is less likely to be killed by a predator when in a group than when alone (e.g. Hamilton, 1971; Kenward, 1978a). While this consequence of group living must be a primary consideration it should also be emphasized that there is no energetic equivalent for *the chance of predation* which could be used in an economic model. This assumes particular importance when the selection pressures that operate on economic costs are compared with those concerned with the chance of death by predation; generally it must be expected that the latter will have a greater effect, a prediction that has been described in a different context as the 'life-dinner' principle (Dawkins & Krebs, 1979).

The habitat in which groups are found has a profound influence on their significance in anti-predator behaviour. We shall distinguish the two broad categories of 'open' and 'closed' habitats and deal with each in turn.

The hypotheses associated with the first three social behaviours listed above would all predict that the chance of predation will decline when an individual belongs to increasingly large groups. The relationship should take the form illustrated in Fig. 13.1 since there will be a relatively large chance of, for example, detecting an approaching predator when one animal joins another than when one joins a large group (Pulliam, 1973). Kenward's (1978) observations of goshawk (*Accipiter gentilis*) predation on woodpigeons (*Columba palumbus*) provide empirical support for this prediction. There might also be a tendency for the chance of predation to increase when group size becomes very large because of interference between potential prey. The curve would then take the form (b) in Fig. 13.1. Crisler's (1956) observations of wolf (*Canis lupus*) predation on caribou (*Rangifer arcticus*) calves in very large groups and those by Schaller (1972) of lion predation on African ungulates are good examples of this effect. However, it should not be assumed that an increased chance of killing can be simply translated to a disadvantage for the individual prey in the group. Although the predator is more likely to obtain a meal, the mathematical chance of the meal consisting of any particular individual will be reduced if the group size is very large.

But as group size increases there may be associated costs. A simple model involving increasing numbers and a finite food supply would predict that the amount of food per individual would decline in the fashion indicated by the depletion curve in Fig. 13.1. A relationship of this type could exist when redshank (*Tringa totanus*) hunt invertebrates by sight: under these circumstances capture rate declines with nearest-neighbour distance (Goss-Custard, 1976). However, a simple relationship between numbers in a group and food depletion is probably rare. Often group members take compensatory action, an example being redshank which enlarge their individual distance when visually selecting prey.

Other costs of belonging to a group are those of interference, or

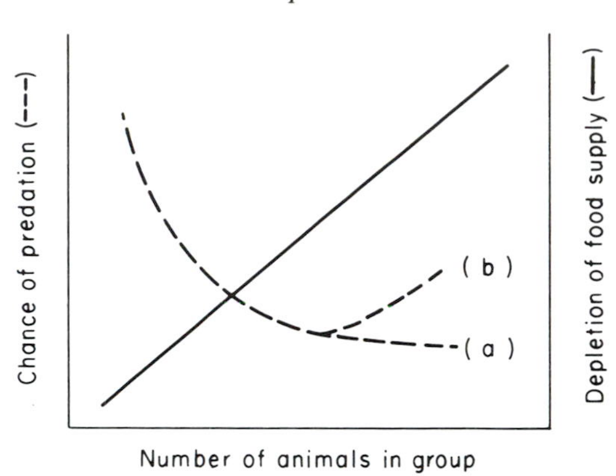

Fig. 13.1 Hypothetical interaction between the chance of predation and depletion of the food supply in groups of increasing size. The curve (b) assumes an interference effect between individuals in large groups.

competition, which directly or indirectly lead to a reduction in the time available for feeding in addition to involving energy expenditure. The food intake of starlings (*Sturnus vulgaris*) is reduced as their number at feeding troughs increases beyond a critical point, due to increased interference and aggression between individuals (Feare & Inglis, 1979) (Fig. 13.2.). However, the feeding rate of individuals in a small group is higher than that of an isolated animal. As might be predicted this is due mainly to a reduced need for anti-predator scanning. Similar conclusions have been reached by Powell (1974), based on experiments in which he flew hawk models over starling groups of various size, and Bertram (1980), for ostrich (*Struthio camelus*).

The time lost by individuals will vary according to their dominance status. Pulliam (1976), in a mathematical model, analysed the varying conditions under which dominant and subordinate birds should form groups. His model suggests that at high food concentrations a dominant bird would reach a maximum feeding rate and then use excess time to chase subordinates. When this harrassment reduced the feeding rate of a subordinate to below that of a solitary bird it would leave the group. In general, dominants should feed more efficiently than subordinates, and empirical data on yellow-eyed juncos (*Junco phaeonotus*) support this prediction (Caraco, 1979).

All animals occasionally make mistakes in predator identification (e.g. finches: Newton, 1972) and the number of false alarms that result might increase with the number of animals in a group. However, this prediction has not been tested.

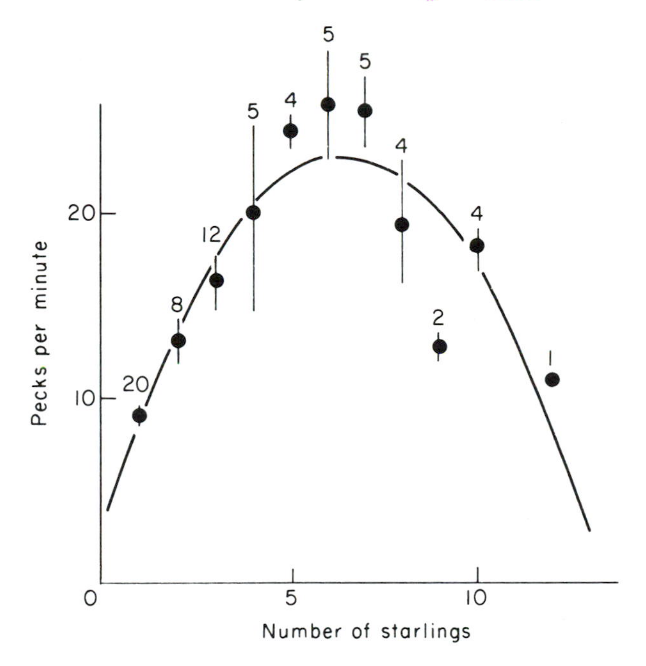

Fig. 13.2 The relationship between feeding rate and the number of starlings (*Sturnus vulgaris*) feeding at a cattle trough. Vertical bars = one standard error; numbers of observations are shown above each point (from Feare & Inglis, 1979).

If group size is limited by the costs of competition between individuals for food then we can predict that very large groups should form if the feeding constraint is removed. Under such circumstances the anti-predator advantages of joining a group should outweigh the effects of interference. A case that lends substance to this point is the formation of very large groups of wildebeest, often numbering tens of thousands, in the Serengeti region of Tanzania. In this instance the food constraint is 'removed' by the rapid production of grasses in the plains habitat and by the migratory habit of the population (Grzimek & Grzimek, 1960; Watson, 1967; Pennycuik, 1975). Moorhens (*Gallinula chloropus*) in England forage in groups of 5–30 during the day in winter, but at night, when there is no feeding, they collect into communal roosts which contain many hundreds of birds (Petrie, unpublished observations). The possibility of feeding interference becomes even less likely in multi-species groups where individuals that feed on subtly different food resources obtain an anti-predator advantage by associating with individuals of other species (Morse, 1970; Gosling, 1980b).

However, even in cases where competition for food is reduced or absent there are energetic costs associated with joining and remaining with the group. Thus the cost to wildebeest in maintaining both high-quality food and group membership is the energy expended during migration. Similarly, moorhens walk large distances each evening in order to join communal roosts.

In addition to obtaining anti-predator advantages by joining spatially discrete groups, animals can also obtain protection by synchronizing their behaviour in time. A well-known example is that of synchronized breeding. Wildebeest produce 80% of their calves in a 3-week period, and calves born outside the main peak are at greatest risk from hyaena (*Crocuta crocuta*) predation (Estes, 1966, 1976; Kruuk, 1972). But such synchronization must have costs and, in the case of wildebeest breeding, the simultaneous increase in food requirements during lactation might lead to a temporary depletion of the food supply.

13.4.2 Anti-predator behaviour in 'closed' habitats

Many of the issues discussed above are significant only in open habitats. In closed habitats, for example scrubland or forest, the anti-predator value of behaviours such as joining groups can be reversed. It is difficult for prey to detect approaching predators, as they can do more readily in open habitats, and so crypsis assumes greater importance. Predictably, groups in closed habitats, for example those of forest antelopes, are usually very small (Jarman, 1974; Estes, 1974). However, models of this strategy have generally employed unrealistic assumptions about the behaviour of prey and predator: the implication of Vine's (1971) model is that group living would be particularly advantageous in a closed habitat because a scanning predator would fail to detect a group as readily as scattered individuals, while Treisman (1975) found it necessary to invoke an inability of the prey to escape following detection to arrive at a prediction where crypsis improves on group living as an anti-predator strategy. These models underestimate the ability of the predator to employ auditory and olfactory information in locating prey and behaviour, such as the regular use of game trails and water-holes by both predator and prey. Taking these factors into account it is intuitively obvious that large groups will be easily detected in closed habitats, hence their rarity. This general conclusion has been documented for a number of vertebrate groups including ungulates (Geist, 1974; Jarman, 1974; Estes, 1974) and primates (Crook & Gartlan, 1966; Clutton-Brock & Harvey, 1977).

As shown experimentally by Tinbergen, Impekoven & Franck (1967) spacing out is an important defence against predators by cryptic prey. Predators usually search more intensively after finding one concealed prey item and so any nearby are more likely to be found (Croze, 1970). But spacing out generally involves a cost. The energy spent in agonistic and demarcation behaviour would generally increase in relation to the area defended. We do not, of course, maintain that the primary function of spacing out is anti-predator behaviour; as discussed below, spacing is frequently linked with the defence of food or mates. But in some situations spacing has clear anti-predator significance. This is particularly obvious when the ability of animals to escape predators is temporarily impaired. An example is the isolation of female antelopes to dense cover when they are about to give birth (Fig. 13.3)

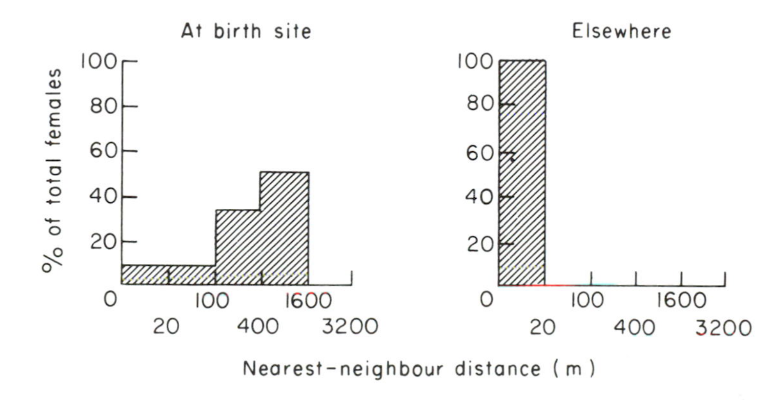

Fig. 13.3 The nearest-neighbour distances of female Coke's hartebeest (*Alcelaphus buselaphus*) when giving birth, compared with those of other females (from Gosling, 1969).

and the subsequent 'lying-out' behaviour of young calves (Gosling, 1969; Lent, 1974).

The differences between the anti-predator strategies of animals in open and closed habitats can be summarized in the model shown in Fig. 13.4. The model employs the universal currency of the contribution to inclusive fitness and can thus yield optima. In practice it is very difficult to obtain such data but there is probably no other way to avoid the problem of incompatible currencies in an economic approach.

13.4.3 Condition and the chance of predation

Intuitively it seems obvious that an animal's condition (nutrient reserves, muscle tone, etc.) will influence the chance that it will be killed by a predator

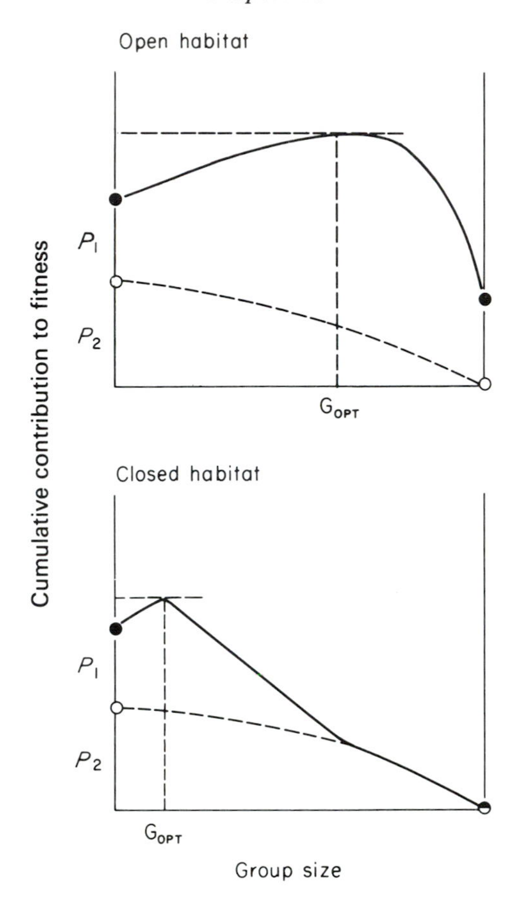

Fig. 13.4 Two phenotypic measures of fitness (P_1 = probability of survival in the face of predation; P_2 = resources gained per unit feeding time) vary in a different manner according to the size of group an animal joins. In theory these measures could be expressed in the same currency (contribution to inclusive fitness) and their contributions added together. The form of the curves allows predictions of the way optimal group size will vary in open and closed habitats.

(see Chapter 8). Even at close quarters the ability of an animal to accelerate and change direction should depend, to some extent, on condition. When long chases are involved, however, this factor seems almost certain to be important. Wild dog (*Lycaon pictus*) chase their chosen prey, often gazelle (*Gazella thompsoni* and *Gazella granti*) or impala (*Aepyceros melampus*) (Kuhme, 1965; Estes & Goddard, 1967; Pienaar, 1969; Schaller, 1972), at speeds of about 55 km/h for between 0.6 and 3.4 km. Towards the end of such chases the prey are clearly exhausted and presumably could run faster or further if they had larger nutrient reserves. However, data on this important point are rare.

The most critical data available are those of Kenward who compared the condition (dry weight of the *pectoralis minor* muscle) of woodpigeons caught by goshawk with a sample that were shot. In all cases the birds killed by the raptor were in poorer condition. Schaller (1972) reached similar conclusions using the appearance of marrow fat in the wildebeest prey of lion (*Panthera leo*) and hyaena compared with that of a sample that died in accidental drowning. Those killed by lion were in slightly worse condition than those that drowned, and the sample killed by hyaena were substantially worse. Perhaps this reflected the hunting techniques of the two predators: hyaena usually run down their prey (Kruuk, 1972) while lion rely mainly on stalking and ambush (Schaller, 1972). Good condition would presumably be important for an animal attempting to flee from hyaena while escape from a lion in an ambush might depend on chance.

Some observations indicate that predators actively select individuals that are in poor condition. Kruuk (1972) noticed that hyaena sometimes ran into a group of animals, then stopped and watched as they scattered. He speculated that this behaviour allowed the hyaena to select physically inferior individuals. Similarly, other running predators such as African wild dog have ample opportunity to select on the basis of condition as they run into fleeing groups of gazelle; wild dog often 'chased prey a few metres and then changed their mind' (Schaller, 1972). From the predators viewpoint such selection of animals in poor condition would clearly minimize energy expenditure on the chase and might also reduce the chance of injury at the kill.

The prey, on the other hand, must attempt to avoid selection by the predator. This can be done in three ways. First, and most obviously, it can maintain good condition. Second, it can reduce the chance of being the animal in poorest condition by joining a group; as it enters groups of increasing size, the probability of joining individuals that are in poorer condition will increase. Third, it can actively demonstrate to the predator that it is in good condition. Zahavi (in Dawkins, 1976) has suggested that this is the function of stotting (Fig. 13.5.), a bounding gait (Estes & Goddard, 1967; Walther, 1969) shown by many antelopes during the early stages of a predation attempt. The potential prey demonstrates that, relative to the other prey, it will be hard to catch. Smythe (1970) has also suggested that stotting and the conspicuous rump patches of many cursorial mammals may elicit pursuit and thus shorten an interaction that might be protracted and energetically expensive.

Fig. 13.5 A stotting Thomson's gazelle (*Gazella thomsoni*) (from Walther, 1969). The height of the bounds might be condition-dependent and, if so, should influence whether or not a predator selects the animal for serious pursuit.

13.5 FEEDING BEHAVIOUR

How do energetic considerations affect the way in which social animals obtain food?

First, what are the social behaviours? These can be grouped under three main headings:

1 defending a food supply;
2 hunting in a group;
3 obtaining information about food from group members.

We will not deal with kleptoparasitism in birds which has been reviewed by Brockman and Barnard (1979) or with the possibility that feeding in groups optimizes the efficient use of food patches (Cody, 1971).

13.5.1 DEFENDING A FOOD SUPPLY

Davies (1978) has reviewed the economics of area defence. He summarized the idea of economic defensibility in the graphical model shown in Fig. 13.6. Many recent attempts to investigate the food value of territories have concentrated on problems posed by the model in Fig. 13.6, that is, when

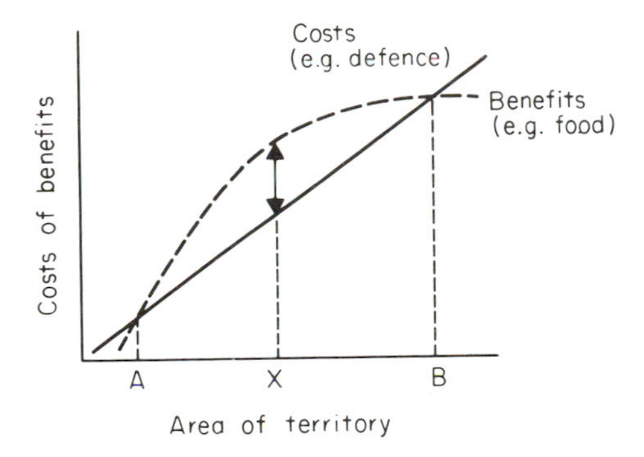

Fig. 13.6 The economics of territorial behaviour. Costs continue to rise as the defended area becomes larger but benefits level off when the resident's requirements are satisfied. Territories should only be defended between sizes A and B and maximum net benefit (benefit minus cost) is obtained at territory size X (from Davies, 1978).

should an animal defend a territory at all, and, assuming that a territory exists, what is its optimum size or shape? Further, what behavioural strategies do territorial animals use to maximize net benefit? Davies (1976) describes a case relevant to the first question where the abundance of wagtail food dictates the amount of time that they spend in winter territories: when feeding conditions are good the wagtails stay in their territories all day and when they are poor they join flocks. However, even when feeding conditions are worse than in the flock area they spend a little time defending territories and may thereby retain territorial rights for the future. Hartebeest (*Alcelaphus buselaphus*) males, similarly, spend time away from their territories at the time of poorest food quality and the frequency of such absences is highest in the territories with the poorest food supply (see Table 13.1, section 13.6.2, and Gosling 1974, 1975).

Hartebeest territories also provide a case where males apparently maximize the diversity of sward types within the defended area. Many territories are oblong in shape and established at right angles to grassland ecotones; although averaging only 0.31 km^2 they sometimes contain five or more major grassland communities. At high population density territories were sometimes divided between an invading male and the previous resident. Such 'splits' always occurred at right angles to the grassland ecotones (Fig. 13.7) and each male thus maintained maximum grassland diversity. The significance of this behaviour appears to be in maintaining an all year round food supply (Gosling, 1974). It is difficult to distinguish the role of such areas as a personal food supply and that of an attraction for mates (females in the hartebeest case,

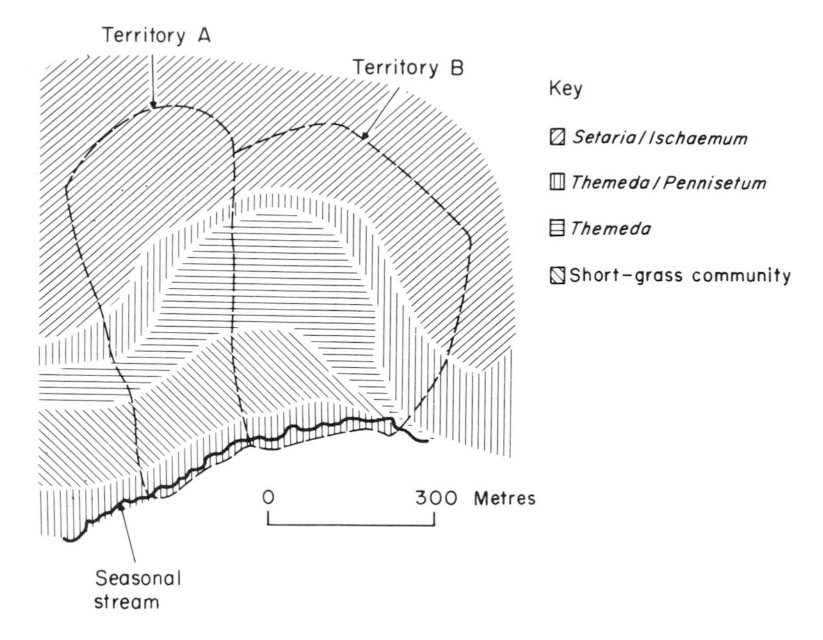

Fig. 13.7 Hartebeest (*Alcelaphus buselaphus*) territories are sometimes divided between an invading male (occupying territory B in this case) and a male that previously occupied the area A + B. In such cases the split is always at right angles to the parallel grass communities; grassland diversity is thus maximized by each male.

see section 13.5.2). In fact, territories with complex functions are in the great majority and their analysis is an outstanding problem in behavioural ecology.

The study of nectar-feeding birds has provided a number of important insights into the economics of defending a food supply. Golden-winged sunbirds (*Nectarinia reichenowi*) defend territories which vary in size, but which each contain about the same number of nectar-producing flowers (Gill & Wolf, 1975). The defended flowers have a higher nectar content than undefended flowers because their defence allows nectar renewal. In an optimization analysis of the data collected by Gill & Wolf, Pyke (1979) considered the following hypotheses about territorial behaviour.

1 Territorial behaviour maximizes net daily energy gain;
2 time spent sitting or resting is maximized;
3 daily energetic cost is minimized;
4 the ratio of gross daily energy gain to daily energy gain must be positive and the total time available each day for feeding, sitting and territory defence is fixed.

Pyke modelled 'optimal' territory size and a time budget for each hypothesis and, comparing his results with the data collected by Gill & Wolf, found best agreement with the hypothesis that daily energy cost is minimized.

13.5.2 Hunting in a group

The efficiency of prey capture (captures per chase) increases with the number of lions hunting (Schaller, 1972). Large groups also make more multiple kills and can cope with larger prey. However, the obvious cost of hunting in a group is that the amount of food available for each individual declines (Caraco & Wolf, 1975). Assuming that an individual should try to maximize its food intake these relationships can be used to calculate the optimum number of animals in a hunting group (Fig. 13.8). When capture efficiency

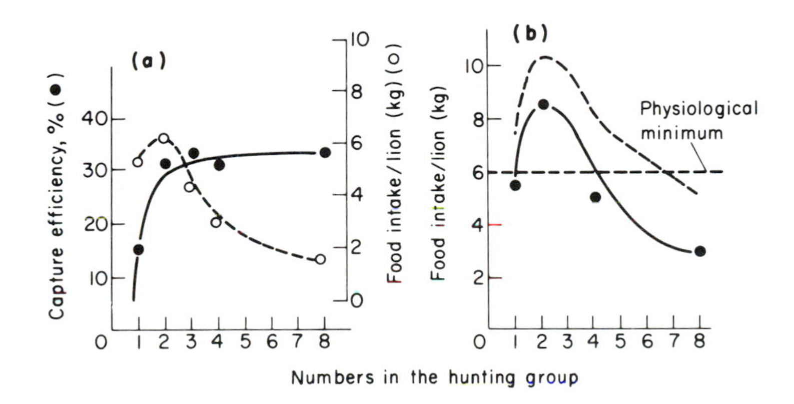

Fig. 13.8 (a) Capture efficiency and daily prey intake (of Thomson's gazelle, *Gazella thomsoni*) per lion (*Panthera leo*) in relation to lion group size; (b) hypothetical prey intake (wildebeest, *Connochaetes taurinus*) in areas of high (– – – –) and low (—) cover (from Caraco & Wolf, 1975).

increases, for example in better cover or when the group contains more females, a curve of the type indicated by the broken line in Fig. 13.7(b) (from Caraco & Wolf) predicts the occurrence of larger groups. Schaller's (1972) data provide empirical support for this prediction.

Other factors may also affect whether or not carnivores hunt co-operatively. Kruuk (1972) and Schaller (1972) both describe interspecific competition over carcasses. The factors influencing the outcome of such encounters are complex but a group of hyaenas, for example, can displace a solitary lion.

13.5.3 Obtaining information about food from group members

The advantages gained by individuals in locating or learning about food are:

1 they can locate ephemeral food sources. Krebs (1974) found that individuals from neighbouring nests in a colony of great blue herons (*Ardea herodias*) tended to leave at the same time and feed at the same (variable) place;
2 they can join animals that are already feeding as shown in woodpigeons (Murton, 1971), hyaenas (Kruuk, 1972) and mixed-species flocks of chickadees (*Parus* spp.) (Krebs, 1973);
3 they can learn a new food type by watching other individuals (Turner, 1964: chaffinches (*Fringilla coelebs*)). In this case the benefits to a younger, inexperienced bird will be greatest.

All of these behaviours have the associated cost of declining available food with increasing group size. Behaviour of type 1 has the additional cost of travelling to a communal roost each night. The benefits obtained from types 1 and 2 would be particularly important when food is dispersed in widely spaced patches.

When there is competition over food items, as in the woodpigeon (Murton, 1971) a dominant animal should gain more by joining a group than one that is subordinate. In such cases the feeding efficiency of subordinates presumably reaches a threshold below which, other factors being equal, it would pay to leave the group.

13.6 MATING BEHAVIOUR

We shall not review the entire range of mating systems, partly because such an approach suffers from the problems discussed at the start of the chapter. Adequate recent reviews have been provided by Wilson (1975) and Emlen & Oring (1977). Instead, we shall explore the role of energy dynamics in the mating strategies of individual males and females.

13.6.1 Competition between males for mates

Trivers (1972) has argued that members of the sex that invests the least in each offspring, usually males, will compete for members of the higher investing sex, usually females.

Males can secure a female:

1 directly, by defending a female against competitors, either by establishing access to the female by competition at the time of mating (e.g. toads, *Bufo bufo*: Davies & Halliday, 1979) or using a previously established individual reference for dominance (e.g. wolves, *Canis lupus*: Woolpy, 1968);
2 indirectly, by defending either a resource, such as a food supply (e.g. hartebeest: Gosling, 1974) or a nest site that is necessary for the female, or by defending a conventional reference for dominance such as a territory on a lek (e.g. Uganda kob, *Adenota kob thomasi*: Buechner, 1961; Leuthold, 1966);
3 in poorly understood permissive mating situations where competition between males appears unimportant in gaining access to females (e.g. chimpanzees, *Pan troglodytes*: Van Lawick-Goodall, 1968).

13.6.1.1 Direct competition between males

Male toads compete directly for access to females, and those that succeed fertilize a batch of eggs laid by a female (Davies & Halliday, 1979). The energetic costs of competing for mates are incurred directly in fighting for receptive females or in closely linked displays such as calling. It can thus be predicted that males will try to avoid competitors in order to reduce these costs. Davies & Halliday (1979) have shown that some toads adopt this alternative. Some males search in areas where both females and competing males are rare so that while competition is reduced there are fewer potential mates. However, Davies & Halliday showed that the strategy can be as successful as searching for a mate in high density areas.

In addition to energetic costs, male toads risk injury during competition for, and displacement from, females. Females can also be killed when large numbers of competing males keep them below the water surface (Davies & Halliday, 1979). Such intense competition also renders the combatants vulnerable to predation: fighting animals are notoriously easy to approach (e.g. Verschuren, 1958).

When the energetic cost of competing for females is high, each time they become receptive there could be selection for a male to establish, and maintain, a dominant role in a social group. Prolonged or violent fighting is generally rare in this context, instead there is a continual low level of reinforcement which incurs low costs but ensures access to receptive females. General accounts of such strategies are well known but the energetic considerations have not been documented. It seems likely that the strategy could only be effective when small groups of relatively long-lived individuals persist for long periods or when dominant males can maintain their status 'cheaply', for example by scent marking. Correlations between dominance, rank and reproductive success have been recorded in, for example, rabbits (*Oryctolagus cuniculus*) (Myers, Hale, Mykytowycz & Hughes, 1971) and mountain sheep (*Ovis* spp.) (Geist, 1971).

The behaviour and reproductive success of dominant males in multi-male groups can only be considered in relation to the functional significance of the behaviour of subordinates. Why do subordinates not simply leave the group? The answer is probably that subordinates derive a number of advantages from group membership (see previous sections) and are also well placed to acquire dominant status when the existing alpha male dies or becomes exhausted and can be easily displaced. Dunbar & Dunbar (1975) found that subordinate male Gelada baboons (*Theropithecus gelada*) could acquire a harem by joining an existing harem as a follower, behaving submissively toward the dominant male and then eventually leaving with a small number of young or peripheral females. In other species subordinates may achieve a low frequency of mating (e.g. bison (*Bison bison*) Lott, 1979).

In some species males try to monopolize many females by acquiring harems. Thus, rutting red deer (*Cervus elaphus*) stags spend most of their time in energetic attempts to acquire females, is preventing them from leaving and in defending them against other males (Darling, 1937). A similar situation exists in the impala (*Aepyceros melampus*) where, although within a large territory, males spend most of their time with a harem of females and try to protect these from non-territorial males that are often found nearby and attempt to enter harems (Jarman, 1979). Stable female groups protected by a single dominant also occur in primates, for example in the hamadryas baboon (*Papio hamadryas*) (Kummer, 1968).

In these situations the costs of herding and defence should increase steadily with the number of females in the harem, while the number of matings that can be achieved will level out at some maximum (Fig. 13.9). There will thus be an optimum harem size where a male can maximize his time as a dominant animal and female numbers in order to achieve the maximum number of matings. However, this simple model does not take into account the timing of female reproduction. A male will be more successful if it times its energy expenditure on harem defence to coincide with a time when many females are receptive.

The mating success of male wild turkey (*Meleogris gallopavo*) depends on the competitive ability of groups of male siblings (Watts & Stokes, 1971). The

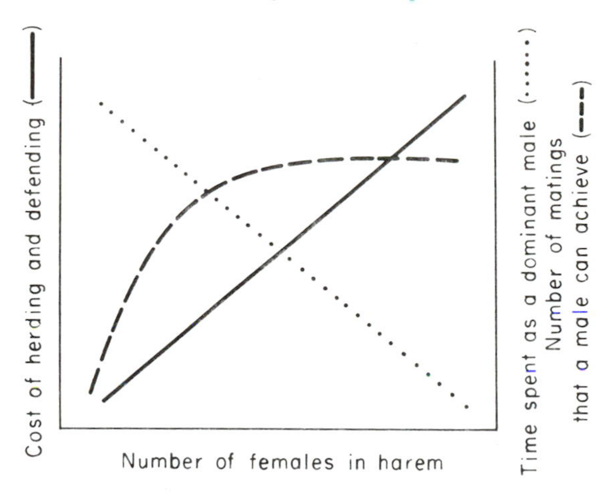

Fig. 13.9 Hypothetical costs and benefits of maintaining a female harem of increasing size. Details in text.

largest groups usually win and a single member of the group, whose dominant status is established by prior within-group agonistic behaviour, mates with all the females thus acquired.

The main energetic interest of this behaviour lies in the way that parental reserves are invested in offspring of varying size and number (see Smith & Fretwell, 1974; section 13.7.1).

13.6.1.2 SECURING A RESOURCE THAT FEMALES NEED

The economic aspects of defending and feeding in a territory that contains a food supply have been briefly introduced in the preceding section and have been reviewed in more detail by Davies (1978). In a more complex case territories are occupied by a resident male whose reproductive success is determined largely by the number of oestrus females that enter the area during his period of tenure: the attractiveness of the territory to females is thus critical.

Good examples of this male strategy are available among the African antelopes whose territories cover most, or all, of the available habitat. Hartebeest males occupy territories that average 0.31 km^2 in area and which extend continuously over large tracts of suitable habitat (Gosling, 1974, 1975). The territories can be classified into six types on the basis of the quality and diversity of grassland within them. Evidence including inter-territory movements, indicate that, of the most common territory types, males showed preference in the sequence numbered in Table 13.1: territory types 1 and 2 were across vegetation ecotones and types 4 and 5 in scrubland. The mean number of males that occupied each type of territory was consistent with this sequence. Changeovers in territory ownership were also more frequent in the ecotone territories (Table 13.1). Males thus competed more intensely for the preferred territories and this was obvious from observations of behaviour: males usually attempted to take over ecotone territories, and fought and were injured more frequently in doing so than for the scrubland territories.

The ranges of females included 20–30 male territories and all of their feeding was within territory boundaries. However, they spent more time in some territories than others and predictably the ecotone territories preferred by the males had the highest frequency of female visits (Table 13.1).

The costs of occupying ecotone territories were thus a high level of competition presumably with a high energy cost; injuries were very common in competition for territories and, rarely, animals were killed during fighting (Gosling, 1974, 1975). Males that acquired ecotone territories were expelled more quickly than those in scrubland but more females visited their territories and, assuming a simple relationship between visits and matings, such males presumably reproduced as least as well as the scubland males. One way to view the scrubland and ecotone males is that they represent alternatives in a mixed strategy: in one case males reproduce at a low rate over a long period with a low rate of energy expenditure on area defence, herding and mating females, and with ample time for feeding. In the other, males invest energy quickly in the expectation of rapid reproductive success.

Table 13.1 Hartebeest territory dynamics in relation to female preference (from Gosling, 1974, 1975).

	Territory type			
	Ecotone		Scrubland	
	2	3	4	5
Mean number males/year	1.1	1.0	0.7	0.7
Mean number changeovers/year	0.9	0.8	0.5	0.7
% observations with territorial male absent	14	16	43	60
% observations with females present	34	29	21	16

The costs of defending a territory increase continuously as it becomes larger and must ultimately place a constraint on the amount of resources that a male can monopolize. In the present context this limits the attractiveness of the territory to females. Thus there should be strong selective advantages for devices that assist in territory defence at lower cost than that involved in chasing out intruders. Demarcation by glandular secretion, urine and faeces may be such a device. Although open to a variety of functional interpretations (Ralls, 1971; Eisenberg & Kleiman, 1972; Johnson, 1973), some form of area defence remains a likely explanation in most cases. Boundary marking should theoretically reduce ambiguity about territory limits and the cost of such ambiguity in competition for dominance in disputed areas. Not surprisingly, marking of this sort, particularly with energetically 'free' substances such as faeces, is very common. Territory defence by bird song, as demonstrated in the great tit (*Parus major*) (Krebs, 1977), may be selected for similar reasons, being energetically cheaper than agonistic behavior and less dangerous.

13.6.1.3 Securing a conventional resource

In some species males defend a conventional resource, usually an area, that is visited by females for mating. The classical case of such a conventional reference for dominance are the clusters of territories, usually lacking any significant food resources, that are referred to as leks. Leks are found in all major vertebrate groups. Emlen & Oring (1977) suggest that they occur when resources are superabundant, unpredictable in time or space, or otherwise undefendable from an economic point of view.

It may be that the sites occupied by leks originated as important food resources in evolutionary time. The intense competition between males at such locations, leading to a hierarchy of male quality, may have provided an incentive for females to visit the area and this function may have prevailed as the food supply became unimportant or secondary.

Females usually select particular territories in a lek and the males in these territories achieve the largest number of matings. In the Uganda kob (*Adenota kob thomasi*) these territories are at the centre of the lek (Buechner & Schloeth, 1965) regardless of ownership changes. Thus males that obtain central territories can achieve a high rate of mating but they do so at the cost of intense competition. In addition the central territories on a kob lek have very little food and so the males presumably lose condition rapidly. A similar case to that of the hartebeest might thus exist with males in the central territories expending energy at a very high rate on defence and reproductive behaviour.

At the other extreme some male kob occupy 'single territories' away from the leks (Leuthold, 1966). These males are involved in a far less competitive system: Leuthold (1966) found that males in single territories spent 1% of their time in territory defence compared with 8% for males on a lek. It can be hypothesised that males must be in better condition to acquire a central lek territory than a single territory but that, as in the hartebeest, these alternatives might form a mixed strategy with the alternative between rapid energy expenditure and a high reproductive rate, and slow energy expenditure and a slow but prolonged rate of reproduction.

13.6.1.4 Economic constraints on male reproductive behaviour

A number of studies have identified male body size as a factor that limits competitive access to females. The experimental study by Davies & Halliday (1979) on the role of body size in mating success by toads (*Bufo bufo*) is perhaps the most elegant demonstration. In sexually dimorphic species where males are larger and polygyny is the rule, it is generally considered that the difference is a consequence of intrasexual competition in which large body size is advantageous. Presumably there are also disadvantages in large size, such as those of obtaining sufficient food during seasons of limited food supply, but relevant data are not available.

Parental investment, in its various forms, is probably the most critical determinant of adult body size. The body size of coypus (*Myocastor coypus*) at 3 months is correlated with size when fully grown (Fig. 13.10). Berg, Simms & Everitt (1963) obtained the same result for the relationship between weight at

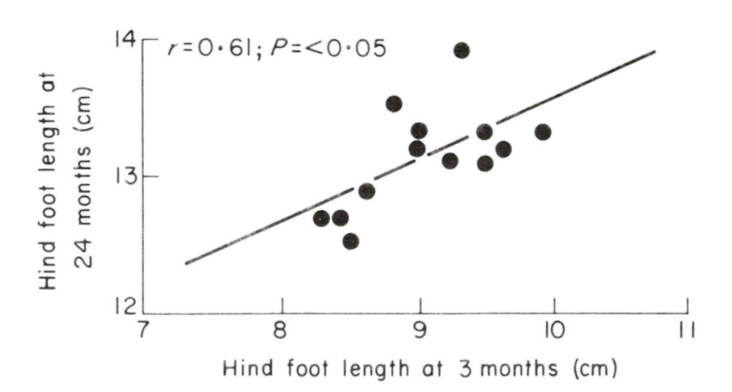

Fig. 13.10 The relation between body size of coypus (*Myocastor coypus*) at 3 and 24 months of age (Gosling, unpublished).

weaning and adult weight in laboratory rats (*Rattus norvegicus*). Bernstein (1978) showed that the quantity of milk ingested during lactation could significantly affect the weight at weaning in male laboratory rats and that the small differences established were then maintained into adult life with *ad lib* feeding. The main effect of post-weaning nutrition is probably only a negative one, that is, poor nutrition can probably prevent an individual from achieving its full potential, but excellent nutrition cannot produce a larger size.

Nutrition may also affect the size of horns and antlers, the weapons used by some artiodactyls in intraspecific combat. Both body and antler size of deer are smaller in areas where the food supply is poor (Mitchell, Staines & Welch, 1977) and Vogt (1947) showed a direct link between antler size and nutritional plane. Such factors might have a direct effect on a male's chance of reproductive success: Schaller (1967), for example, showed a correlation between the antler size of Axis deer (*Axis axis*) and dominance status and Geist (1966) found that those male Stone's sheep (*Ovis dalli*) with the largest horns achieved most mounts. However, the effect of age differences are rarely taken into account in such analyses (Rowell, 1974), and Clutton-Brock, Albon, Gibson & Guinness (1979) found no correlation between the antler lengths of red deer stags of similar age and either dominance or reproductive success.

There are few data on the relationship between male condition, or energy reserves, and reproductive status. In discussing competition between red deer stags Clutton-Brock & Albon (1979) make the general point that in many larger animals changes in fighting ability within breeding seasons or across the lifetime of individuals are related to changes in body condition but not to obvious changes in size. A conclusion of this type seems inescapable in view of the dramatic depletion of fat reserves in rutting stags. For example, kidney fat declines by 87% between September and October (Mitchell, McCowan & Nicholson, 1976).

13.6.1.5 Assessment

The costs of securing a mate are sometimes purely energetic but in many cases competitors risk injury or death (e.g. hartebeest, Gosling, 1974). It can be predicted that an animal could reduce these costs by prior assessment of the likely outcome of any interaction (Maynard Smith & Price, 1973; Parker, 1974; Maynard Smith & Parker, 1976; Zahavi 1975, 1977). Evidence for such assessment has been found in the agonistic behaviour of toads (Davies & Halliday, 1978) and red deer (Clutton-Brock & Albon, 1979). Both assess competitors from the quality of their calls and Clutton-Brock & Albon suggest that the length of a roaring bout by red deer stags might be condition dependent. In general, selection should favour 'cheat-proof' assessment criteria that closely reflect individual differences in fighting ability.

13.6.2 Female strategies

It has often been suggested that females could gain a long-term genetic advantage and/or short-term phenotypic advantage by mating with a high-quality partner. For example, a female that mates with a male holding a large territory may produce offspring that inherit the ability to acquire large territories and, at the same time, the female and her offspring will gain the advantage of living in that territory (Halliday, 1978).

Females will tend to acquire a high-quality mate as a consequence of male intrasexual competition. Thus in the Uganda kob females can hypothetically ensure high mate quality simply by visiting the central lek territories where intrasexual male competition is most intense. In elephant seals (Cox & le Boeuf, 1977) females protest loudly when males attempt to mount them. These calls attract other males in the vicinity who directly compete for the female: the male that wins, i.e. in intrasexual terms the highest quality male, then mates with the female.

In the hartebeest (Gosling, 1974, 1975) males that acquire high-quality territories attract, on average, more females. But an individual female might sometimes obtain a better food supply by visiting low-quality territories where the grassland is not depleted. This general point was emphasized by Orians (1969). In other words the strategy of the female is frequency dependent and, in economic terms, a mixed strategy might be evolutionarily stable.

There is very little evidence in the vertebrates for active choice by females of individual males. Gosling (1975) noticed a tendency for individual female hartebeest to associate with individual males but, as in most field studies, it was difficult to disentangle the influence of other attractive features of the environment. Females could exert choice by escaping the attention of males. Female hartebeest drive away young, but sexually mature, males when these attempt to mount (Gosling, 1975). Some female primates, for example chimpanzees (*Pan troglodytes*) (successful Van Lawick-Goodall, 1968), allow a number of males to copulate in succession but the chance of successful fertilization by each male is not currently known.

In a few bird species males perform most parental care, an example being the American jacana (*Jacana spinosa*) (Jenni & Collier, 1972). In this species females compete for males and sexual selection has operated in such a way that females are larger and more brightly coloured than males. Male moorhens (*Gallinula chloropus*) also perform the majority of parental care (Siegfried & Frost, 1975) and Petrie (unpublished) has shown that the females that win most of their interactions (the heaviest), when competing with other females for male partners, eventually pair with males that are in the best

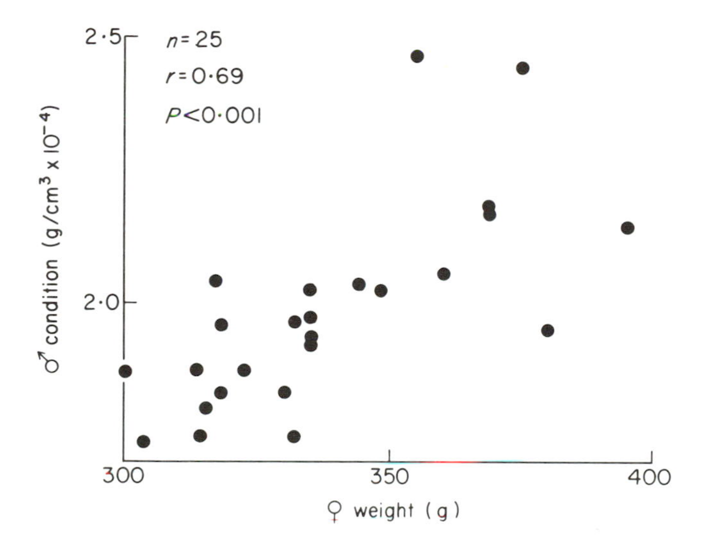

Fig. 13.11 The relationship between body weight of female moorhens (*Gallinula chloropus*) and the condition of their mates. Females may select males on the basis of their ability to incubate (Petrie, unpublished).

condition (Fig. 13.11). Males in very poor condition usually fail to form pairs. It appears that female moorhens are prepared to expand energy, and risk injury, in competition for males whose energy reserves are needed for parental care.

13.7 PRODUCING AND REARING OFFSPRING

In this section we consider both the social behaviour of the parent, and any kin that participate in rearing, and also the social environment of the young. We will not consider reproduction as a facet of life history tactics, as in the reviews by Stearns (1977) and Horn (1978), but rather will deal with a few economic considerations from the viewpoint of individual investment.

13.7.1 Offspring number, quality and sex

First, we can ask how the social environment affects the number, quality and sex of offspring. A simple case is that one or both parents could defend a food supply, such as a territory, and that the usual economics of territory defence (Davies, 1978) would dictate the amount of food available for parental investment. The best-documented example of this sort of relationship is the red grouse (*Lagopus lagopus scoticus*), where males defend territories which contain the food supply for themselves and the females that they mate with and which nest there. The quality of the female's food before incubation determines egg quality (Moss, Watson & Parr, 1975) and this in turn determines the quality of the hatchlings as measured by chick mortality (Jenkins, Watson & Picozzi, 1965). Artificial fertilization of heather, the main food supply, allows the production of larger broods and also, presumably because of a higher quality food supply, allows males to defend smaller territories (Millar, Watson & Jenkins, 1970). Similarly, Southern (1970) showed that pairs of tawny owls (*Strix aluco*), whose food supply is the rodent population within their woodland territory, may fail to lay any eggs at all when rodents are rare.

An obvious constraint on the production of young is the risk of impairing future reproduction or the survival of the parent. Tinkle & Hadley (1975) found that reproductive effort (ratio of clutch calories to body calories) was inversely correlated with adult survival in lizards (see Chapter 10).

Spacing patterns, and their associated energy costs, have important consequences for the chance of predation on the young. Krebs (1971) demonstrated a higher chance of predation when great tit (*Parus major*) nests were close together, that is, when territories were small, while Birkhead (1977) showed the opposite effect in guillemots (*Uria aalge*) apparently because of co-operative defence against potential predators. Female hartebeest, and other antelopes similarly, reduce the chance of predation on their newborn offspring by enlarging their nearest-neighbour distance (Gosling, 1969); in this case the costs are mainly in feeding time lost in anti-predator scanning, while those to the tits are in energy expended and feeding time lost in agonistic behaviour. There is no feeding cost to the colonial guillemots because their distant food supply is undefendable.

Trivers & Willard (1973) discussed the importance of parental investment in influencing the chance of reproductive success by offspring. They hypothesized that:

1 the condition of the offspring at the end of the period of parental investment will be correlated with that of the mother during parental investment;
2 differences in condition of young at the end of the period of parental investment will be maintained into adulthood;
3 in a polygynous mating system males which have a slight advantage in condition will achieve a greater differential reproductive success over other males than would females having the same advantage of condition over other females. Thus females in poor condition should produce female offspring, each with a moderate chance of reproductive success, while those in good condition should produce males.

Myers (1978) has pointed out a number of defects in the data presented by Trivers & Willard in support of this hypothesis and Williams (1979), while accepting the theoretical advantages, was unable to find any case of parental control over the sex ratio of their progeny.

However, the absence of convincing reports on parental control of sex ratio may be partly due to the problems in obtaining large samples and data are now available to show that female coypus (*Myocastor coypus*) can apparently control the sex ratio of their offspring (Gosling, unpublished). This was initially detected by the observation, in very large post-mortem samples, that the sex composition of litters changes at around week 14 of the 19-week gestation period. The proportion of females in litters decreases by about 10% while that of males *increases* to a similar extent. The change is due to the selective abortion of predominantly female litters and the females that abort these litters are those with the largest amounts of stored fat per embryo (Fig. 13.12).

Large amounts of stored fat per embryo are generally found in small litters and the embryos of such litters are significantly larger than those in large litters at full term. Thus there is a tendency for females with small litters of large embryos to abort unless these are predominantly male. Large fat deposits per embryo are also characteristic of seasons when large litters are conceived and so females, which can conceive at a post-partum oestrus (Gosling, 1980a), trade a small litter of large, predominantly female embryos for a large litter that will inevitably result in smaller neonates. At this stage females employ another physiological device: they show a slight, but significant, tendency to selectively resorb individual male embryos. Both sorts of parental control thus tend to produce small numbers of large male neonates and large numbers of small female neonates, as indeed might be

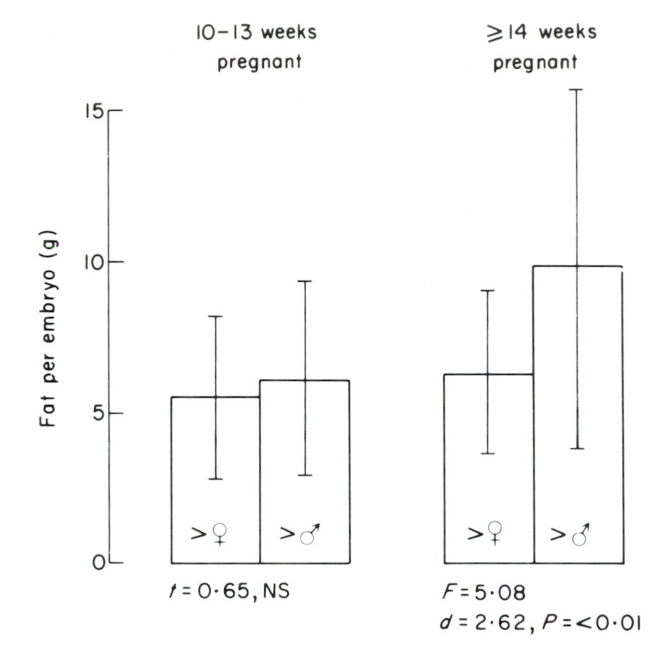

Fig. 13.12 The fat reserves of female coypus (*Myocastor coypus*) with predominantly female litters (> ♀) compared with those that are predominantly male (> ♂) before and after week 14 of gestation. After week 14 females bearing predominantly female litters are thinner because the fattest females have aborted. The general increase in fat reserves between early and late stages occurs because all females accumulate fat throughout pregnancy. See text for interpretation. (Gosling, unpublished). Vertical bars represent two standard deviations.

predicted in this species where a male's reproductive strategy is to defend a number of females (Gosling, 1977).

Parental control of this kind would be irrelevant if the differential size of neonates was not translated into similarly different adult sizes. However, as described in section 13.6.4, there is a significant correlation between the size of both males and females at 3 and 24 months of age ($r_{13} = 0.61$ and $r_{14} = 0.72$, respectively).

13.7.2 Rearing offspring

In the previous section we concluded that the reproductive success of progeny is determined to a large extent by events before they are born. However, there will obviously be some relative adjustment of prospects depending on the differential success of individual offspring in acquiring resources. At the start of life altricial birds and mammals are entirely dependent on a food supply provided by the parents either in the form of solid food, brought to the nest by one or both parent birds, or by lactation. An exceptional case in birds is the production of crop milk by pigeons, which can be considered as an analogue of lactation in mammals. In precocial birds, reptiles and amphibians young do not depend directly on their parents for food, but there is often a degree of dependence, for example, in learning food types by imitation. In general the amount of energy spent by a parent on its progeny should increase the chance of their survival. The general pattern of such expediture is illustrated in female mammals by the large increase in food requirements (Sadlier, 1969), and the rapid depletion in fat reserves during lactation. However, there are obvious constraints on this expenditure and on the corresponding investment by birds: if parents expend too much energy on their progeny they put themselves at risk.

The amount of investment by female mammals in particular offspring may influence their future reproductive performance. Thus female red deer (*Cervus elaphus*) that do not produce calves in a particular year continue to suckle their previous offspring and thus enhance its survival chances (Guiness, Hall & Cockerill, 1979). Similar observations were made by Spinage (1969) in the case of waterbuck (*Kobus defassa*). In other artiodactyls investment in a previous offspring can also occur opportunistically when a calf dies: at these times some females allow previous calves to suck (Altman, 1960; Gosling, 1975). However, when calves are produced and survive, older offspring are kept away by the mother which thus switches its main investment to the new offspring. The breakdown of mother – offspring bonds in ungulates has been reviewed by Lent (1974).

There is an extensive literature on co-operative behaviour by adults in the rearing of young, Emlen (1978), for example, has reviewed its occurrence in birds. The African wild dog (*Lycaon pictus*) provides a good mammalian example which is unusual because, unlike most mammals (eg. Eisenberg, 1977), the social group consists primarily of genetically related males; females leave the group in which they are born and form, or join, other groups (Frame & Frame, 1976). Only one female breeds successfully and all adult members of the pack help to rear the pups by regurgitating meat. Estes & Goddard (1967) observed a case where an orphaned litter was successfully reared in a group consisting only of adult males. Such cases, once believed to be cases of real altruism, are now believed to have evolved through kin selection (Maynard-Smith, 1964). The males that contribute to rearing the offspring of other males increase their own inclusive fitness (Hamilton, 1964) in the same way that the male turkey siblings mentioned earlier increase their own fitness by helping their dominant brother.

The Florida scrub jay (*Aphelocoma coerulescens*) (Woolfenden, in Emlen, 1978) shows some similarities to the wild dog case with a number of closely related jays assisting the parents to feed nestlings. These helpers provide about 30% of the food and thus provide a substantial economic benefit to the parents while boosting their own inclusive fitness. However, there is a sting in the tail because the degree to which relatives help is closely related to the chance that they may be able to take over the breeding territory. The parent owners thus trade economic assistance for an increased risk of losing breeding status.

13.8 CONCLUDING REMARKS

We have discussed some main categories of social behaviour in a series of discrete accounts. This approach is simplistic, not only because in the real world these behaviours may follow each other in a rapid and complex sequence, but also because many occur simultaneously. Thus an animal may bite a mouthful of food, lift its head and inspect its surroundings (antipredator scanning) while chewing and, by its presence in a territory, be deterring potential intruders. Behaviour may also be mutually exclusive in a complex fashion: a male antelope might for example be isolated in order to maintain an area of undisputed dominance for future mating and/or an exclusive food supply and thus be unable to enjoy the anti-predator advantages of group membership. This lack is sometimes demonstrated by association with similarly isolated territorial males of other species (Gosling, 1980b).

Any attempt to integrate the behaviours that we have reviewed into an optimum life-time strategy will face the problem of all optimization models, namely choosing an appropriate maximizing assumption. Even within the categories of behaviour that have been covered it has generally been difficult to employ the same currency in evaluating the trade-off between costs and benefits. An exception is in feeding where social animals attempt to maximize energy gain or minimize time spent feeding (Schoener, 1971): the attribute of a universal currency accounts, to some extent, for the outstanding success of optimal foraging theory in understanding feeding behaviour (see Chapter 4).

The maximizing assumption of ergonomic efficiency goes some way to an understanding of anti-predator or reproductive behaviour, but often such considerations are trivial. For example, an animal may spend energy and lose feeding time while inspecting its surroundings for predators, but the primary consideration in such behaviour is the effect it has on the probability that the animal will detect a predator and the consequences for its own survival chances. But in these terms the costs and benefits are confused by expression in incompatible currencies. In general, any optimization model of social behaviour that aims to be both comprehensive and quantitative must employ inclusive fitness as a maximizing assumption.

References

Abeloos, M. (1930) Recherches expérimentales sur la croissance et la régeneration chez les planaires. *Bulletin biologique de la France et de la Belgique*, **64,** 1–140.
9.3.2

Abeloos, M. (1944) Recherches expérimentales sur la croissance. La croissance des mollusques Aronidés. *Bulletin biologique de la France et de la Belgique*, **78,** 215–216.
9.3.2

Abrahamson, W. G. & Gadgil, M. (1973) Growth form and reproductive effort in goldenrods (Solidago, Compositae). *Amer. Natur.*, **107,** 651–661.
12.4

Adolph, E. F. & Heggeness, F. W. (1971) Age changes in body weight and fat in fetal and infant mammals. *Growth*, **35,** 55–63.
8.5

Akre, R. D. & Davis, H. G. (1978) Biology and pest status of venomous wasps. *Ann. Rev. Ent.*, **23,** 215–238.
6.11

Alexander, R. McN. (1970) *Functional Design in Fishes*. 160, pp. Hutchinson, London.
8.7

Alexander, R. McN. (1971) *Size and shape*. Edward Arnold, London.
9.7.2

Al-Joborae, F. F. (1980) The influence of diet on the gut morphology of the starling (*Sturnus vulgaris* L. 1758). Unpubl. D. Phil. thesis, University of Oxford.
5.3.1, 5.3.3

Allen, W. V. (1976) Biochemical aspects of lipid storage and utilization in animals. *Am. Zool.*, **16,** 631–647.
8.4

Altenberg, E. (1934) A theory of hermaphroditism. *Amer. Natur.*, **68,** 88–91.
10.6

Altmann, M. (1960) The role of juvenile elk and moose in the social dynamics of their species. *Zoologica*, **45,** 35–39.
13.7.2

Amir, Sh. (1979) On the optimal timing of reproduction. *Amer. Natur.*, **114,** 461–466.
10.2.1

Andrews, J. T. & Lorimer, G. H. (1978) Photorespiration—still unavoidable? *FEBS Lett.*, **90,** 1–9.
3.5

Ankney, C. D. (1977) Feeding and digestive organ size in breeding Lesser Snow Geese. *Auk*, **94,** 275–282.
5.3.1

Anderson, C. W. (1942) *Thoroughbreds*. 72. pp. Macmillan Co., New York.
8.4

Anderson, M. & Krebs, J. R. (1978) On the evolution of hoarding behaviour. *Anim. Behav.*, **26,** 707–711.
8.5

Argenzio, R. A., Lowe, J. E., Pickard, D. W. & Stevens, C. E. (1974) Digesta passage and water exchange in the equine large intestine. *Am. J. Physiol.*, **226,** 1035–1042.
5.5.4.

Askenmo, C. (1979) Reproductive effort and return rate of male pied flycatchers. *Amer. Natur.*, **114,** 748–752.
10.4

Atsatt, P. R. & O'Dowd, D. J. (1976) Plant defense guilds. *Science*, **193,** 24–29
5.1

Avery, R. A. (1973) Morphometric and functional studies on the stomach of the lizard *Lacerta vivipara*. *J. Zool., (Lond.)*, **169,** 157–167.
5.5.2 (Fig 5.9 legend)

Badger, M. R. & Andrews, T. J. (1974) Effects of CO_2, O_2 and temperature on a high

affinity form of ribulose diphosphate carboxylase-oxygenase from spinach. *Biophys. Biochem. Res. Commun.*, **60,** 204–210.
3.4

Badger, M. R. & Collatz, G. J. (1977) Studies on the kinetic mechanism of ribulose-1,5-bisphosphate carboxylase and oxygenase reactions with particular reference to the effect of temperature on kinetic parameters. *Carnegie Inst. Washington Yearbook*, **76,** 355–361.
3.5

Baile C. A. (1975) Control of feed intake in ruminants. In *Digestion and Metabolism in the Ruminant*. (Eds. McDonald, I. W. & Warner, A. C. I. pp. 333–350. University of New England Publishing Unit, Australia.
5.4.4

Baile, C. A. & Forbes, J. M. (1974) Control of feed intake and regulation of energy balance in ruminants. *Physiol. Rev.*, **54,** 160–214.
5.4.4

Baker, H. G. (1970) *Plants and Civilization*. 2nd edn. 194. pp. Wadsworth Publ. Co., Belmont, Ca.
8.2

Balch, C. C. (1950) Factors affecting the utilization of food by dairy cows. *J. Dairy Sci.*, **4,** 361–388.
5.5.4 (Fig 5.10 legend)

Balch, C. C. & Campling, R. C. (1965) Rate of passage of digesta through the ruminant digestive tract. In *Physiology of Digestion in the Ruminant*. (Eds. Dougerty, R. W., Allen, R. S., Burrough, W., Jacobson, N. L. & McGillard, A. D.) Butterworths, London.
5.5.4

Baldwin, R. L. (1968) Estimation of theoretical calorific relationships as a teaching technique. A review. *J. Dairy Sci.*, **51,** 104–111.
10.4

Ballinger, R. (1977) Reproductive strategies: food availability as a source of proximal variation in a lizard. *Ecology*, **58,** 628–635.
10.2.2

Barnes, H. & Barnes, M. (1964) Egg size, nauplius size and their variation with local, geographic and specific factors in some common cirripedes. *J. Anim. Ecol.*, **33,** 391–402.
10.3

Barr, T. C. (1968) Cave ecology and the evolution of troglodites. In *Evolutionary Biology*. (Eds. Dobzhansky, T., Hecht, M. K. & Steere, W. C.) Appleton-Century, Crofts, New York.
9.7.3

Bartholomew, G. A. (1972) Body temperature and energy metabolism. In *Animal Physiology: Principles and Adaptations*. (Ed. Gordon, M. S.) (pp. 298–368.) Macmillan, New York.
12.2.8

Bazzaz, F. A., Carlson, R. W. & Harper, J. L. (1979) Contribution to reproductive effort by photosynthesis of flowers and fruits. *Nature (London)*, **279,** 554–555.
10.2.1

Beall, G. (1948) The fat content of a butterfly *Danaus plexippus* Linn., as affected by migration. *Ecology*, **29,** 80–94.
8.3

Beer, J. R. & Richards, A. G. (1956) Hibernation of the big brown bat. *J. Mammal.*, **37,** 31–41.
8.6

Belkin, D. A. (1961) The running speeds of *Dipsosaurus dorsalis* and *Callisaurus draconoides*. *Copeia*, 1961, 223–224.
8.4

Bell, R. H. V. (1971) A grazing ecosystem in the Serengeti. *Sci. Am.*, **224,** (1), 86–93.
5.4.2

Bell, G. (1978) The evolution of anisogamy. *J. theor. Biol.*, **73,** 247–270.
10.3

Belovsky, G. E. (1978) Diet optimization in a generalist herbivore: the moose. *Theor. Popul. Biol.*, **14,** 105–134.
4.3

Bennett, A. R. & Dawson, W. R. (1976) Metabolism. In *Biology of the Reptilia*. Vol. 5 (Eds. Gans, C. & Dawson W. R.) pp. 127–223. Academic Press, London.
2.7

Bennett, C. H. (1979) Dissipation—error tradeoff in proofreading. *Bio-Systems*, **11,** 85–91.
7.3

Benson, A. A. & Lee, R. F. (1975). The role of wax in oceanic food chains. *Sci. Am.*, **232,** 76–86.
8.2

Benton, M. J. (1979) Ectothermy and the success of dinosaurs. *Evolution*, **33,** 983–997.
2.6, 2.7

Berg, B. N., Simms, H. S. & Everitt, A. V. (1963). Nutrition and longevity in the rat. V.

Weaning weight, adult size, and onset of disease. *J. Nutr.*, **80,** 255–262.
13.6.4

Berman, M. (1963) A postulate to aid in model building. *J. theor. Biol.*, **4,** 229–236.
5.5.4

Bernstein, H. (1977) Germ line recombination may be primarily a manifestation of DNA repair processes. *J. theor. Biol.*, **69,** 371–380.
7.6

Bernstein, I. L. (1978). Re-evaluation of the effect of early nutritional experience on body weight. *Physiol. Behav.*, **21,** 821–823.
13.6.4

Berry, J. & Björkman, O. (1980) Photosynthetic response and adaptation to temperature in higher plants. *Ann. Rev. Plant Physiol.*, **31,** 491–543.
3.9

Bertram, B. C. R. (1978) Living in Groups: Predators and Prey. In *Behavioural Ecology: an Evolutionary Approach.* (Eds. Krebs, J. R. & Davies, N. B.) pp. 64–96. Blackwell Scientific Publications, Oxford.
13.4

Bertram, B. C. R. (1980) Vigilance and group size in ostriches. *Anim. Behav.*, **28,** 278–286.
13.4.1, 13.4.2

Bidder, G. P. (1932) Senescence. *Br. med. J.*, **ii,** 583–585.
7.3

Birkebak, R. & Birkebak, R. (1964) Solar radiation characteristics of tree leaves. *Ecology*, **45,** 646–649
12.3

Birkhead, T. R. (1977) The effect of habitat and density on breeding success in common guillemot (*Uris aalge*). *J. Anim. Ecol.*, **46,** 751–764.
13.7.1

Bishop, G. H. (1922) Cell metabolism in the insect fat body. 1. Cytological changes accompanying growth and histolysis. *J. Morph.*, **26,** 567–594.
8.3

Bjorksten, J. (1968) The crosslinkage theory of ageing. *J. Am. Ger. Soc.*, **16,** 408–427.
7.8

Blaxter, K. L., Graham, N. McC. & Wainman, F. W. (1956) Some observations on the digestibility of food by sheep, and on related problems. *Br. J. Nutr.*, **10,** 69–91.
5.5.4

Blaxter, K. L., Kay, R. N. B., Sharman, G. A. M., Cunningham, J. M. M. & Hamilton W. J. (1974) *Farming the Red Deer.* H.M.S.O., Edinburgh.
5.5.4 (Fig 5.10 caption)

Blem, C. R. (1976) Patterns of lipid storage and utilization in birds. *Am. Zool.*, **16,** 671–684.
8.3

Bligh, J. (1973) *Temperature Regulation in Mammals and other Vertebrates.* 436, pp. Elsevier North Holland, Amsterdam.
2.7

Bligh, J. (1976) Temperature regulation. In *Environmental Physiology of Animals.* (Eds. Bligh, J., Cloudsley-Thompson, J. L. & Macdonald, A. G.) pp. 415–430. Blackwell Scientific Publications, Oxford.
2.6, 2.7

Bligh, J., Cloudsley-Thompson, J. L. & Macdonald, A. G. (1976) *Environmental Physiology of Animals.* Blackwell Scientific Publications, Oxford.
1.4.1

Bliss, L. C. (1971) Arctic and alpine plant life cycles. *Ann. Rev. Ecol. Syst.*, **2,** 405–438.
8.4

Boardman, N. K. (1977) Comparative photosynthesis of sun and shade plants. *Ann. Rev. Plant Physiol.*, **28,** 353–357.
3.9

Booth, D. A. (1978a) Prediction of feeding behaviour from energy flows in the rat. In *Hunger Models* (Ed. Booth, D. A.) pp. 227–278. Academic Press, London.
5.5.2, 5.5.3

Booth D. A. (Ed.) (1978b) *Hunger Models.* Academic Press, London.
5.6.2

Boucher-Rodoni, R. (1973) Vitesse de digestion d'*Octopus cyanea* (Cephalopoda: Octopoda). *Mar. Biol.*, **18,** 237–242.
5.5.2 (Fig 5.9 caption)

Boucher-Rodoni, R. & Mangold, K. (1977) Experimental study of digestion in *Octopus vulgaris* (Cephalopoda: Octopoda). *J. Zool., (London)*, **183,** 505–515.
5.6.1

Boucot, A. J. (1976) Rates of size increase and phyletic evolution. *Nature (London)*, **261,** 694–695.
2.2

Bounoure, L. (1919) *Aliments, Chitine et Tube digestif chez les Coleopteres.* A. Hermann et Fils, Paris.
5.6.3

Bowes, G., Ogren, W. L. & Hageman, R. H. (1971) Phosphoglycollate production catalysed by ribulose diphosphate carboxylase.

Biochem. Biophys. Res. Commun., **45,** 716–722.
3.4

Boyer, J. S. (1976) Water deficits and photosynthesis. In *Water Deficits and Plant Growth.* Vol. IV. (Ed. Kozlowski, T. T.) Academic Press, New York.
12.3

Boyer, J. S. & Bowen, B. L. (1970) Inhibition of oxygen evolution in chloroplasts isolated from leaves with low water potentials. *Plant Physiol.*, **45,** 612–615.
12.3

Bradbury, S. (1957) A histochemical study of pigment cells of the leech, *Glossosiphonia complanata. Quart. J. Micros. Sci.*, **98,** 301–314.
8.3

Bradbury, S. (1958) A cytological and histochemical study of the connective tissue fibres of the leech, *Hirudo medicinalis. Quart. J. Micros. Sci.*, **99,** 130–142.
8.3

Bradford, D. F. & Smith, C. C. (1977) Seed predation and seed number in *Scheelea rostrata* palm fruits. *Ecology*, **58,** 667–673.
6.6

Braekkan, O. R. (1977) Nutrition in fish. In *Recent Advances in Animal Nutrition* (Eds. Haresign, W. & Lewis, D.) Butterworths, London.
5.6.3

Brandon, R. N. (1978) Evolution. *Philos. Sci.*, **45,** 96–109.
1.3

Brandt, C. S. & Thacker, E. J. (1958) A concept of rate of food passage through the gastro-intestinal tract. *J. Anim. Sci.*, **17,** 218–223.
5.1, 5.5.4

Branscomb, E. W. & Galas, D. J. (1975) Progressive decrease in protein synthesis accuracy induced by streptomycin in *Escherichia coli. Nature (London)*, **254,** 161–163.
7.3

Brattsten, L. B., Wilkinson, C. F. & Eisner, T. (1977) Herbivore-plant interactions: mixed-function oxidases and secondary plant substances. *Science*, **196,** 1349–1352
5.5

Bremermann, H. J. (1979) Theory of spontaneous cell fusion. Sexuality in cell populations as an evolutionarily stable strategy. Applications to immunology and cancer. *J. theor. Biol.*, **76,** 311–334.
7.6

Brockelman, W. Y. (1975) Competition, the fitness of offspring, and optimal clutch size. *Amer. Natur.*, **109,** 677–699.
12.1.2

Brockman, H. J. & Barnard, C. J. (1979) Kleptoparasitism in Birds. *Anim. Behav.*, **27,** 487–515.
13.5

Brown, A. V. & Fitzpatrick, L. C. (1978) Life history and population energetics of the Dobson fly, *Corydalus cornutus. Ecology*, **59,** 1091–1108.
12.3

Brown, J. H. & Lasiewski, R. C. (1972) Metabolism of weasels: the cost of being long and thin. *Ecology*, **53,** 939–943.
12.2.4

Brulfert, J., Guerrier, D. & Quieroz, O. (1975) Photoperiodism and enzyme rhythms: kinetic characteristics of the photoperiodic induction of crassulacean acid metabolism. *Planta*, **125,** 33–34.
3.7.3.

Bryant, D. M. (1979) Reproductive costs in the housemartin. *J. Anim. Ecol.*, **48,** 655–75.
10.4

Broda, E. & Peschek, G. A. (1979) Did respiration or photosynthesis come first? *J. theor. Biol.*, **81,** 201–212.
2.2

Buechner, H. K. (1961) Territorial behaviour in Uganda Kob. *Science*, **133,** 698–699.
13.6.1

Buechner, H. K. & Schloeth, R. (1965) Ceremonial mating behaviour in Uganda Kob (*Adenota kob thomasi* Newmann). *Z. Tierpsychol.*, **22,** 209–225.
13.6.3

Bukhari, A. I. & Zipser, D. (1973) Mutants of *Escherichia coli* with a defect in the degradation of nonsense fragments. *Nature New Biol.*, **243,** 238–241.
7.3

Bull, J. J. & Shine, R. (1979) Iteroparous animals that skip opportunities for reproduction. *Amer. Natur.*, **114,** 296–303.
10.2.2

Bullough, W. S. (1952) The energy relations of mitotic activity. *Biol. Rev.*, **27,** 133–168.
9.7.1

Bullough, W. S. (1967) *The Evolution of Differentiation.* Academic Press, London.
9.7.4

Burnet, F. M. (1974) *Intrinsic Mutagenesis: a Genetic Approach to Ageing.* John Wiley,

New York.
7.3

Burnet, F. M. (1978) *Endurance of Life: The Implications of Genetics for Human Life*. Cambridge University Press, Cambridge.
7.10

Cairns, J. (1975) Mutation selection and the natural history of cancer. *Nature (London)*, **225,** 197–200.
7.4.1

Caldwell, M. M., White, R. S., Moore, R. T. & Camp, L. B. (1977) Carbon balance, productivity and water use of cold desert shrub communities dominated by C_3 and C_4 species. *Oecologia*, **29,** 275–300.
3.8.2

Calow, P. (1973) The relationship between fecundity, phenology and longevity: a systems approach. *Amer. Natur.*, **107,** 559–574.
10.2.2

Calow, P. (1975) The feeding strategies of two freshwater gastropods, *Ancylus fluviatilis* Müll and *Planorbis contortus* Linn. (Pulmonata), in terms of ingestion rates and absorption efficiencies. *Oecologia*, **20,** 33–49.
10.2.1

Calow, P. (1976) *Biological Machines: a Cybernetic Approach to Life*. Edward Arnold, London
9.5, 9.6

Calow, P. (1977a) Ecology, Evolution and Energetics: A Study in Metabolic Adaptation. In *Advances in Ecological Research*. Vol. 10. (Ed. Macfadyen, A.) pp. 1–62. Academic Press, London.
2.2, 5.6.3, 7.5, 9.1, 9.6, 10.2.1

Calow, P. (1977b) Irradiation studies on rejuvenation in triclads. *Exp. Geront.*, **12,** 173–179.
7.8, 7.10

Calow, P. (1977c) Conversion efficiencies in heterotrophic organisms. *Biol. Rev.*, **52,** 385–409.
9.2.3, 10.5

Calow, P. (1978a) Bidder's hypothesis revisited: solutions to some key problems associated with general molecular theory of ageing. *Gerontology*, **24,** 448–458.
1.4.1, 7.3, 7.6, 7.7, 7.8, 10.2.2

Calow, P. (1978b) *Life Cycles*. Chapman Hall, London.
9.7.4, 10.5

Calow, P. (1978c) The evolution of life-cycle strategies in freshwater gastropods. *Malacologia*, **13,** 351–364.
10.2.2, 10.3

Calow, P. (1979) The cost of reproduction—a physiological approach. *Biol. Rev.*, **54,** 23–40.
1.4.2, 7.7, 7.8, 8.8, 10.2.1, 10.2.2

Calow, P. (*in press*). Growth in lower invertebrates. In *Comparative Animal Nutrition*. (Ed. Rechigl, M.) Karger, Basle.
9.5

Calow, P., Beveridge, M. & Sibly, R. (1979) Adaptational aspects of asexual reproduction in freshwater triclads. *Amer. Zool.*, **19,** 715–727.
10.2.1, 10.7

Calow, P., Beveridge, M. & Sibly, R. (1980) Heads and tails: Adaptational aspects of asexual reproduction in freshwater triclads. *Amer. Zool.*, **19,** 715–728.
7.6

Calow, P. & Jennings, J. B. (1974) Calorific values in the phylum Platyhelminthes. The relationship between potential energy, mode of life and evolution of entoparasitism. *Biol. Bull.*, **147,** 81–94.
8.2

Calow, P. & Jennings, J. B. (1977). Optimal strategies for the metabolism of reserve materials in Microbes and Metazoa. *J. theor. Biol.* 65: 601–603
8.5

Calow, P. & Woollhead, A. S. (1977) The relation between ration, reproductive effort and age-specific mortality in the evolution of life-history strategies—some observations on freshwater triclads. *J. Anim. Ecol.*, **46,** 765–781.
10.2.2

Campbell, A., Singh, N. B. & Sinha, R. N. (1976) Bioenergetics of the granary weevil, *Sitophilus granarius* (L.) (Coleoptera: Curculionidae). *Can. J. Zool.*, **54,** 786–798.
12.3

Campling, R. C. (1970) Physical regulation of voluntary intake. In *Physiology of Digestion and Metabolism in Ruminant*. (Ed. Phillipson A. T.) Oriel Press, Newcastle upon Tyne.
5.4.4

Caraco, T. (1979) Time budgeting and group size: A test of theory. *Ecology*, **60,** 618–627.
13.4.1

Caraco, T. & Wolf, L. L. (1975) Ecological determinants of group sizes of foraging lions. *Amer. Natur.*, **109,** 343–352.
13.5.2

Case, T. J. (1978) On the evolution and adaptive significance of postnatal growth rates in terrestrial vertebrates. *Q. Rev. Biol.* **53,** 243–281
9.3.1, 9.4.1, 9.4.2, 9.5

Castle, E. J. & Castle, M. E. (1956) The rate of passage of food through the alimentary tract of pigs. *J. agric. Sci., Camb.*, **47,** 196–204.
5.5.4 (Fig 5.10 caption)

Charlesworth, B. (1980) *Evolution in Age-Structured Populations.* Cambridge University Press, Cambridge.
7.5, 7.8

Charlesworth, B. & Williamson, J. A. (1975) The probability of survival of a mutant gene in an age-structured population and implications for the evolution of life-histories. *Genet. Res.*, **26,** 1–10.
7.5, 7.8

Charnov, E. L. (1976) Optimal foraging: the marginal value theorem. *Theor. Popul. Biol.*, **9,** 129–136.
4.2

Charnov, E. L. & Krebs, J. R. (1973) On clutch size and fitness. *Ibis*, **116,** 217–219.
10.2.2, 10.4

Charnov, E. L., Maynard Smith, J. & Bull, J. J. (1976) Why be an hermaphrodite? *Nature (London)*, **263,** 125–126.
10.6

Charnov, E. L. & Schaffer, W. M. (1973) Life history consequences of natural selection: Cole's result revisited. *Amer. Natur.*, **107,** 791–793.
7.7

Chinnici, J. P. (1971) Modification of recombination frequency in *Drosophila*. I. Selection for increased and decreased crossing-over. *Genetics*, **69,** 71–83.
12.1

Chollet, R. & Ogren, W. L. (1972) Oxygen inhibits maize bundle sheath photosynthesis. *Bioch. Biophys. Res. Commun.*, **46,** 2062–2066.
3.7.2

Church D. C. (1975) *Digestive Physiology and Nutrition of Ruminants.* 2nd edn. O & B Books, Corvallif Oregon
5.1, 5.5.4

Clausen, J., Keck, D. D. & Hiesy, W. M. (1948) Experimental studies on the nature of species II. Environmental responses of climatic races of Achillea. Carnegie Inst. Washington. Pub. 581, iii + 129 pp.
3.1

Clegg, D. O., Conn, E. E. & Janzen, D. H. (1979) Developmental fate of the cyanogenic glucoside linamarin in Costa Rican wild lima bean seeds. *Nature (London)*, **278,** 343–344.
6.2

Clemens, E. T., Stevens, C. E. & Southworth, M. (1975) Sites of organic acid production and pattern of digesta movement in the gastrointestinal tract of geese. *J. Nutr.*, **105,** 1341–1350.
5.6.1

Clutton-Brock, T. H. & Albon, S. D. (1979) The roaring of red deer and the evolution of honest advertisement. *Behaviour*, **69,** 145–170.
13.6.1.

Clutton-Brock, T. H., Albon, S. D., Gibson, R. M. & Guinness, F. E. (1979) The logical stag: Adaptive aspects of fighting in red deer (*Cervus elaphus* L.). *Anim. Behav.*, **27,** 211–225.
13.6.1.

Clutton-Brock, T. H. & Harvey, P. H. (1977) Primate ecology and social organisation. *J. Zool. (Lond.)*, **183,** 1–39.
13.2., 13.4.2

Clutton-Brock, T. H. & Harvey, P. H. (1979) Comparison and adaptation. *Proc. R. Soc. Lond. B.*, **205,** 547–565.
1.4.1

Cody, M. L. (1966) A generalised theory of clutch size. *Evolution*, **20,** 174–184.
12.4

Cody, M. L. (1973) Coexistence, coevolution and convergent evolution in seabird communities. *Ecology*, **54,** 31–44.
9.4.1, 9.4.2

Cody, M. L. (1971) Finch flocks in the Mohave desert. *Theor. Popul. Biol.*, **2,** 142–158.
13.5.

Coe, M. J., Bourn, D. & Swingland, I. R. (1979) The biomass production and carrying capacity of giant tortoises on Aldabra. *Phil. Trans. Roy. Soc. Lond. B*, **286,** 163–176.
5.6.3

Cohen, D. (1971) Maximising final yield when

growth is limited by time or by limiting resources. *J. theor. Biol.*, **33,** 299–307.
10.2.1

Cohen, D. & Parnas, H. (1976) An optimum policy for the metabolism of storage materials in unicellular algae. *J. theor. Biol.*, **56,** 1–19.
8.5

Cohen, J. (1977) *Reproduction.* Butterworths, London.
7.6

Cole, L. C. (1954) The population consequences of life history phenomena. *Quart. Rev. Biol.*, **29,** 103–137.
1.6, 7.7, 10.1, 12.4

Coleman, J. D. (1974) Breakdown rates of foods ingested by starlings. *J. Wildl. Mgmt*, **38**, 910–912.
5.5.2 (Fig 5.9 caption)

Comfort, A. (1979) *The Biology of Senescence.* 3rd edn. Churchill Livingstone, Edinburgh.
7.8

Constantz, G. D. (1979) Life history patterns of a livebearing fish in contrasting environments. *Oecologia*, **40,** 189–206.
12.3

Cook, R. E. (1979) Patterns of juvenile mortality and recruitment in plants. In *Topics in Plant Population Biology*. (Eds. Solbrig, O. T. *et al.*) Columbia University Press, New York.
12.4

Cornic, G. (1976) Physiologie végétale: Effet exercé sur l'activité photosynthétique du *Sinapsis alba* L. par une inhibition temporaire de la photorespiration se deroulant dansunair sans CO_2 *C. R. Acad. Sci., Paris*, **282,** 1955–57.
3.5

Cowie, R. J. (1977) Optimal foraging in Great Tits (*Parus major*). *Nature*, **268,** 137–139.
4.2

Cox, C. R. & Le Boeuf, B. J. (1977) Female incitation of male competition: a mechanism of mate selection. *Amer. Natur.*, **111,** 317–335.
13.6.6

Crampton, E. W. & Harris, L. E. (1969) *Applied Animal Nutrition.* 2nd edn. Freeman, San Francisco.
5.4.2

Cremer, T. & Kirkwood, T. B. L. (1981) Cytogerontology since 1881: a reappraisal of August Weisman and a review of modern progress. *Human Genetics, (in press).*
7.8

Crick, F. (1979) Split genes and RNA splicing. *Science*, **204,** 264–271.
12.1

Crisler, L. (1956) Observations of wolves hunting caribou. *J. Mammal.*, **37,** 337–346.
13.4.1

Crompton, A. W., Taylor, C. R. & Jagger, J. A. (1978) Evolution of homeothermy in mammals. *Nature (London)*, **272,** 333–336.
2.7

Crook, J. H. (1964) The evolution of social organisation and visual communication in the weaver birds (Ploceinae). *Behaviour*, **10,** (Suppl.) 1–178.
13.2

Crook, J. H. (1965) The adaptive significance of avian social organisation. *Symp. zool. Soc. Lond.*, **14,** 181–218.
13.2

Crook, J. H. (1970) The socio-ecology of primates. In: *Social Behaviour in Birds and Mammals* (Ed. Crook, J. H.) pp. 103–166. Academic Press, London.
13.2

Crook, J. H. & Gartlan, J. S. (1966) Evolution or primate societies. *Nature (London)* **210,** 1200–1203.
13.4.2.

Croze, H. J. (1970) Searching image in carrion crows. *Z. Tierpsychol.*, **5,** (Suppl.) 1–86.
13.4.2.

Curtis, H. J. (1966) *Biological Mechanisms of Ageing.* Charles C. Thomas, Springfield, Illinois.
7.3

Cys, J. M. (1967) The inability of dinosaurs to hibernate as a possible key factor in their extinction. *J. Paleontol.*, **41,** 266.
2.6

Daly, M. (1979) Why don't male mammals lactate? *J. theor. Biol.*, **78,** 325–345.
10.4

Darling, F. F. (1937) *A Herd of Red Deer.* Oxford University Press, Oxford
13.6.1.

Davenport, H. W. (1977) *Physiology of the Digestive Tract.* 4th edn. Year Book Medical Publishers, Chicago.
5.1, 5.3.3

Davenport, H. W. (1978) *A Digest of Digestion.* 2nd edn. Year Book Medical Publishers, Chicago.
5.1, 5.3.3

Davies, D. D. (1979) The central role of

phosphoenol-pyruvate in plant metabolism. *Ann. Rev. Plant Physiol.*, **30,** 131–158.
3.7.1

Davies, N. B. (1976) Food, flocking and territorial behaviour of the pied wagtail (*Motacilla alba yarellii* Gould) in winter, *J. Anim. Ecol.*, **45,** 215–235.
13.5.1

Davies, N. B. (1978) Ecological questions about territorial behaviour. In *Behavioural Ecology: an Evolutionary Approach.* (Eds. Krebs, J. R. & Davies N. B.) pp. 317–350. Blackwell Scientific Publications, Oxford.
13.5.1., 13.6.2., 13.7.1.

Davies, N. B. & Halliday, T. R. (1978) Deep croaks and fighting assessment in toads, *Bufo bufo. Nature (London)*, **391,** 56–58.
13.6.5.

Davies, N. B. & Halliday, T. R. (1979) Competitive mate searching in male common toads, *Bufo bufo. Anim. Behav.*, **27,** 1253–1268.
13.6.1., 13.6.4.

Davies, N. B. & Krebs, J. R. (1978) Introduction: Ecology, Natural Selection and Social Behaviour. In *Behavioural Ecology: an Evolutionary Approach.* (Eds. Krebs J. R. & Davies N. B.) pp. 1–18. Blackwell Scientific Publications, Oxford.
13.2

Davis, J. (1961) Some seasonal changes in morphology of the rufous-sided towhee. *Condor*, **63,** 313–321.
5.3.1

Davis, W. H. (1970) Hibernation: ecology and physiological ecology. In *Biology of Bats.* Vol. 1. (Ed. Wimsatt W. A.) pp. 265–300. Academic Press, New York.
8.5

Dawkins, R. (1976) *The Selfish Gene.* Oxford University Press, Oxford.
1.3, 13.2, 13.4.3

Dawkins, R. & Carlisle, T. R. (1976) Parental investment, mate desertion and a fallacy. *Nature (London)*, **262,** 131–133.
10.4

Dawkins, R. & Krebs, J. R. (1979) Arms races between and within species. *Proc. R. Soc. Lond., B*, **205,** 489–511.
13.4.1

Delong, R. & Poplin, L. (1977) On the etiology of ageing. *J. theor. Biol.*, **67,** 111–120.
7.8

Derickson, W. K. (1976) Lipid storage and utilization in reptiles. *Am. Zool.*, **16,** 711–723
8.3

Diefenbach, C. O. da C. (1975) Gastric function in *Caiman crocodilus* (Crocodylia: Reptilia) 1. Rate of gastric motility as a function of temperature. *Comp. Biochem. Physiol.*, **51A,** 259–265.
5.5.2 (Fig. 5.9 caption)

Dijk, Th. van. (1979) On the relationship between reproduction, age and survival of the carabid beetles: *Calathus melanocephatus* L. and *Pterostichus coerulescens* L. (Coleoptera, Carabidae). *Oecologia,* **40,** 63–80.
10.2.2

Dinius, D. A. & Baumgardt, B. R. (1970) Regulation of food intake in ruminants. 6. Influence of caloric density of pelleted rations. *J. Dairy Sci.*, **53,** 311–316.
5.4.4 (Fig. 5.7 caption)

Dittrich, P. (1976) Equilibration of label in malate during dark fixation of CO_2 in *Kalanchoë fedtochenkoi. Plant Physiol.*, **58,** 288–291.
3.7.1

Dittrich, P., Campbell, W. H. & Black, C. C. (1973) Phosphoenolpyruvate carboxykinase in plants exhibiting crassulacean acid metabolism. *Plant Physiol.*, **52,** 357–361.
3.7.1

Doliner, L. H. & Joliffe, R. A. (1979) Ecological evidence concerning the adaptive significance of the C_4 dicarboxylic acid pathway of photosynthesis. *Oecologia*, **38,** 23–34.
3.8.1

Donart, G. B. (1969) Carbohydrate reserves of six mountain range plants as related to growth. *J. range Manag.*, **22,** 411–415.
8.5

Downer, R. G. H. & Matthews, J. R. (1976) Patterns of lipid distribution and utilisation in insects. *Am. Zool.*, **16,** 733–745.
8.2

Doyle, R. W. (1979) Ingestion rate of a selective deposit feeder in a complex mixture of particles. Testing the energy maximization hypothesis. *Limnol. Oceanogr.*, **24,** 867–874.
4.3.2

Doyle, W. L. & Watterson, R. L. (1949) The accumulaion of glycogen in the "glycogen body" of the nerve cord of the developing chick. *J. Morph.*, **85,** 391–403.
8.3

DuMond, F. V. & Hutchinson, T. C. (1967) Squirrel monkey reproduction: the "fatted" male phenomenon and seasonal spermato-

genesis. *Science*, **158,** 1067–1070.
8.5

Dunbar, R. I. M. & Dunbar, P. (1975) Social dynamics of gelada baboons. *Contrib. Primatol.*, **6,** (Karger: Basle).
13.6.1

Dunn, E. H. (1975) Growth, body composition and energy content of nestling double-crested cormorants *Condor*, **77,** 431–438.
8.5

Eckhardt, R. C. (1979) The adaptive syndromes of two guilds of insectivorous birds in the Colorado Rocky Mountains. *Ecol. Monogr.*, **49,** 129–149.
12.2.7

Ehleringer, J. R. & Björkman, O. (1977) Quantum yields for CO_2 uptake in C_3 and C_4 plants. Dependence on temperature, CO_2 and O_2 concentration. *Plant Physiol.*, **59,** 85–90.
3.4, 3.7.2

Eisenberg, J. F. (1977) The evolution of the reproductive unit in the Class Mammalia. In *Reproductive Behaviour and Evolution*. (Eds. Rosenblatt, J. S. & Komisaruk, B. R.). Plenum, New York.
13.7.2

Eisenberg, J. F. & Kleiman, D. G. (1972) Olfactory communication in mammals. *Ann. Rev. Ecol. Syst.*, **3,** 1–32.
13.6.2

Eisner, T. & Adams, P. A. (1975) Startle behavior in an ascalaphid (Neuroptera). *Psyche*, **82,** 304–305.
6.18

Eisner, T. & Dean, J. (1976) Ploy and counterploy in predator-prey interactions: orb-weaving spiders versus bombardier beetles. *Proc. Nat. Acad. Sci. USA*, **73,** 1365–1367.
6.8, 6.18

Eisner, T., Kriston, I. & Aneshansley, D. J. (1976) Defensive behavior of a termite (*Nasutitermes exitiosus*). *Behav. Ecol. Sociobiol.*, **1,** 83–125.
6.18

Eisner, T. & Shepherd, J. (1966) Defense mechanisms of arthropods. XIX. Inability of sundew plants to capture insects with detachable integumental outgrowths. *Ann. Ent. Soc. Amer.*, **59,** 868–870.
6.18

Eisner, T., van Tassell, E. & Carrel, J. E. (1967) Defensive use of a "fecal shield" by a beetle larva. *Science*, **158,** 1471–1473.
6.18

Elder, D. (1979) Why is regenerative capacity restricted in higher organisms? *J. theor. Biol.*, **81,** 563–568.
7.4.1

Elliott, J. M. (1976) Energy losses in the waste products of brown trout (*Salmo trutta* L.). *J. Anim. Ecol.*, **45,** 561–580.
5.6.3

Elliott, J. M. & Persson, L. (1978) The estimation of daily rates of food consumption for fish. *J. Anim. Ecol.*, **47,** 977–991.
5.5.2, 5.5.3

Elner, R. W. & Hughes, R. N. (1978) Energy maximization in the diet of the shore crab, *Carcinus maenas*. *J. Anim. Ecol.*, **47,** 103–116.
4.3.1

Emlen, J. M. (1973) *Ecology: An Evolutionary Approach*. Addison-Wesley Publishing Company, Massachusetts.
10.3, 10.4

Emlen, S. T. (1978) Co-operative breeding in birds. In *Behavioural Ecology: an Evolutionary Approach*. (Eds. Krebs, J. R. & Davies, N. B.) pp. 245–281. Blackwell Scientific Publications, Oxford.
13.7.2

Emlen, S. T. & Oring, L. W. (1977) Ecology, sexual selection, and the evolution of mating systems. *Science*, **197**, 215–223.
13.6., 13.6.3

Engelmann, M. D. (1966) Energetics, terrestrial field studies and animal productivity. In: *Advances in Ecological Research*. Vol. 3 (Ed. Cragg, J. B.) pp. 73–115. Academic Press, London.
2.8

Estes, R. D. (1966) Behaviour and life history of the wildebeest. *Nature*, **212,** 999–1000.
13.4.1

Estes, R. D. (1974) Social organisation of the African Bovidae. In *The Behaviour of Ungulates and its Relation to Management*. (Eds. Geist, V. Walther, F. R.) pp. 166–205. I.U.C.N. publication, Morges.
13.4.2

Estes, R. D. (1976) The significance of breeding synchrony in the wildebeest. *E. Afr. Wildl. J.*, **14,** 135–152.
13.4.1

Estes, R. D. & Goddard, J. (1967) Prey selection and hunting behaviour in the Af-

rican wild dog. *J. Wildl. Mngt.*, **31,** 52–70.
13.4.3., 13.7.2

Evans, C. H. (1977) Further evidence against the accumulation of altered enzymes in late passage embryonic mouse fibroblasts *in vitro. Exp. Geront.*, **12,** 169–171.
7.3

Evans, G. C. (1972) *The Quantitative Analysis of Plant Growth.* Blackwell Scientific Publications, Oxford.
10.2.1

Evans, P. G. (1969) Winter fat depostion and overnight survival of yellow buntings (*Emberiza citrinella* L.). *J. Anim. Ecol.*, **38,** 415–423.
8.5

Fábry, P (1969) Feeding pattern and nutritional adaptations. Butterworths, London. (Translated by K. Osancova.)
5.6.3

Farlow, J. O., Thompson, C. V. & Rosner, D. E. (1976) Plates of the dinosaur *Stegosaurus*: forced convection heat loss fins? *Science*, **192**, 1123–1125.
2.6

Farner, D. S. (1960) Digestion and the digestive system. In *Biology and Comparative Physiology of Birds* (Ed. Marshall, A. J.) Academic Press, New York.
5.6.3

Farquhar, G. C., van Caemmerer, S. & Berry, J. A. (1980) A biochemical model of photosynthetic CO_2 assimilation in leaves of C_3 species. *Planta*, **149,** 78–90.
3.4

Feare, C. J. & Inglis, I. R. (1979) The effects of reduction of feeding space on the behaviour of captive starlings (*Sturnus vulgaris*). *Ornis Scandinavica*, **10,** 42–47.
13.4.1

Feeny, P. P. (1975) Biochemical coevolution between plants and their insect herbivores. In *Coevolution of Animals and Plants.* (Ed. Gilbert, L. E. & Raven, P. H.) pp. 3–19. University of Texas Press, Austin.
1.6

Feeny, P. (1976) Plant apparency and chemical defense. In *Biochemical Interaction between Plants and Insects* (Eds. Wallace, J. W. & Mansell, R. L.) pp. 1–40. Plenum, New York.
6.10

Feinsinger, P. (1978) Ecological interactions between plants and hummingbirds in a successional tropical community. *Ecol. Monogr.*, **48,** 269–287.
12.2.5

Fell, B. F. & Weekes, T. E. C. (1975) Food intake as a mediator of adaptation in the ruminal epithelium. In *Digestion and Metabolism in the Ruminant*. (Eds. McDonald, I. W. & Warner, A. C. I.) pp. 101–118 University of New England Publishing Unit, Australia.
5.6.3

Fernando, W. (1945) The storage of glycogen in the Temnocephaloidea. *J. Parasitol.*, **31,** 185–190.
8.3

Finkelstein, A., Rubin, L. L. & Tzeng, M. (1976) Black widow spider venom: effect of purified toxin on lipid bilayer membranes. *Science,* **193,** 1009–1011.
6.18

Fischer, H., Hollands, K. G. & Weiss, H. S. (1922) Environmental temperature and composition of body fat. *Proc. Soc. Exp. Biol. Med.*, **110,** 832–833.
8.2

Fisher, R. A. (1930) *The Genetical Theory of Natural Selection.* Clarendon Press, Oxford.
1.2, 1.3, 1.6, 7.5, 7.8, 8.7

Fitzpatrick, L. C. (1973) Energy allocation in the Allegheny Mountain Salamander, *Desmognathus ochrophaeus. Ecol. Monogr.*, **43,** 43–58.
12.3

Fletcher, D. J. C. (1978) The African bee, *Apis mellifera adansonii*, in Africa. *Ann. Rev. Ent.*, **23,** 151–171.
6.6

Flux, J. E. C. (1970) Lifehistory of the mountain hare (*Lepus timidus scotius*) in Northeast Scotland. *J. Zool.*, **161,** 75–123
8.5

Forbes, J. M. (1978) Models of the control of food intake and energy balance in ruminants. In *Hunger Models* (Ed. Booth, D. A.) Academic Press, London.
5.4.4

Forrester, M. L., Krotkov, G. & Nelson, C. D. (1966) Effect of oxygen on photosynthesis, photorespiration and respiration in detached leaves I. Soybean. *Plant Physiol.*, **41,** 422–427.
3.4

Frame, L. H. & Frame, G. W. (1976) Female African wild dogs emigrate. *Nature (London)*, **263,** 227–229.
13.7.2

Freeland, W. J. & Janzen, D. H. (1974) Strategies in herbivory by mammals: the role of plant secondary compounds. *Amer. Natur.*, **108,** 269–289.
4.3

Frisch, R. E. & McArthur, J. W. (1974) Menstrual cycles: fatness as a determinant of minimum weight for height necessary for their onset. *Science*, **185**, 949–951.
8.5

Fry, C. H., Ash, J. S. & Ferguson-Lee, I. J. (1970) Spring weights of Palearctic migrants at Lake Chad. *Ibis*, **112,** 58–62.
8.5

Gadgil, M. & Bossert, W. H. (1970) Life historical consequences of natural selection. *Amer. Natur.*, **104,** 1–24.
7.5, 10.1, 10.2.2

Gadgil, M. & Solbrig, O. T. (1972). The concept of r and K selection: evidence from wild flowers and some theoretical considerations. *Amer. Natur.*, **106,** 14–31.
1.6, 7.5, 12.4

Galas, D. J. & Branscomb, E. W. (1976) Ribosomes slowed by mutation to streptomycin resistance. *Nature (London.)* **262,** 617–619.
7.3

Gallant, J. & Palmer, L. (1979) Error propagation in viable cells. *Mech. Ageing Dev.*, **10,** 27–38.
7.3

Gasaway, W. C. (1976) Cellulose digestion and metabolism by captive rock ptarmigan. *Comp. Biochem. Physiol.*, **54A,** 179–182.
5.3.5

Gates, D. M. (1962) *Energy Exchange in the Biosphere*. Harper & Row, New York.
12.3

Gates, D. M. (1968) Transpiration and leaf temperature. *Ann. Rev. Plant. Physiol.*, **19,** 211–238.
12.3

Gates, D. M., Adlerfer, R. & Taylor, S. E. (1968) Leaf temperature of desert plants. *Science*, **159,** 994–995.
12.3

Geist, V. (1966) The evolution of horn-like organs. *Behaviour*, **27,** 175–214.
13.6.1, 13.6.4

Geist, V. (1971) *Mountain Sheep: a Study in Behaviour and Evolution*. University of Chicago Press, Chicago.
13.6.2

Geist, V. (1974) On the relationship of social evolution and ecology in ungulates. *Am. Zool.*, **14,** 205–220.
13.4.2

Gibb, J. A. (1956) Food, feeding habits, and territory of the Rock Pipit, *Anthus spinoletta*. *Ibis*, **98,** 506–530.
12.2

Gibb, J. A. (1960) Populations of tits and goldcrests and their food supply in pine plantations. *Ibis*, **102,** 163–208.
12.2

Gilbert, L. E. & Raven, P. H. (Eds.) (1975) *Coevolution of Animals and Plants*. 246 pp. University of Texas Press, Austin.
6.1

Gilbert, W. (1978) Why genes in pieces? *Nature (London)*, **271,** 501.
12.1

Gill, F. B. & Wolf, L. L. (1975) Economics of feeding territoriality in the golden-winged sunbird. *Ecology*, **56,** 333–345.
13.5.1

Givnish, T. (1979) On the adaptive significance of leaf form. In *Topics in Plant Population Biology*. (eds Solbrig, O. T. *et al.*) Columbia University Press, New York.
12.3

Givnish, T. J. & Vermeij, G. J. (1976) Sizes and shapes of liana leaves. *Amer. Natur.*, **110,** 743–776.
12.3

Glesener, R. R. & Tilman, D. (1978) Sexuality and the components of environmental uncertainty. *Amer. Natur.*, **112,** 659–673.
10.5

Goddard, D. R. (1947) The respiration of cells and tissues. In *Physical Chemistry of Cells and tissues*. (Ed. Höber R.) pp. 371–444. J. & A. Churchill, London.
2.6

Goldberg, A. L. (1972) Degradation of abnormal proteins in *Escherichia coli*. *Proc. Nat. Acad. Sci. USA*, **69,** 422–426.
7.3

Goldberg, A. L. & Dice, J. F. (1974) Intracellular protein degradation in mammalian and bacterial cells. *Ann. Rev. Biochem.*, **43,** 835–869.
7.3

Goldberg, A. L. & St. John, A. C. (1975) Intracellular protein degradation in mammalian and bacterial cells, Part 2. *Ann. Rev. Biochem.*, **45,** 747–803.
7.3

Goldstein, R. A. & Elwood, J. W. (1971) A two-compartment three-parameter model for the absorption and retention of ingested

elements by animals. *Ecology*, **52,** 935–939.
5.5.4

Goldsworthy, A. & Day, P. R. (1970) Further evidence for reduced role of photorespiration in low compensation point species. *Nature (London)*, **228,** 687–688.
3.7.2

Goodman, D. (1979) Regulatory reproductive effort in a changing environment. *Amer. Natur.*, **113,** 735–748.
10.2.2

Goodson, W. H. & Hunt, T. K. (1979) Wound healing and ageing. *J. Invest. Dermatol.*, **73,** 88–91.
7.9

Gosling, L. M. (1969) Parturition and related behaviour in Coke's hartebeest. *Alcelaphus buselaphus cokei* Gunther. *J. Reprod. Fert.*, **6,** (Suppl.) 265–286.
13.4.2, 13.7.1

Gosling, L. M. (1974) The social behaviour of Coke's hartebeest. *Alcelaphus buselaphus cokei*. In *The Behaviour of Ungulates and its Relation to Management* (Eds. Geist, V. & Walther, F. R.) pp. 488–511. I.U.C.N., Morges.
13.5.1, 13.6.2, 13.6.5, 13.6.6

Gosling, L. M. (1975) The ecological significance of male behaviour in Coke's hartebeest. Unpublished Ph.D. thesis, University of Nairobi.
13.5.1, 13.6.2 13.6.6, 13.7.2

Gosling, L. M. (1977) Coypu *Myocastor coypus*. In *The Handbook of British Mammals*. (Eds. Corbet, G. B. & Southern H. N.) pp. 256–265. Blackwell Scientific Publications, Oxford.
13.6.5, 13.7.1

Gosling, L. M. (1980a) The duration of lactation in feral coypus (*Myocastor coypus*). *J. Zool., (Lond.)*, **191,** 461–474.
13.7.1

Gosling, L. M. (1980b) Defence guilds of savannah ungulates as a context for scent communication. *Symp. Zool. Soc. Lond.*, **45,** 195–212.
13.4.1, 13.8

Goss, R. J. (1964) *Adaptive Growth*. Lagos Press, London.
9.7.4

Goss, R. J. (1967) *Principles of Regeneration*. Academic Press, New York.
7.4.1

Goss-Custard, J. D. (1976) Variation in the dispersion of redshank (*Tringa totanus*) on their winter feeding grounds. *Ibis*, **118,** 257–263.
13.4.1

Gotto, R. V. (1962) Egg number and ecology in commensal and parasitic copepods. *Ann. Mag. Nat. Hist.*, **13,** 97–104.
10.3

Gould, S. J. & Lewontin, R. C. (1979) The spandrels of San Marco and the Panglossian paradigm: a critique of the adaptationist's programme. *Proc. R. Soc. Lond., B*, **205,** 581–598.
1.4.1, 1.4.2

Gradmann, H. (1926) Untersuchungen über die wasserverhältrisse des bodens als grundlage des pflanzenwachstums. *Gahrb. Wiss. Botan.*, **69,** 1–100.
3.6.1

Grime, J. P. (1977) Evidence for the existence of three primary strategies in plants and its relevance to ecological and evolutionary theory. *Amer. Natur.*, **111,** 1169–1194.
12.4

Grime, J. P. (1978) Interpretation of small-scale patterns in the distribution of plant species in space and time. In *Structure and Functioning of Plant Populations*. pp. 101–124. Verhandelingen der Koninklijke Nederlandse Akademie van Wetenschappen Afdeling Natuurkunde, Tweede Reeks, deel 70.
1.4.1

Grime, J. P. (1979) *Plant Strategies and Vegetation Processes*. John Wiley & Sons, Chichester.
1.4.1, 1.6, 3.1, 7.4.2, 9.4.1, 9.4.2, 12.4

Grime, J. P. & Hunt, R. H. (1975) Relative growth rate: its range and adaptive significance in a local flora. *J. Ecol.*, **63,** 393–422.
9.3.3, 9.4.2, 9.5

Grodzinski, B. & Butt, V. S. (1976) Hydrogen peroxide production and the release of carbon dioxide during glycolate oxidation in leaf peroxisomes. *Planta*, **128,** 225–231.
3.4

Grovum W. L. & Williams V. J. (1973) Rate of passage of digesta in sheep. 4. Passage of marker through the alimentary tract and the biological relevance of rate-constants derived from the changes in concentration of marker in faeces. *Br. J. Nutr.*, **30,** 313–329.
5.5.4

Grovum W. L. & Williams V. J. (1977). Rate of passage of digesta in sheep. 6. The effect of level of food intake on mathematical predictions of the kinetics of digesta in the reticulorumen and intestines. *Br. J. Nutr.*, **38,**

425–436.
5.5.4

Grzimek, M. & Grzimek, B. (1960) A study of the game of the Serengeti Plains. *Z. Saugetierk.*, **25,** 1–61.
13.4.1

Guinness, F. E., Hall, M. J. & Cockerill, R. A. (1979) Mother–offspring association in red deer (*Cervus elaphus* L.) on Rhum. *Anim. Behav.*, **27,** 536–544.
13.7.2

Gurdon, J. S. (1974) *Control of Gene Expression in Animal Development*. Oxford University Press, Oxford.
7.2.3

Haberlandt, B. (1904) *Physiologische Pflanzanatomie*. 616 pp. Engelmann, Leipzig.
3.7.2

Hails, C. J. & Bryant, D. M. (1979) Reproductive energetics of a free-living bird. *J. Anim. Ecol.*, **48,** 471–482.
10.4

Hairston, N. G., Tinkle, D. W. & Wilbur, H. M. (1970) Natural selection and the parameters of population growth. *J. Wildlife Mngt*, **34,** 681–690.
12.4

Haldane, J. B. S. (1924) A mathematical theory of natural and artificial selection. *Trans. Camb. Philos. Soc.*, **23,** 19–40.
1.2, 1.3

Halliday, T. R. (1978) Sexual selection and mate choice. In *Behavioural Ecology: an Evolutionary Approach*. (Eds Krebs, J. R. & Davies N. B.) pp. 180–213. Blackwell Scientific Publications, Oxford.
13.6.6.

Halliwell, B. & Butt, V. S. (1974) Oxidative decarboxylation of glycolate and glyoxylate by leaf peroxisomes. *Biochem. J.*, **138,** 217–224.
3.4

Hallwachs, W. & Janzen, D. H. Guanacaste tree seeds (Leguminosae: *Enterolobium cyclocarpum*) as food for Costa Rican spiny pocket mice (Heteromyidae: *Liomys salvini*). *Ecology* (*submitted for publication*).
6.13

Hamilton, W. D. (1964) The genetical theory of social behaviour (I & II). *J. theor. Biol.*, **7,** 1–16; 17–32.
13.2, 13.7.2

Hamilton W. D. (1966) The moulding of senescence by natural selection. *J. theor. Biol.*, **12,** 12–45.
7.8

Hamilton, W. D. (1971) Geometry for the selfish herd. *J. theor. Biol.*, **31,** 295–311.
13.4, 13.4.1

Hammond, J., Mason, I. L. & Robinson, T. J. (1971) *Hammond's Farm Animals*. 4th edn. 293 pp. Edward Arnold, London.
8.5

Hanawalt, P. C., Cooper, P. K., Ganesan, A. K. & Smith, C. A. (1979) DNA repair in bacteria and mammalian cells. *Ann. Rev. Biochem.*, **48,** 783–836
7.3

Harborne, J. B. (1978a) *Introduction to Ecological Biochemistry*, 243 pp. Academic Press, New York.
6.1

Harborne, J. B. (Ed.) (1978b) *Biochemical Aspects of Plant and Animal Coevolution*, 435 pp. Phytochemical Society of Europe Symposia Series, No. 15. Academic Press, London.
6.1

Harborne, J. B. (Ed.) (1972) *Phytochemical Ecology*. 272 pp. Annual Proceedings of the Phytochemical Society, No. 8. Academic Press, London.
6.1

Harper, J. L. (1977) *Population Biology of Plants*. Academic Press, London.
1.5, 9.3.3, 9.4.1, 12.4

Harper, J. L., Lovell, P. H. & Moore, K. G. (1970) The shapes and sizes of seeds. *Ann. Rev. Ecol. Syst.*, **1,** 327–356.
10.3

Haukioja, E. & Hakala, T. (1978) Life history evolution in *Anodonta piscinalis* (Mollusca, Pelecyopoda). *Oecologia*, **35,** 253–266.
10.2.2

Harman, D. (1956) Ageing: a theory based on free radical and radiation chemistry. *J. Gerontol.*, **11,** 298–300.
7.8

Hart, R. W., D'Ambrosio, S. M., Ng, K. J. & Modak, S. P. (1979) Longevity, stability and DNA repair. *Mech. Ageing Dev.*, **9,** 203–223.
7.8

Hart, R. W. & Setlow, R. B. (1974) Correlation between deoxyribonucleic acid excision-repair and lifespan in a number of mammalian species. *Proc. Nat. Acad. Sci. USA*, **71,** 2169–2173.
7.1, 7.8

Hartnell, G. F. & Salter, L. D. (1979) Determination of rumen fill, retention and ruminal turnover rates of ingesta at different stages of lactation in dairy cows. *J. Anim. Sci.*, **48,** 381–392.
5.5.4

Harwood, J. (1975) The grazing strategies of blue geese, *Anser caerulescens*. Unpublished Ph.D. thesis, University of Western Ontario.
5.6.1

Hassell, M. P. & May, R. M. (1974) Aggregation of predators and insect parasites and its effect on stability. *J. Anim. Ecol.*, **43**, 567–594.
4.2.1

Hatch, M. D. (1970) Chemical energy costs for CO_2 fixation by plants with differing photosynthetic pathways. In *Prediction and Measurement of Photosynthetic Productivity*. pp. 215–220. Centre for Agric. Publ. Doc., Wageningen.
3.6.1

Hatch, M. D. (1979) Regulation of C_4 photosynthesis: factors affecting cold-mediated inactivation and reactivation of pyruvate, Pi dikinase. *Aust. J. Plant Physiol.*, **6**, 607–619.
3.8.3

Hatch, M. D., Kagawa, T. & Craig, S. (1975) Subdivision of C_4 pathway species based on different C_4 acid decarboxylating systems and ultrastructural features. *Aust. J. Plant Physiol.*, **2**, 111–128.
3.7.2

Hatch, M. D. & Slack, C. R. (1968) A new enzyme for the interconversion of pyruvate and phosphopyruvate and its role in the C_4 dicarboxylic acid pathway of photosynthesis. *Biochem. J.*, **106**, 141–146.
3.8.3

Hayflick, L. (1965) The limited *in vitro* lifetime of human diploid cell strains. *Expl. Cell Res.*, **37**, 614–636.
7.8

Hayflick, L. (1973) The biology of human ageing. *Amer. J. Med. Sci.*, **265**, 432–445.
7.8

Hayflick, L. (1977) The cellular basis for biological ageing. In *Handbook of the Biology of Ageing* (Eds. Finch, C. E. & Hayflick, L.) pp. 159–186. Van Nostrand-Reinhold, New York.
7.8

Hayflick, L. & Moorhead, P. S. (1961) The serial cultivation of human diploid cell strains. *Expl. Cell Res.*, **25**, 585–621.
7.8

Heath, D. J. (1977) Simultaneous hermaphroditism; cost and benefit. *J. theor. Biol.*, **64**, 363–373.
10.6

Heath, D. J. (1979) Brooding and the evolution of hermaphroditism. *J. theor. Biol.*, **81**, 151–155.
10.6

Hegsted, D. M., Jack, C. W. & Stare, F. J. (1962) The composition of human adipose tissue from several parts of the world. *Am. J. clin. Nutr.*, **10**, 11–18.
8.2

Heinrich, B. (1975) Energetics of pollination. *Ann. Rev. Ecol. Syst.*, **6**, 139–170.
12.2.5

Heinrich, B. (1977) Why have some animals evolved to regulate a high body temperature? *Amer. Natur.*, **111**, 623–640.
2.7

Heinrich, B. (1979) Majoring and minoring by foraging bumblebees, *Bombus vagans*: an experimental analysis. *Ecology*, **60**, 245–255.
4.3.1.1

Heinrich, B. (1979) *Bumblebee Economics*. Harvard University Press, Cambridge, Mass.
12.2.5

Heuts, M. (1953) Regressive evolution in cave animals. In *Evolution*. pp. 290–309. Society for Experimental Biology, Symposium 7. Academic Press, New York.
9.7.3

Hemmingsen, A. M. (1950) The relation of standard (basal) energy metabolism to total fresh weight of living organisms. *Rep. Steno Memo. Hosp. Nord. Insulin Lab.*, **4**, 1–58.
2.4, 2.6

Hemmingsen, A. M. (1960) Energy metabolism as related to body size and respiratory surfaces, and its evolution. *Rep. Steno Memo. Hosp. Nord. Insulin Lab.*, **9**, 1–110.
2.4, 2.5

Hew, C. S., Krotkov, G. & Canvin, D. T. (1969) Effects of temperature on photosynthesis and CO_2 evolution in light and darkness by green leaves. *Plant Physiol.*, **44**, 671–677.
3.4

Hickling, C. F. 1930. A contribution towards the lifehistory of the spur-dog. *J. mar. Biol. Ass., U.K.*, **16**, 529–576.
8.7

Hickman, J. C. (1975) Environmental unpredictability and plastic energy allocation strategies in the annual *Polygonum cascadense* (Polygonaceae). *J. Ecol.*, **63**, 689–701.
12.4

Hickman, J. C. (1977) Energy allocation and niche differentiation of four co-existing annual species of *Polygonum* in western North America. *J. Ecol.*, **65**, 317–326.
12.4

Hickman, J. C. (1979) The biology of plant numbers. In *Topics in Plant Population Biology*. (eds. Solbrig, O. T. *et al.*) Columbia University Press, New York.
12.4

Hilderbrand, M. (1974) *Analysis of Vertebrate Structure*. 710 pp. John Wiley & Sons, New York.
8.3

Hildrew, A. G. & Townsend, C. R. (1977), The influence of substrate on the functional response of *Plectrocnemia conspersa* (Curtis) larvae (Trichoptera: Polycentropodidae). *Oecologia*, **31,** 21–26.
4.2

Hildrew, A. G. & Townsend, C. R. (1980) Aggregation, interference and foraging by larvae of *Plectrocnemia conspersa* (Trichoptera: Polycentropodidae). *Anim. Behav.*, **28,** 553–560.
4.2

Hirsch, H. R. (1978) The waste-product theory of ageing: waste dilution by cell division. *Mech. Ageing Dev.*, **8,** 51–62.
7.3, 7.8

Hirshfield, M. F. & Tinkle, D. W. (1975) Natural selection and the evolution of reproductive effort. *Proc. Nat. Acad. Sci. U.S.A.*, **72,** 2227–2231.
10.2.2

Hochachka, P. W. (1976) *Design of metabolic and enzyme machinery to fit lifestyle and environment*. pp. 3–31. Biochemical Society Symposium 41.
9.2.2

Hochachka, P. W. & Somero, G. N. (1973) *Strategies of Biochemical Adaptation*. Saunders, Philadelphia.
1.1, 9.2.2, 12.1

Hoffman, G. W. (1974) On the origin of the genetic code and the stability of the translation apparatus. *J. Mol. Biol.*, **86,** 349–362.
7.3

Hofmann R. R. (1973) *The Ruminant Stomach*. East African Literature Bureau, Nairobi, Kenya.
5.1, 5.4.1

Holaday, A. S. & Bowes, G. (1980) C_4 acid metabolism and dark CO_2 fixation in a submersed aquatic macrophyte (*Hydrilla verticillata*). *Plant Physiol*, **65,** 331–335.
3.6.1

Holliday, R. & Tarrant, G. M. (1972) Altered enzymes in ageing human fibroblasts. *Nature (London)*, **238,** 26–30.
7.3

Hopfield, J. J. (1974) Kinetic proofreading: a new mechanism for reducing errors in biosynthetic processes requiring high specificity. *Proc. Nat. Acad. Sci., USA* **71,** 4135–4139.
7.3

Hopkins, A. (1966) The pattern of gastric emptying: a new view of old results. *J. Physiol., Lond.*, **182,** 144–149.
5.5.2, 5.5.3

Hopkins, C. A., Subramanian, G. & Stallard, H. E. (1972) The development of *Hymenolepis diminuta* in primary and secondary infections in mice. *Parasitology*, **64,** 401–412.
9.3.2

Hoppe P. P. (1977) Rumen fermentation and body weight in African ruminants. *Proceedings of XIIIth Congress of Game Biologists*. pp. 141–150. Atlanta, USA.
5.4.1

Horn, H. S. (1978) Optimal tactics of reproduction and Life History. In *Behavioural Ecology: an Evolutionary Approach* (Eds. Krebs, J. R. & Davies, N. B.) pp. 411–429. Blackwell Scientific Publications, Oxford.
10.1, 13.7

Hubbard, S. F. & Cook, R. M. (1978) Optimal foraging by parasitoid wasps. *J. Anim. Ecol.*, **47,** 593–604.
4.2

Hubbell, S. P. (1971) Of sowbugs and systems: the ecological energetics of a terrestrial isopod. In *Systems Analysis and Simulation in Ecology*. (Ed. Pattern, B. C.) Academic Press, New York.
9.6

Hubbell, S. P., Sikora, A. & Paris, O. H. (1965) Radiotracer, gravimetric and calorimetric studies of ingestion and assimilation rates of an isopod. *Health Physics*, **11,** 1485–1501.
5.6.3

Huey, R. B. & Pianka, E. R. (1981) Ecological Consequences of foraging mode. *Ecology*, (*in press*).
12.2.7

Hughes, R. N. (1979) Optimal diets under the energy maximization premise: the effects of recognition time and learning. *Amer. Natur.*, **113,** 209–22.
4.3.1.1

Hughes, R. N. & Elner, R. W. (1979) Tactics of a predator, *Carcinus maenas*, and morphological responses of the prey, *Nucella lapillus*. *J. Anim. Ecol.*, **48,** 65–78.
4.3.1.1

Humphreys, W. F. (1979) Production and

respiration in animal populations. *J. Anim. Ecol.*, **48,** 427–453.
2.8

Hungate, R. E. (1966) *The Rumen and its Microbes,* Academic Press, New York.
5.1, 5.5.4, 5.6.3 (Fig 5.11 legend)

Hungate, R. E. (1975) The rumen microbial ecosystem. *Ann. Rev. Ecol. Syst.*, **6,** 39–66.
5.5.4

Hunt, J. N. & Knox, M. T. (1968) Regulation of gastric emptying. In *Handbook of Physiology. Section 6. Alimentary Canal.* American Physiological Society, Washington.
5.5.2 (Fig 5.9 caption)

Hunt, J. N. & Stubbs, D. F. (1975) The volume and energy content of meals as determinants of gastric emptying. *J. Physiol.*, **245,** 209–225.
5.5.3

Hussell, D. T. J. (1972) Factors affecting clutch size in arctic passerines. *Ecol. Monogr.*, **42,** 317–64.
10.4

Huxley, J. (1932) *Problems of Relative Growth.* Methuen, London.
9.7

Hyman, L. H. (1951) *The Invertebrates: Platyhelminthes and Rhynchocoela. The Acoelomate Bilateria.* Vol II. McGraw Hill Book Co., New York.
10.5, 10.6

Jagendorf, A. T. & Uribe, E. (1967) ATP formation caused by acid-base transition of spinach chloroplasts. *Proc. Nat. Acad. Sci. USA*, **55,** 170–177.
3.2

Janis, C. (1976) The evolutionary strategy of the Equidae and the origins of rumen and cecal digestion. *Evolution*, **30,** 757–774.
5.4.2, 5.4.3

Janzen, D. H. (1966) Coevolution of mutualism between ants and acacias in Central America. *Evolution*, **20,** 249–275.
6.12

Janzen, D. H. (1969) Seed-eaters versus seed size, number, toxicity and dispersal. *Evolution*, **23,** 1–27.
10.3

Janzen, D. H. (1971) Seed predation by animals. *Ann. Rev. Ecol. Syst.*, **2,** 465–492.
6.10

Janzen, D. H. (1973a) Dissolution of mutualism between *Cecropia* and its *Azteca* ants. *Biotropica*, **5,** 15–28.
6.6

Janzen, D. H. (1973b) Host plants as islands. II. Competition in evolutionary and contemporary time. *Amer. Natur.*, **107,** 786–790.
6.15

Janzen, D. H. (1974) Tropical blackwater rivers, animals, and mast fruiting by the Dipterocarpaceae. *Biotropica,* **6,** 69–103.
6.10

Janzen, D. H. (1976a) Why bamboos wait so long to flower. *Ann. Rev. Ecol. Syst.*, **7,** 347–391.
6.10

Janzen, D. H. (1976b) Why tropical trees have rotten cores. *Biotropica*, **8,** 110.
6.16

Janzen, D. H. (1976c) The depression of reptile biomass by large herbivores. *Amer. Natur.*, **110,** 371–400.
6.2, 6.17

Janzen, D. H. (1977a) What are dandelions and aphids? *Amer. Natur.*, **111,** 586–589.
6.10, 10.5

Janzen, D. H. (1977b) Why fruits rot, seeds mold, and meat spoils. *Amer. Natur.*, **111,** 691–713.
6.16

Janzen, D. H. (1979a) How many babies do figs pay for babies? *Biotropica*, **11,** 48–50.
6.9

Janzen, D. H. (1979b) How to be a fig. *Ann. Rev. Ecol. Syst.*, **10,** 13–51.
6.16

Janzen, D. H. (1980a) Specificity of seed-attacking beetles in deciduous forests of Costa Rica. *J. Ecol.*, **68,** 929–952.
6.5

Janzen, D. H. (1980b) When is it coevolution? *Evolution*, **34,** 611–612.
6.7

Janzen, D. H., Juster, H. B. & Bell, E. A. (1977) Toxicity of secondary compounds to the seed-eating larvae of the bruchid beetle *Callosobruchus maculatus. Phytochemistry*, **16,** 223–227.
6.14

Janzen, D. H. & Martin, P. S. (1981) Neotropical anachronisms: the fruits the mastodons left behind. *Science (in press)*
6.6

Jarman, P. J. (1974) The social organisation of antelope in relation to their ecology. *Behaviour*, **48,** 215–267.
13.2, 13.4.2

Jarman, M. V. (1979) Impala social behaviour:

territory, hierarchy, mating and the use of space *Z. Tierpsychol.*, **21,** (Suppl.) 1–92.
13.6.1

Jeanrenaud, B. (1965) Lipid components of adipose tissue. In *Handbook of Physiology. Section 5: Adipose Tissue.* pp. 169–176. American Physiological Society, Washington, D.C.
8.2

Jenkin, F. (1867) The origin of species. *The North British Review*, **46,** 277–318.
12.1

Jenkins, D., Watson, A. & Miller, G. R. (1963) Population studies on red grouse, *Lagopus lagopus scoticus* (Lath.), in north-east Scotland. *J.Anim.Ecol.*, **32,** 317–375.
5.3.5

Jenkins, D., Watson, A. & Picozzi, N. (1965) Red grouse chick survival in captivity and in the wild. *Trans. Congr. int. Un. Game Biol.*, **6,** 63–70.
13.7.1

Jenni, D. A. & Collier, G. (1972). Polyandry in the American Jacana (*Jacana spinosa*). *Auk*, **89,** 743–765.
13.6.6

Jennings, J. B. & Colam, J. B. (1970) Gut structure, digestive physiology and food storage in *Pontobena vulgaris* (Nematoda: Enoplida). *J. Zool.* (*Lond.*), **161,** 211–221.
8.3

Jennings, J. B. & Mettrick, D. F. (1968) Observation on the ecology, morphology and nutrition of the rhabdocoel turbellarian, *Syndesmis fraciscana* (Lehman, 1946) in Jamaica. *Caribb. J. Sci.*, **8,** 57–67.
8.2

Jobling, M. (1978) Aspects of food consumption and energy metabolism of farmed plaice, *Pleuronectes platessa* L. Unpublished Ph.D. thesis, University of Glasgow.
5.6.3

Jobling, M & Davies, P. S. (1979) Gastric evacuation in plaice, *Pleuronectes platessa* L.: effects of temperature and meal size. *J. Fish Biol.*, **14,** 539–546.
5.5.2 (Fig. 5.9 legend)

Johnson, R. P. (1973) Scent marking in mammals. *Anim. Behav.*, **21,** 521–535.
13.6.2

Jolliffe, P. A. & Tregunna, E. B. (1968) Effect of temperature, CO_2 concentration and light intensity on oxygen inhibition of photosynthesis in wheat leaves. *Plant Physiol.*, **43,** 902–906.
3.4

Jolly, A. (1972) *The evolution of Primate Behaviour.* 397 pp. Macmillan Co., New York.
8.7

Jones, D. A. & Wilkins, D. A. (1971) *Variation and Adaptation in Plant Species.* Heinemann, London.
3.1

Kacser, H. & Burns, J. A. (1973) The control of flux. *Symp. Soc. Exp. Biol.*, **27,** 65–104.
3.7.1

Kaplan, R. H. & Salthe, S. N. (1979) The allometry of reproduction: an empirical view in salamanders. *Amer. Natur.*, **113,** 671–689.
10.2.1

Karlin, S. & Nevo, E. (Eds) (1976) *Population Genetics and Ecology.* Academic Press, New York.
12.1

Kay, R. N. B., Engelhardt, W. V. & White, R. G. (in press) The digestive physiology of wild ruminants. In *Digestive Physiology and Metabolism in Ruminants* (Eds. Ruckebusch, Y. & Thivend, P.) M.T.P. Press, Lancaster, UK.
5.4.1, 5.6.3

Kelly, G. J., Latzko, E. & Gibbs, M. (1976) Regulatory aspects of photosynthetic carbon metabolism. *Ann. Rev. Plant Physiol.*, **27,** 181–205.
3.4

Kennedy, R. A. & Laetsch, W. M. (1974) Plant species intermediate for C_3, C_4 photosynthesis. *Science*, **184,** 1087–1089.
3.7.3

Kenward, R. E. (1978a) Hawks and doves: attack success and selection in goshawk flights at woodpigeons. *J. Anim. Ecol.*, **47,** 449–460.
13.4.1

Kenward, R. E. (1978b) Hawks and doves: factors affecting success and selection in goshawk attacks on woodpigeons. *J. Anim. Ecol.*, **47,** 491–502.
5.3.3

Kenward, R. E. & Sibly, R. M. (1977) A woodpigeon (*Columba palumbus*) feeding preference explained by a digestive bottleneck. *J. appl. Ecol.*, **14,** 815–826.
5.3.1, 5.3.2, 5.3.3

Kenward, R. E. & Sibly, R. M. (1978) Woodpigeon feeding behaviour at *Brassica* sites. A

field and laboratory investigation of wood-pigeon feeding behaviour during adoption and maintenance of a *Brassica* diet. *Anim. Behav.*, **26**, 778–790.
5.3.2, 5.3.3

Kidder, D. E. & Manners, M. J. (1978) *Digestion in the Pig*. Scientechnica, Bristol.
5.5.4 (Fig 5.10 legend)

Kidwell, M. (1972) Genetic change of recombination value in *Drosophila melanogaster* II. Simulated natural selection. *Genetics*, **70**, 433–443.
12.1

Kiltie R. A. (1980) Seed predation and group size in rain forest peccaries. 170 pp. Ph.D. thesis, Princeton University, Princeton, New Jersey.
6.6

Kinne, O. (1964) The effects of temperature and salinity on marine and brackish water animals. II Salinity and temperature salinity relations. *Oceanogr. Mar. Biol. A. Rev.*, **2**, 281–339.
2.3

Kirkwood, T. B. L. (1977) Evolution of ageing. *Nature (London)*, **270**, 301–304.
7.8

Kirkwood, T. B. L. (1980) Error propagation in intracellular information transfer. *J. theor. Biol.*, **82**, 363–382.
7.3, 7.8

Kirkwood, T. B. L. & Holliday, R. (1975) The stability of the translation apparatus. *J. Mol. Biol.*, **97**, 257–265.
7.3

Kirkwood, T. B. L. & Holliday, R. (1979) The evolution of ageing and longevity. *Proc. Roy Soc. Lond. B*, **205**, 531–546.
7.8

Kleiber, M. (1961) *The Fire of Life*. 454 pp. Wiley, New York.
2.5

Klekowski, R. Z. & Duncan, A. (1975) Physiological approach to ecological energetics. In *Methods for Ecological Bioenergetics*, (Eds Grodzinski, W., Klekowski, R. Z. & Duncan A.) pp. 15–64. Blackwell Scientific Publications, Oxford.
2.8

Klekowski, R. Z., Prus, T. & Zyromska-Rudzka, H. (1967) Elements of energy budget of *Tribolium castaneum* (Hbst) in its developmental cycle. *Secondary Productivity of Terrestrial Ecosystems*. Vol. II. (Ed. Petrusewicz, K.) pp. 859–879. Polish Academy of Science, Institute of Ecology, I. B. P., Warszawa, Krakow.
12.3

Kluge, M., Lange, O. L., von Eichmann, M. & Schmid, R. (1973) Diurnaler Saurerhythmus bei *Tillandsia usneoides*: Untersuchungen über den Weg des Kohlenstoffs Sowie die Abhängigkeit des CO_2-Gaswechsels von Lichtintensität, Temperatur und Wassergehalt der Pflance. *Planta*, **112**, 357–372.
3.7.1

Kluge, M. & Osmond, C. B. (1972) Studies on PEP carboxylase and other enzymes of crassulacean acid metabolism of *Bryophyllum tubiflorum* and *Sedum praealtum*. *Z. Pflanzenphysiol.*, **66**, 97–105.
3.7.1

Kluijver, H. N. (1952) Notes on body weight and time of breeding in the great tit (*Parus m. major* L.). *Ardea*, **40**, 123–141.
10.4

Kluijver, H. N. (1970) Regulation of numbers in populations of great tits (*Parus m. major*). In (Eds. der Boer, P. J. & Gradwell, G. R.) Proceedings of the Advanced Study Institute (Oysterbech) Netherlands. pp. 507–52, PUDOC, Wagenigen.
10.4

Knittle, J. L. (1978) Adipose tissue development in Man. In *Human Growth*. Vol. 2, Postnatal growth (Eds. Falkner, F. & Tanner, J. M.) pp. 295–315. Plenum, London.
8.3

Knowlton, N. (1974) A note on the evolution of gamete dimorphisms. *J. theor. Biol.*, **46**, 283–285.
10.3

Koch, K. & Kennedy, R. A. (1980) Characteristics of crassulacean acid metabolism in the succulent C_4 dicot, *Portulaca oleracea* L. *Plant Physiol.*, **56**, 193–197.
3.7.3

Krause, G. H., Lorimer, G. H., Heber, U. & Kirk, M. R. (1978) Photorespiratory energy dissipation in leaves and chloroplasts. *Proc. Fourth Int. Congress on Photosynthesis*. (Eds. Hall, D. O., Coombs, J. & Goodwin, T. W.) pp. 299–310. The Biochemical Society, London.
3.5

Krebs, J. R. (1971) Territory and breeding density in the great tit, *Parus major* L. *Ecology*, **52**, 2–22.
13.7.1

Krebs, J. R. (1973) Social learning and the

significance of mixed-species flocks of chickadees. *Can. J. Zool.*, **52**, 1275–1288.
13.5.3

Krebs, J. R. (1974) Colonial nesting and social feeding as strategies for exploiting food resources in the Great Blue Heron (*Ardea herodias*). *Behaviour*, **51**, 99–134.
13.5.3.

Krebs, J. R. (1977) Song and territory in the great tit. In *Evolutionary Ecology* (Eds. Stonehouse, B. & Perrins, C. M.) pp. 47–62. Macmillan, London.
13.6.2

Krebs, J. R. (1978) Optimal foraging: decision rules for predators. In *Behavioural Ecology: an Evolutionary Approach.* (Eds. Krebs, J. R. & Davies, N. B.) Blackwell Scientific Publications, Oxford.
4.4

Krebs, J. R. & Davies, N. B. (1978) *Behavioural Ecology: an Evolutionary Approach.* Blackwell Scientific Publications, Oxford.
1.4.2

Krebs, J. R., Erichson, J. T., Webber, M. I. & Charnov, E. L (1977) Optimal prey selection in the great tit (*Parus major*). *Anim. Behav.*, **25**, 30–38.
4.3.1

Krebs, J. R., Kacelnik, A. & Taylor, P. (1978) Optimal sampling by foraging birds: an experiment with great tits (*Parus major*). *Nature (London)*, **275**, 27–31.
4.3.1.1

Krebs, J. R., Ryan, J. C. & Charnov, E. L. (1974) Hunting by expectation or optimal foraging? A study of patch use by chickadees. *Anim. Behav.*, **22**, 953–964.
4.2

Krogh, A. (1914) The quantitative relation between temperature and standard metabolism in animals. *Internationale Zeitschrift fur physikalisch-chemische Biologie*, **1**, 491–508.
9.3.2

Krogh, A. (1941) *The Comparative Physiology of Respiratory Mechanisms*. 172 pp. University of Philadelphia Press, Philadelphia.
2.5, 2.6

Kruuk, H. (1972) *The Spotted Hyaena.* University of Chicago Press, Chicago.
13.4.1, 13.4.3, 13.5.2, 13.5.3

Ku, S. B. & Edwards, G. E. (1975) Photosynthesis in mesophyll protoplasts and bundle sheath cells of various types of C_4 plants. IV Enzymes of respiratory metabolism and energy-utilising enzymes of photosynthetic pathways. *Z. Pflanzenphysiol.*, **77**, 16–32.
3.7.2

Ku, S. B. & Edwards, G. E. (1977) Oxygen inhibition of photosynthesis. I. Temperature dependence and relation to O_2/CO_2 solubility ratio. *Plant Physiol.*, **59**, 986–990.
3.4, 3.5

Kuhme, W. (1965) Freilandstudien zur Soziologie des Hyanenhundes (*Lycaon pictus lupinus* Thomas 1902). *Z. Tierpsychol.*, **22**, 495–541.
13.4.3

Kummer, H. (1968) *Social Organization of Hamadryas Baboons: a Field Study*. University of Chicago Press, Chicago.
13.6.1

Lack, D. (1948) The significance of clutch size. *Ibis*, **90**, 25–45.
1.6

Lack, D. (1954) *The Natural Regulation of Animal Numbers*. Clarendon Press, Oxford.
7.8, 10.4

Lack, D. (1968) *Ecological Adaptations for Breeding in Birds*. Methuen, London.
9.4.1, 9.4.2

Laetsch, W. M. (1974) The C_4 syndrome: a structural analysis. *Ann. Rev. Plant Physiol.*, **25**, 27–52.
3.7.2

Laing, W. A., Ogren, W. L. & Hageman, R. H. (1974) Regulation of soybean net photosynthetic CO_2 fixation by the interaction of CO_2, O_2 and ribulose-1, 5-diphosphate carboxylase. *Plant Physiol.*, **54**, 678–685.
3.4

Lam, R. K. & Frost, B. W. (1976) Model of copepod filtering response to changes in size and concentration of food. *Limnol. Oceanogr.*, **21**, 490–500.
4.3.2

Lange, O. L. & Zuber, M. (1977) *Frerea indica*, a stem succulent CAM plant with deciduous C_3 leaves. *Oecologia*, **31**, 67–72.
3.7.3

Lanham, U. (1964) *The Insects*. 298 pp. Columbia University Press, New York.
8.5

Laplace, J. P. & Lebas, F. (1975) Le transit digestif chez le lapin III.—Influence de l'heure et du mode d'administration sur

l'excrétion fécale du césium-141 chez le lapin alimenté *ad libitum*. *Ann. Zootech.*, **24,** 255–265
5.1, 5.5.4

Law, R. (1979a) Optimal life history under age-specific predation. *Amer. Natur.*, **114,** 399–417.
10.2.2

Law, R. (1979b) Ecological determinants in the evaluation of life histories. In *Population Dynamics* (Eds. Anderson, R. M., Turner, B. D. & Taylor, L. R.) pp. 81–103. B.E.S. Symposium 20. Blackwell Scientific Publications, Oxford.
1.6

Law, R., Bradshaw, A. D. & Putwain, P. D. (1977) Life history variation in *Poa annua*. *Evolution*, **31,** 233–246.
1.4.2

Lawlor, L. R. (1976) Molting, growth and reproductive strategies in the terrestrial isopod, *Armadillidium vulgare*. *Ecology*, **57,** 1179–1194.
10.2.1, 12.2.3

Laws, R. M., Parker, I.C.S. & Johnstone, R. C. B. (1975) *Elephants and their Habitat*. 276pp. Clarendon Press, Oxford.
8.5

Lawton, J. H. (1970) Feeding and food energy assimilation in larvae of the damselfly *Pyrrhosoma nymphula* (Sulz.) (Odonata: Zygoptera). *J. Anim. Ecol.*, **39,** 669–689.
5.6.3

Layzer, D. (1980) Genetic variation and progressive evolution. *Amer. Natur.*, **115,** 809–826.
12.1

Lee, R. F. & Barnes, A. T. (1975) Lipids in the mesopelagic copepod, *Gaussia princeps*: wax ester utilization during starvation. *Comp. Biochem. Physiol.*, **52B,** 265–268.
8.2

Lehman, J. T. (1976) The filter feeder as an optimal forager, and the predicted shapes of feeding curves. *Limnol. Oceanogr*, **21,** 501–516.
4.3.2, 5.6.3

Lehninger, A. L. (1973) *Bioenergetics*. W. A. Benjamin, California.
9.2.2

Lent, P. C. (1974) Mother–infant relationships in ungulates. In *The Behaviour of Ungulates and its Relation to Management* (Eds Geist, V. & Walther, F. R.) pp. 14–55. I.U.C.N., Morges.
13.4.2, 13.7.2

Leon, J. A. (1976) Life histories as adaptive strategies. *J. theor Biol.*, **60,** 301–335.
7.5

Leopold, A. S. (1953) Intestinal morphology of gallinaceous birds in relation to food habits. *J. Wildl. Mgmt*, **17,** 197–203.
5.3.5

Leuthold, W. (1966) Variations in territorial behaviour of Uganda kob. *Behaviour*, **27,** 214–257.
13.6.1, 13.6.3

Levin, D. A. (1976a) The chemical defenses of plants to pathogens and herbivores. *Ann. Rev. Ecol. Syst.*, **7,** 121–159.
6.1

Levin, D. A. (1976b) Alkaloid-bearing plants: an ecogeographic perspective. *Amer. Natur.*, **110,** 261–284.
6.17

Levin, V. (1963) Reproduction and development of young in a population of California quail. *Condor*, **65,** 249–278.
5.3.1

Lewis, C. M. & Holliday, R. (1970) Mistranslation and ageing in *Neurospora*. *Nature (London)*, **228,** 877–880.
7.3

Lewontin, R. C. (1965) Selection for colonizing ability. In *The Genetics of Colonizing Species*. (Eds. Baker, H. G. & Stebbins, G. L.) pp. 79–94. Academic Press, New York.
7.8, 10.2.1

Lewontin, R. C. (1974) *The Genetic Basis of Evolutionary Change*. Columbia University Press, New York.
1.4.2

Lewontin, R. C. (1978) Fitness, survival and optimality. In *Analysis of Ecological Systems*, (Eds. Horn, D. H., Mitchell, R. & Stairs, G. R.) Ohio State University Press, Columbus, Ohio.
13.3

Lifson, N. & McClintoch, R. M. (1966) Theory and use of the turnover rates of body water for measuring energy and material balance. *J. theor. Biol.*, **12,** 46–74.
10.4

Linn S., Kairis, M. & Holliday, R. (1976) Decreased fidelity of DNA polymerase activity isolated from ageing human fibroblasts. *Proc. Nat. Acad. Sci. USA*, **73,** 2818–2822.
7.3

Long, S. P. & Incoll, L. D. (1979) The prediction and measurement of photosynthetic rate of *Spartina* x *townsendii* (*sensu lato*) in

the field. *J. appl. Ecol.*, **16,** 879–891.
3.8.4

Long, S. J., Incoll, L. D. & Woolhouse, H. W. (1975) C_4 photosynthesis in plants from cool temperate regions with particular reference to *Spartina townsendii. Nature (London)*, **257,** 622–624.
3.8.4

Long, S. P. & Woolhouse, H. W. (1978a) The responses of net photosynthesis to vapour pressure deficit and CO_2 concentration in *Spartina* × *townsendii (sensu lato). J. exp. Bot.*, **29,** 567–577.
3.8.4

Long, S. P. & Woolhouse, H. W. (1978b) The responses of net photosynthesis to light and temperature in *Spartina* × *townsendii* (*sensu lato*). *J. Exp. Bot.*, **29,** 803–814.
3.8.4

Long, S. P. & Woolhouse, H. W. (1979) *Primary production of* Spartina *marshes.* Proceedings of the 1st European Ecological Symposium. Blackwell Scientific Publications, Oxford.
3.8.4

Lorimer, G. H., Woo, K. C., Berry, J. A. & Osmond, C. B. (1978) *The C_2 photorespiratory carbon oxidation cycle in leaves of higher plants: pathway and consequences.* Proc. of the Fourth Int. Congress on Photosynthesis. (Eds. Hall, D. O., Coombs, J. & Goodwin, T. W.) Biochemical Society London.
3.4, 3.5

Lotka, A. J. (1928) *Elements of Mathematical Biology.* Republished by Dover, New York (1958).
7.5

Lott, D. F. (1979) Dominance relations and breeding rate in mature male American bison. *Z. Tierpsychol.*, **49,** 418–432.
13.6.1

Lynch, M. (1980) The evolution of cladoceran life histories. *Q. Rev. Biol.*, **55,** 23–42.
10.2.1

MacArthur, R. H. (1962) Some generalised theorems of natural selection. *Proc. Nat. Acad. Sci., U.S.A.*, **48,** 1893–1897.
1.1, 12.4

MacArthur, R. H. & Pianka, E. R. (1966) On the optimal use of a patchy environment. *Amer. Natur.*, **100,** 603–609.
13.3

MacArthur, R. H. (1972) *Geographical Ecology.* Harper & Row, New York.
1.1, 1.6, 4.4

MacArthur, R. H. & Wilson, E. O. (1967) *The Theory of Island Biogeography.* Princeton University Press, Princeton, N. J.
1.6, 7.5, 12.4

MacConnell, J. G., Blum, M. S. & Fales, H. M. (1970) Alkaloid from fire ant venom: identification and synthesis. *Science*, **168,** 840–841.
6.18

MacDonald, D. W. (1976) Food caching by red foxes and some other carnivores. *Z. Tierpsychol.*, **42,** 170–185.
8.5

Mahmoud, A., Grime, J. P. & Furness, S. B. (1975) Polymorphism in *Arrhenathenum elatius* (L.) Beauv. ex J. & C. Presl. *New Phytol.*, **75,** 269–276.
7.4.2

Malins, D. C. & Barone, A. (1970) Glyceryl ether metabolism: regulation of buoyancy in the dogfish, *Squalus acanthias. Science*, **167,** 79–80.
8.7

Malpas, R. C. (1977) Diet and the condition and growth of elephants in Uganda. *J. appl. Ecol.*, **14,** 489–504.
8.5

Marcus, J. H., Sutcliffe, D. W. & Willoughby, L. G. (1978) Feeding and growth of *Asellus aquaticus* (Isopoda) on food items from the littoral of Windermere including green leaves of *Elodea canadensis. Freshwat. Biol.*, **8,** 505–519.
9.3.2

Marestom, V. (1966) Mallard ducklings (*Anas platyrhynchos* L.) during the first days after hatching. *Viltrevy*, **4,** 343–370.
8.3

Marriott, R. W. & Forbes, Dorothy K. (1970) The digestion of Lucerne chaff by Cape Barren geese, *Cereopsis novaehollandiae* Latham. *Aust. J. Zool.*, **18,** 257–263.
5.3.4

Martin, G. M., Sprague, C. A. & Epstein, C. J. (1970) Replicative life–span of cultivated human cells: effects of donor's age, tissue and genotype. *Lab. Invest.*, **23,** 86–92.
7.8

Martin, P. S. (1973) The discovery of America. *Science*, **179,** 969–974.
6.10

Martin, R. (1977) A possible genetic mechanism of ageing, rejuvenation and recombination in germinal cells. In *Molecular*

Human Cytogenetics. (Eds. Sparkes, R. S., Comings, D. E. & Fox, C. F.) pp. 355–373. Academic Press, New York.
7.6

Mason, H. L. & Langenheim, J. H. (1957) Language analysis and the concept environment. *Ecology*, **38,** 325–340.
12.1

Mattocks, J. G. (1971) Goose feeding and cellulose digestion. *Wildfowl*, **22,** 107–113.
5.3.4

May, R. M. (1977). Optimal life-history strategies. *Nature (London)*, **267,** 394–395.
10.1

Mayer, W. V. & Roche, E. T. (1954) Developmental patterns in the Barrow ground squirrel, *Spermophilus undulatus barrowensis*. *Growth*, **18,** 53–69
9.4.1

Maynard Smith, J. (1962) Review lectures on senescence. I. The causes of ageing. *Proc. Roy. Soc. Lond. B*, **157,** 115–127.
7.3

Maynard Smith, J. (1964) Group selection and kin selection. *Nature (London)*, **201,** 1145–1147.
7.8, 13.7.2

Maynard Smith, J. (1969) The status of neo-Darwinism. In *Towards a Theoretical Biology* (Ed. Waddington, C. H.), pp. 82–89. Edinburgh University Press, Edinburgh.
1.2

Maynard Smith, J. (1974) *Models in Ecology*. 146 pp. Cambridge University Press, Cambridge.
3.1

Maynard Smith, J. (1977) Parental investment—a prospective analysis. *Anim. Behav.*, **25,** 1–9.
10.4

Maynard Smith, J. (1978a) Optimization theory in evolution. *Ann. Rev. Ecol. Syst.*, **9,** 31–56.
1.2, 1.4.1, 13.3

Maynard Smith, J. (1978b) *The Evolution of Sex*. Cambridge University Press, Cambridge.
7.6, 10.3, 10.4, 10.5, 10.6

Maynard Smith, J. & Parker, G. A. (1976). The logic of asymmetric contests. *Anim. Behav.*, **24,** 159–175.
13.3, 13.6.5

Maynard Smith, J. & Price, G. R. (1973) The logic of animal conflict. *Nature, (London)*, **246,** 15–18.
1.4.2, 10.3, 13.6.5

Mayr, E. (1956) Geographic character displacement and climatic adaptation. *Evolution*, **10,** 105–108.

Mayr, E. (1963) *Population, Species and Evolution*. Harvard University Press, Cambridge, Mass.
1.2

Mazanov X. Y. & Nolan J. V. (1976) Simulation of the dynamics of nitrogen metabolism in sheep. *Br. J. Nutr.*, **35,** 149–174.
5.5.4

McCleery, R. H. (1977) On satiation curves. *Anim. Behav.*, **25,** 1005–1015.
5.6.1

McDougall, P. & Milne, H. (1978) The anti-predator function of defecation on their own eggs by female Eiders. *Wildfowl*, **29,** 55–60.
5.1

McKey, D., Waterman, P. J., Mbi, C. N., Gartlan, J. S., & Struhsaker, T. T. (1978) Phenolic content of vegetation in two African rain forests: ecological implications. *Science*, **202,** 61–64.
6.13

McKillup, S. C. & Butler, A. J. (1979) Modification of egg production and packaging in response to food availability by *Nassarius pauperatus*. *Oecologia*, **43,** 221–223.
10.2.2

McMeekan, C. P. (1940) Growth and development of the pig, with special reference to carcass quality characters. *J. Agr. Sci.*, **30,** 276–344.
9.7.2

McNab, B. K. (1978) The evolution of endothermy in the phylogeny of mammals. *Amer. Natur.*, **112,** 1–21.
2.6

McNab, B. K. (1980) Food habits, energetics, and the population biology of mammals. *Amer. Natur.*, **116,** 106–124.
12.2.6

McNaughton, S. J. (1975) *r*- and *K*-selection in *Typha*. *Amer. Natur.*, **109,** 251–261.
1.6

McNeill, S. & Lawton, J. H. (1970) Annual production and respiration in animal populations. *Nature (London)*, **225,** 472–474.
2.8

Medawar, P. B. (1952) *An Unsolved Problem in Biology*. H. K. Lewis, London. (Reprinted in *The Uniqueness of the Individual*. Methuen, London (1957)).
7.8

Medvedev, Zh. A. (1962) Ageing at the mole-

cular level. In *Biological Aspects of Ageing*. (Ed. Shock, N. W.) pp. 255–266. Columbia University Press, New York.
7.3

Meier, A. H., Trobec, T. N., Haymaker, H. G., MacGregor, R. III & Russo, A. C. (1973) Daily variations in the effects of handling on fat storage and testicular weights in several vertebrates. *J. exp. Zool.*, **184,** 281–288.
8.6

Menge, J. L. (1974) Prey selection and foraging period of the predaceous rocky intertidal snail, *Acanthina punctulata. Oecologia*, **17,** 293–316.
4.3.1

Metz, J. H. M. (1975) Time patterns of feeding and rumination in domestic cattle. *Meded. LandbHoogesch., Wageningen*, 75–12, 1–166.
5.4.4

Michael, E. D. (1972) Growth rates in *Anolis carolinensis. Copeia*, 1972, 575–577.
8.5

Michon, J. (1954) Influence de l'isolement à partir de la maturité sexuelle sur la biologie des Lumbricidae. *Compte rendues hebdomadaire des seances de l'Academie des Sciences, Paris,* **238,** 2457–2458.
9.3.2

Miflin, B. J. & Lea, P. J. (1976) The pathway of nitrogen assimilation in plants. *Phytochemistry*, **15,** 873–885.
3.5.

Milinski, M. & Heller, R. (1978) Influence of a predator on the optimal foraging behaviour of sticklebacks (*Gasterosteus aculeatus* L.). *Nature* (*London*), **275,** 642–644.
5.6.2

Miller, G. R., Watson, A. & Jenkins, D. (1970) Responses of red grouse populations to experimental improvement of their food. In *Animal Populations in Relation to their Food Resources* (Ed. Watson, A.) pp. 323–335. Blackwell Scientific Publications, Oxford.
13.7.1

Millar, J. S. (1977) Adaptive features of mammalian reproduction. *Evolution*, **31,** 370–386.
10.4

Milne, J. A., Macrae, J. C., Spence, A. M. & Wilson, S. (1978) A comparison of the voluntary intake and digestion of a range of forages at different times of the year by the sheep and the red deer (*Cervus elaphus*). *Br. J. Nutr.*, **40,** 347–357.
5.5.4

Minot, C. S. (1908) *The Problem of Age, Growth and Death*. Putnam, New York.
9.5

Mitchell, B., McCowan, D. & Nicholson, I. A. (1976) Annual cycles of body weight and condition in Scottish Red deer, *Cervus elaphus. J. Zool. (Lond.)*, **180,** 107–128.
13.6.4

Mitchell, B., Staines, B. W. & Welch, D. (1977) *Ecology of Red Deer*. ITE, Cambridge.
13.6.4

Mitchell, P. (1961) Coupling of phosphorylation to electron and hydrogen transfer by a chemi-osmotic type of mechanism. *Nature* (*London*), **191,** 144–148.
3.2

Mitchell, R. (1973) Growth and population dynamics of a spider mite (*Tetranychus urticae* K., Acarina: Tetranychidae). *Ecology*, **54,** 1349–1355.
9.3.2

Monteith, J. L. (1973) *Principles of Environmental Physics*. American Elsevier Publishing Co., New York.
12.3

Mooney, H. A. & Gulmon, S. L. (1979) Environmental and evolutionary constraints on the photosynthetic characteristics of higher plants. In *Topics in Plant Population Biology*. (Eds. Solbrig, O. T. *et al.* Columbia University Press, New York.
12.3

Mooney, H. A. & Hays, R. I. (1973) Carbohydrate storage cycles in two Californian Mediterranean-climate trees. *Flora*, **162,** 295–304.
8.4

Moore, I. A. (1942) The role of temperature in speciation of frogs. *Biol. Symp.*, **6,** 189–213.
10.3

Moore, I. A. (1949) Geographic variation of adaptive characters in *Rana pipiens*. *Evolution*, **3,** 1–24.
10.3

Moran, J. B., Norton, B. W. & Nolan, J. V. (1979) The intake, digestibility and utilization of a low-quality roughage by Braham Cross, buffalo, Banteng and Shorthorn steers. *Aust. J. agric. Res.*, **30,** 333–340.
5.6.1

Morgan, P. R. (1972) The influence of prey availability on the distribution and predatory behaviour of *Nucella lapillus* (L.). J. Anim. Ecol., **41**, 257–274
4.3.1

Morris, D. (1977) *Manwatching*. 320 pp. Cape,

London.
8.4

Morrison, P. (1960) Some interrelations between weight and hibernation function. *Bull. Mus. Comp. Zool.*, **124,** 75–91.
8.5, 8.6

Morse, D. H. (1970) Ecological aspects of some mixed species foraging flocks of birds. *Ecol. Monogr.*, **40,** 119–168.
13.4.1

Morton, J. (1979) *Guts.* 2nd edn. Edward Arnold, London.
5.1

Moss, R. (1972) Effects of captivity on gut lengths in red grouse. *J. Wildl. Mgmt.*, **36,** 99–104.
5.3.1 (Table 5.1)

Moss, R. (1974) Winter diets, gut lengths, and interspecific competition in Alaskan ptarmigan. *Auk*, **91,** 737–746.
5.3.1

Moss, R. (1977) The digestion of heather by red grouse during the spring. *Condor*, **79,** 471–477.
5.3.1, 5.3.5

Moss, R. & Parkinson, J. A. (1972) The digestion of heather (*Calluna vulgaris*) by red grouse (*Lagopus lagopus scoticus*). *Br. J. Nutr.*, **27,** 285–298.
5.3.5

Moss, R., Watson, A. & Parr, P. (1975) Maternal nutrition and breeding success in red grouse (*Lagopus lagopus scoticus*). *J. Anim. Ecol.*, **44,** 233–244.
13.7.1

Mrosovsky, N. (1976) Lipid programmes and life strategies in hibernators. *Am. Zool.*, **16,** 685–697.
8.6

Mrosovsky, N. & Fisher, K. C. (1970) Sliding set point for body weight in ground squirrels during the hibernating season. *Can. J. Zool.*, **48,** 241–247.
8.6

Muller, H. J. (1964) The relation of recombination to mutational advance. *Mutat. Res.*, **1,** 2–9.
7.6

Murphy, G. I. (1968) Patterns in life history and the environment. *Amer. Natur.*, **102,** 391–403.
10.2.2

Murton, R. (1971) The significance of a specific search image in the feeding behaviour of the woodpigeon. *Behaviour*, **40,** 10–40.
13.5.3

Myers, J. H. (1978) Sex ratio adjustment under food stress: maximization of quality or numbers of offspring. *Amer. Natur.*, **112,** 381–388.
13.7.1

Myers, K., Hale, C. S., Mykytowycz, R. & Hughes, R. L. (1971) The effects of varying density and space on sociality and health in animals. In *Behaviour and Environment: The Use of Space by Animals and Men.* (Ed. Esser, A. H.) pp. 148–187, Plenum, New York.
13.6.1

Nalborczyk, E., La Croix, L. J. & Hill, R. D. (1974) Environmental influences on light and dark CO_2 fixation by *Kalanchöe daigremontianum. Can. J. Bot.*, **53,** 1132–1138.
3.7.3

Nambudiri, E. M. V., Tidwell, W. D., Smith, B. N. & Hebbert, N. P. (1978) A C_4 plant from the Pliocene. *Nature (London)*, **276,** 21.
3.7.3

Neals, T. F., Patterson, A. A. & Hartney, V. J. (1968) Physiological adaptation to drought in the carbon assimilation and water loss of xerophytes. *Nature (London)*, **219,** 469–472.
3.7.3

Needham, A. E. (1965) *The Uniqueness of Biological Materials.* 593 pp. Pergamon Press, Oxford.
8.2

Nelson, J. S. (1969) The breeding ecology of the red-footed booby in the Galapagos. *J. Anim. Ecol.*, **38,** 181–198.
9.4.2

Nelson, W. G. (1979) Experimental studies of selective predation on amphipods: consequences for amphipod distribution and abundance. *J. exp. mar. Biol. Ecol.*, **38,** 225–245.
4.2

Nelson, R. A., Jones, J. D., Wahner, H. W., McGill, D. B. & Code, C. R. (1975) Nitrogen metabolism in bears: urea metabolism in summer starvation and in winter sleep and role of urinary bladder in water and nitrogen conservation. *Mayo Clin. Proc.*, **50,** 141–146.
8.6

Nevenzel, J. C. (1970) Occurrence, function and biosynthesis of wax esters in marine organisms. *Lipids*, **5,** 308–319.
8.2

Newell, N. D. (1949) Phyletic size increase—an

important trend illustrated by fossil invertebrates. *Evolution*, **3,** 103–124.
2.6

Newell, S. J., Solbrig, O. T. & Kincaid, D. T. (1981) Studies on the population biology of the genus *Viola.* III. The demography of *V. blanda. J. Ecol.*, **69,** (*in press*)
11.4

Newell, S. & Tramer, J. (1978) Reproductive strategies in herbaceous plant communities during succession. *Ecology*, **59,** 228–234.
11.4

Newsholme, E. A. & Start, C. (1973) *Regulation in Metabolism.* 349 pp. John Wiley & Sons, London.
8.4

Newton, I. (1966) Fluctuations in the weights of bullfinches. *British Birds*, **19,** 89–100.
10.4

Newton, I. (1972) *Finches.* Collins, London.
13.4.1

Nichols, J. D., Conley, W., Batt, B. & Tipton, A. R. (1976) Temporally dynamic reproductive strategies and the concept of *r*- and *K*-selection. *Amer. Natur.*, **110,** 995–1005.
10.2.2

Nikolskii, G. V. (1969) *Fish Population Dynamics.* Oliver & Boyd, Edinburgh.
10.3

Ninio, J. (1975) Kinetic amplification of enzyme discrimination. *Biochimie*, **57,** 587–595.
7.3

Nobel, P. S. (1976) Water relations and photosynthesis of a desert CAM plant *Agave deserti. Plant Physiol.*, **58,** 576–582.
3.7.1

Norberg, R. A. (1977) An ecological theory on foraging time and energetics and choice of optimal food searching method. *J. Anim. Ecol.*, **46,** 511–529.
4.4

O'Brian, W. J., Slade, N. A. & Vinyard, G. L. (1976) Apparent size as the determinant of prey selection by bluegill sunfish (*Lepomis macrochirus*). *Ecology*, **57,** 1304–1310.
4.3.2

O'Dor, R. K. & Wells, M. J. (1978) Reproduction versus somatic growth: hormonal control in *Octopus vulgaris. J. exp. Biol.*, **77,** 15–31.
10.2.2

Odum, E. P. (1960) Premigratory hyperphagia in birds. *Am. J. clin. Nutr.*, **8,** 621–629.
8.3, 8.5

Ogren, W. L. & Hunt, L. D. (1978) Comparative biochemistry of ribulose bisphosphate carboxylase in higher plants. In *Photosynthetic Carbon Assimilation.* (Eds. Siegelman, H. W. & Hind, G.) pp. 179–208. Plenum, New York.
3.5

Ollason, J. G. (1980) Learning to forage—Optimally? *Theor. Popul. Biol.*, **18,** 44–56.
4.2.1

Orgel, L. E. (1963) The maintenance of the accuracy of protein synthesis and its relevance to ageing. *Proc. Nat. Acad. Sci. USA*, **49,** 517–521.
7.3

Orgel, L. E. (1973) Ageing of clones of mammalian cells. *Nature* (*London*), **243,** 441–445.
7.3, 7.8

Orgel, L. E. & Crick, F. H. C. (1980) Selfish DNA: the ultimate parasite. *Nature* (*London*), **284,** 604–607.
12.1

Orians, G. H. (1969) On the evolution of mating systems in birds and mammals. *Amer. Natur.*, **103,** 589–603.
13.6.6

Orians, G. H. & Janzen, D. H. (1974) Why are embryos so tasty? *Amer. Natur.*, **108,** 581–592.
6.8

Orians, G. H. & Solbrig, O. T. (1977) A cost–income model of leaves and roots with special reference to arid and semi-arid areas. *Amer. Natur.*, **111,** 677–690.
12.3

Orton, J. H. (1929) Reproduction and death in the invertebrates and fishes. *Nature* (*London*), **123,** 4–5.
10.2.2

Osmond, C. B. (1978) Crassulacean acid metabolism: a curiosity in context. *Ann. Rev. Plant Physiol.*, **29,** 379–414.
3.8.5

Oster, G. & Wilson, E. O. (1978) *Caste and Ecology in the Social Insects.* Princeton University Press, Princeton, N. J.
1.4.2, 12.1

Owen, M. (1972) Some factors affecting food intake and selection in white-fronted geese. *J. Anim. Ecol.*, **41,** 79–92.
5.3.3, 5.3.4

Owen, M. (1975) An assessment of fecal analysis technique in waterfowl feeding studies. *J.*

Wildl. Mgmt., **39,** 271–279.
5.3.1, 5.3.4

Owen, M. (1976) The selection of winter food by white-fronted geese. *J. appl. Ecol.*, **13,** 715–729.
5.3.4

Owen, M., Nugent, M. & Davies, N. (1977) Discrimination between grass species and nitrogen-fertilized vegetation by young Barnacle Geese. *Wildfowl*, **28,** 21–26.
5.3.4

Oxnard, C. E. (1973) *Form and Pattern in Human Evolution.* 2118 pp. University of Chicago Press, Chicago.
8.7

Paigen, K. (1979) Acid hydrolases as models of genetic control. *Ann. Rev. Genet.*, **13,** 417–466.
12.1

Parke, D. V. & Smith, R. L. (1977) *Drug Metabolism—from Microbe to Man.* Taylor & Francis, London.
7.3

Parker, G. A. (1974) Assessment strategy and the evolution of fighting behaviour. *J. theor. Biol.*, **47,** 223–243.
13.6.5

Parker, G. A., Baker, R. R. & Smith, V. G. F. (1972) The origin and evolution of gamete dimorphism and the male–female phenomenon. *J. theor. Biol.*, **36,** 529–553.
10.3

Parkhurst, D. F. & Loucks, D. L. (1972) Optimal leaf size in relation to environment. *J. Ecol.*, **60,** 505–537.
12.3

Parnas, H. & Cohen, D. (1976) The optimal strategy for the metabolism of reserve materials in micro-organisms. *J. theor. Biol.*, **56,** 19–55.
8.5

Patton, R. A. & Krause, G. F. (1972) A maximum-likelihood estimator of food retention time in ruminants *Br. J. Nutr.*, **28,** 19–22.
5.5.4

Pearson, O. P. (1954) The daily energy requirements of a wild anna hummingbird. *Condor*, **56,** 317–322.
12.2.5

Pendergast, B. A. & Boag, D. A. (1973) Seasonal changes in the internal anatomy of spruce grouse in Alberta. *Auk*, **90,** 307–317.
5.3.1

Penning de Vries, F. W. T., Brunsting, A. H. M. & Laar, H. H. van (1974) Products, requirements and efficiency of biosynthesis: a quantitative approach. *J. theor. Biol.*, **45,** 339–357.
9.2.4

Pennycuick, C. J. (1972) *Animal Flight.* 68 pp. Edward Arnold, London.
8.5

Pennycuick, L. (1975) Movements of the migratory Wildebeest population in the Serengeti area between 1960 and 1973. *E. Afr. Wildl. J.*, **13,** 65–87.
13.4.1

Peters, R. H. (1976) Tautology in evolution and ecology. *Amer. Natur.*, **110,** 1–12.
1.2

Phillips, A. M. (1969) Nutrition, digestion and energy utilization. In *Fish Physiology*. Vol. 1 (Eds. Hoar, W. S. & Randall, D. J.) pp. 391–432. Academic Press, New York.
8.2

Phillipson, J. (1966) *Ecological Energetics.* Edward Arnold, London.
1.1

Phillipson, J. (1973) The biological efficiency of protein production by grazing and other land-based systems. In *The Biological Efficiency of Protein Production* (Ed. Jones, J. G. W.) pp. 217–235. Cambridge University Press, Cambridge.
2.8

Pianka, E. R. (1966) Convexity, desert lizards, and spatial heterogeneity. *Ecology*, **47,** 1055–1059.
12.2.7

Pianka, E. R. (1970) On *r*- and *K*-selection. *Amer. Natur.*, **104,** 592–597.
1.6, 10.2.2, 12.4

Pianka, E. R. (1976) Natural selection of optimal reproductive tactics. *Am. Zool.*, **16,** 775–784.
10.1.2

Pianka, E. R., Huey, R. B. & Lawlor, L. R. (1979) Niche segregation of desert lizards. In *Analysis of Ecological Systems.* (Eds. Horn. D. J., Stairs, G. R. & Mitchell, R.) Ohio State University Press, Columbus.
12.2.7

Pianka, E. R. & Parker, W. S. (1975a) Age-specific reproductive tactics. *Amer. Natur.*, **109,** 453–464.
7.8, 10.2.2, 10.3

Pianka, E. R. & Parker, W. S. (1975b) Ecology of horned lizards: a review with special reference to *Phrynosoma platyrhinos.*

Copeia, 1975, 141–162.
12.2.8

Pickard, D. W. & Stevens, C. E. (1972) Digesta flow through the rabbit large intestine. *Am. J. Physiol.*, **222,** 1161–1166.
5.1, 5.5.4

Pienaar, U. de. (1969) Predator-prey relations amongst the larger mammals of the Kruger National Park. *Koedoe*, **12,** 108–761.
13.4.3

Pitts, G. C. & Bullard, T. R. (1968) Some interspecific aspects of body composition in mammals. In *Body Composition in Animals and Man*. pp. 45–70. Nat. Acad. Sci. publ. 1598. Washington, D. C.
8.5

Pond, C. M. (1977) The significance of lactation in the evolution of mammals. *Evolution*, **31,** 177–199.
8.3, 8.5

Pond, C. M. (1978) Morphological aspects and the ecological and mechanical consequences of fat deposition in wild vertebrates. *Ann. Rev. Ecol. Syst.*, **9,** 519–570.
8.3, 8.4, 8.5, 8.6, 8.7

Popper, K. R. (1959) *The Logic of Scientific Discovery*. Hutchinson, London.
1.4.1, 1.4.2, 3.1

Popper, K. R. (1972) *Objective Knowledge*. Oxford University Press, Oxford.
1.4.2

Potapov, R. L. & Andreev, A. V. (1973) Bioenergetics of Heath-Cock *Lyrurus tetrix* (L.) in winter period *Dokl. Akad. Nauk SSSR*, **210,** 499–500.
5.3.3

Powell, G. V. N. (1974) Experimental analysis of the social value of flocking by starlings (*Sturnus vulgaris*) in relation to predation and foraging. *Anim. Behav.*, **22,** 501–505.
13.4.1

Prus, M. & Prus, T. B. (1977) Energy budget of *Tribolium castaneum* (Hbst) at the population level. *Ekologia Polska*, **25,** 115–134.
12.3

Prys-Jones, R. P. (1977) Aspects of Reed Bunting ecology with comparisons with the Yellowhammer. Unpublished D. Phil. thesis, University of Oxford.
5.3.1

Pulliam, H. R. (1973) On the advantages of flocking. *J. theor. Biol.*, **38,** 419–422.
13.4, 13.4.1

Pulliam, H. R. (1976) The principle of optimal behaviour and the theory of communities. *Perspectives in Ethology*, **3,** 311–312.
13.4.1

Pyke, G. H. (1979) The economics of territory size and time budget in the golden-winged sunbird. *Amer. Natur.*, **114,** 131–145.
13.5.1

Pyke, G. H., Pulliam, H. R. & Charnov, E. L. (1977) Optimal foraging: a selective review of theory and tests. *Q. Rev. Biol.*, **52,** 137–154.
4.2.1, 4.4

Quieroz, O. (1974) Circadian rhythms and metabolic patterns. *Ann. Rev. Plant Physiol.*, **25,** 115–134.
3.1.3

Quieroz, O. (1978) CAM; rhythms of enzyme capacity and activity as adaptive mechanisms in photosynthesis. VII. In *Encyclopedia of Plant Physiology*. (New Ser.) (Eds. Gibbs, M. & Latzko, E.). Springer Verlag, Berlin.
3.7.1

Ralls, K. (1971) Mammalian scent marking. *Science*, **171,** 443–449.
13.6.2

Randolph (McClure), P. A. & Randolph, J. C. (1980) Relative allocation of energy to growth and development of homeothermy in the Eastern Wood Rat (*Neotoma floridana*) and Hispid Cotton Rat (*Sigmodon hispidus*). *Ecol. Monogr.*, **50,** 199–219.
12.3

Randolph (McClure), P. A., Randolph, J. C. & Barlow, C. A. (1975) Age-specific energetics of the pea aphid, *Acyrthosiphon pisum*. *Ecology*, **56,** 359–369.
12.2.1, 12.3

Randolph (McClure), P. A., Randolph, J. C. Mattingly, K. & Foster, M. M. (1977) Energy costs of reproduction in the Cotton Rat, *Sigmodon hispidus*. *Ecology*, **58,** 31–45.
12.1.2, 12.3

Ranwell, D. S. (1972) *Ecology of Salt Marshes and Sand Dunes*. 258 pp. Chapman and Hall, London.
3.8.4

Raschke, K. (1975) Stomatal action. *Ann. Rev. Plant Physiol.*, **26,** 309–390.
3.6.1

Raschke, K. (1956) Über die physikalischen Beziehungen zwischen Warmeubergangszahl, Strahlungsuastausch, Temperatur und Transpiration eines Blattes, *Planta*, **48,** 200–238.
12.3

Rathnam, C. K. M. & Chollet, R. (1979) Phosphoenol-pyruvate carboxylase reduces photorespiration in *Panicum milioides*, a C_3–C_4 intermediate species. *Arch. Biochem. Biophys.*, **193,** 346–354.
3.7.3

Rehfeld, D. W., Randall, D. D. & Tolbert, N. E. (1970) Enzymes of the glycolate pathway in plants without CO_2-respiration. *Can. J. Bot.*, **48,** 1219–1226.
3.7.2

Reichl, J. R. & Baldwin, R. L. (1975) Rumen modelling: rumen input–output balance models. *J. Dairy Sci.*, **58,** 879–890.
5.5.4

Reynoldson, T. B. (1961) Environment and reproduction in freshwater triclads. *Nature (London)*, **189,** 329–330.
10.5

Rhoades, D. D. & Duke, G. E. (1975) Gastric function in a captive American Bittern. *Auk*, **92,** 786–792.
5.1

Rhoades, D. F. & Cates, R. G. (1976) Toward a general theory of plant antiherbivore chemistry. In *Biochemical Interaction between Plants and Insects*. (Eds. Wallace, J. W. & Mansell, R. L.) pp. 168–213. Plenum, New York.
6.10

Ricklefs, R. E. (1968) Patterns of growth in birds. *Ibis*, **110,** 419–451
9.5

Ricklefs, R. E. (1969) Preliminary models for growth rates of altricial birds. *Ecology*, **50,** 1031–1039.
9.4.1

Ricklefs, R. E. (1977) A note on the evolution of clutch size in altricial birds. In *Evolutionary Ecology*, (Eds. Stonehouse, B. & Perrins, C.) pp. 193–214. Macmillan, London.
10.4

Ringler, N. H. (1979) Selective predation by drift-feeding brown trout *(Salmo trutta)*. *J. Fish. Res. Board Can.*, **26,** 392–403.
4.3.1

Robertson, A. (1960) A theory of limits in artificial selection. *Proc. Roy. Soc. Lond. B*, **153,** 234–249.
7.5

Robinson, D. W. & Slade, L. M. (1974) The current status of knowledge on the nutrition of equines. *J. Anim. Sci.*, **39,** 1045–1066.
5.4.2

Robinson, M. H. (1969) Defenses against visually orienting predators. *Evolutionary Biology*, **3,** 225–259.
6.1

Rockwood, L. L. & Glander, K. E. (1979) Howling monkeys and leaf-cutting ants: comparative foraging in a tropical deciduous forest. *Biotropica*, **11,** 1–10.
6.5

Rosen, R. (1967) *Optimality Principles in Biology*. Butterworths, London.
1.4.2, 3.1

Rosenthal, G. A. (1977) The biological effects and mode of action of L-canavanine, a structural analogue of L-arginine. *Q. Rev. Biol.*, **52,** 155–178.
6.2

Rosenthal, G. A., Dahlman, D. L. & Janzen, D. H. (1976) A novel means for dealing with L-canavanine, a toxic metabolite. *Science*, **192,** 256–258.
6.13

Rosenthal, G. A. & Janzen, D. H. (Eds.) (1979) *Herbivores, their Interaction with Secondary Plant Metabolites*. 718 pp. Academic Press. New York.
6.1

Rosenthal, G. A., Janzen, D. H. & Dahlman, D. L. (1977) Degradation and detoxification of canavanine by a specialized seed predator. *Science*, **196,** 658–660.
6.13

Rosenthal, G. A., Dahlman, D. L. & Janzen, D. H. (1978) L-canaline detoxification: a seed predator's biochemical mechanism. *Science*, **202,** 528–529.
6.13

Rothschild, M., Keutmann, H., Lane, N. J., Parsons, J., Prince, W. & Swales, L. S. (1979) A study on the mode of action and composition of a toxin from the female abdomen and eggs of *Arctia caja* (L.) (Lep. Arctiidae): an electrophysiological, ultrastructural and biochemical analysis. *Toxicon*, **17,** 285–306.
6.18

Roughgarden, J. (1971) Density-dependent natural selection. *Ecology*, **52,** 453–468.
12.4

Roughgarden, J. (1979) *Theory of Population Genetics and Evolutionary Ecology: An Introduction*. Macmillan, New York.
12.1

Rowell, T. E. (1974) The concept of social dominance *Behav. Biol.*, **11,** 131–154.
13.6.1.5

Rubenstein, D. I. (1980) On the evolution of alternative mating strategies. In *Limits to Action: the Allocation of Individual Behaviour* (Ed. Staddon, J. E. R.) Academic Press, New York.
13.3

Ruggieri, G. D. (1976) Drugs from the sea. *Science*, **194,** 491–497.
6.1, 6.11

Ryan, C. A. (1978) Proteinase inhibitors in plant leaves: a biochemical model for pest-induced natural plant protection. *TIBS*, **3** (no. 3, December), 148–150.
6.3

Sacher, G. A. (1978) Evolution of longevity and survival characteristics in mammals. In *The Genetics of Ageing*. (Ed. Schneider, E. L.) pp. 151–167. Plenum, New York.
7.8

Sacher, G. A. & Hart, R. W. (1978) Longevity, ageing and comparative cellular and molecular biology of the house mouse, *Mus musculus*, and the white-footed mouse, *Peromyscus leucopus*. In *Genetic Effects on Ageing* (*Birth defects: Original Article Series No. 14*). (Eds. Bergsma, D. & Harrison D.E.) Vol. 1, pp. 71–96. Alan R. Liss, New York.
7.8

Sadlier, R.M.F.S. (1969) *The Ecology of Reproduction in Wild and Domestic Mammals*. Methuen, London.
13.7.2

Salisbury, F. B. & Ross, C. (1969) *Plant Physiology*. 747 pp. Wadsworth Publ. Co., Belmont, California.
8.2

Sargent, J. R., Gatten, R. R. & McIntosh, R. (1973) The distribution of neutral lipids in shark tissues. *J. mar. Biol. Ass. UK*, **53,** 649–656.
8.7

Satchell, J. E. (1967) Lumbricidae. In *Soil Biology*. (Eds. Burgess, A. & Raw, F.) pp. 259–322. Academic Press, London.
9.3.2

Savory, C. J. (1978) Food consumption of red grouse in relation to the age and productivity of heather. *J. Anim. Ecol.*, **47,** 269–282
5.3.5

Savory, C. J. & Gentle, M. J. (1976) Changes in food intake and gut size in Japanese quail in response to manipulation of dietary fibre content. *Br. Poult. Sci.*, **17,** 571–580.
5.3.1, 5.3.3

Sayre, R. T., Kennedy, R. A. & Pringnitz, D. J. (1979) Photosynthetic enzyme activities and localisation in *Mollugo verticillata* populations differing in the levels of C_3 and C_4 cycle operation. *Plant Physiol.*, **64,** 293–299.
3.7.3

Scadding, S. R. (1977) Phylogenic distribution of limb regeneration capacity in adult Amphibia. *J. exp. Zool.*, **202,** 57–68.
7.1, 7.4, 7.9

Schaffer, W. M. (1974) Optimal reproductive effort in fluctuating environments. *Amer. Natur.*, **108,** 783–790.
12.4

Schaffer, W. M. & Rosenzweig, M. L. (1977) Selection for optimal life histories. II. Multiple equilibria and the evolution of alternative reproductive strategies. *Ecology*, **58,** 60–72.
12.4

Schaller, G. B. (1967) *The Deer and the Tiger: a Study of Wildlife in India*. University of Chicago Press, Chicago.
13.6.4

Schaller, G. B. (1972) *The Serengeti Lion*. University of Chicago Press, Chicago.
13.4.1, 13.4.3, 13.5.2

Scheerboom, J. E. M. (1978) The influence of food quantity and food quality on assimilation, body growth and egg production in the pond snail, *Lymnaea stagnalis* (L.) with particular reference to the haemolymph-glucose concentration. *Proc. Koninklijke Nederlandse Akademie van Wetenschappen*, **81,** 184–197.
12.3

Schiemer, F., Duncan, A. & Klekowski, R. Z. (1980) A bioenergetic study of the benthic nematode *Plectus palustris* de Man 1880 throughout its life cycle. II. Growth, fecundity and energy budgets at different densities of bacterial food and general ecological considerations. *Oecologia*, **44,** 205–212.
9.3.2

Schimper, A. F. S. (1898) *Pflanzengeographie auf Physiologischer Grundlage*. G. Fisher, Jena.
12.5

Schneider, E. L. & Mitsui, Y. (1976) The relationship between *in vitro* cellular ageing and *in vivo* human age. *Proc. Nat. Acad. Sci. USA*, **73,** 3584–3588.
7.8

Schmidt-Nielsen, K. (1946) Melting points of human fats as related to their location in the body. *Acta Physiol. Scand*, **12,** 123–129.
8.2

Schmidt-Nielsen, K. (1975) *Animal Phys-*

iology. 699 pp. Cambridge University Press, London.
8.2, 8.5

Schoener, T. W. (1971) Theory of feeding strategies. *Ann. Rev. Ecol. Syst.*, **2,** 369–404.
1.6, 13.8

Schultz, A. H. (1969) *The Life of Primates*. 281 pp. Weidenfeld & Nicolson, London.
8.5, 8.7

Schusterman, R. J. & Gentry, R. L. (1971) Development of a fatted male phenomenon in California sea lions. *Devel. psychobiol.*, **4,** 333–338.
8.5

Sebens, K. P. (1979) The energetics of asexual reproduction and colony formation in benthic marine invertebrates. *Amer. Zool.*, **19,** 683–697.
10.2.1

Serres, M. de (1813) Observations sur les usages des diverses parties du tude intestinal des Insectes. *Ann. Mus. Hist. Nat.*, **20,** 48–114.
5.6.3

Shattock, S. G. (1909) On normal tumour-like formations of fat in Man and the lower animals. *Proc. R. Soc. Med. Path.*, Section **2,** 207–270.
8.4, 8.7

Sheldrake, A. R. (1974) The ageing, growth and death of cells. *Nature* (*London*), **250,** 381–385.
7.3, 7.6, 7.8

Shire, R. (1980) Cost of reproduction in reptiles. *Oecologia*, **46,** 92–100.
10.2.2.

Shorland, F. B. (1962) The comparative aspects of fatty acid occurrence and distribution. In *Comparative Biochemistry*. Vol. 3 (Eds. Florkin, M. & Mason, H. S.) pp. 1–102. Academic Press, New York.
8.2, 8.3

Sibbald, I. R. (1979) Passage of feed through the adult rooster. *Poult. Sci.*, **58,** 446–459.
5.5.2 (Fig 5.9 caption)

Sibbald, I. R., Slinger, S. J. & Ashton, G. C. (1960) The weight gain and feed intake of chicks fed a ration diluted with cellulose or kaolin. *J. Nutr.*, **72,** 441–446.
5.4.4 (Fig 5.7 caption)

Sibly, R. M. & McFarland, D. (1976) On the fitness of behaviour sequences. *Amer. Natur.*, **110,** 601–617.
1.4.2, 1.6, 5.6.2

Siegfried, W. R. & Frost, P. G. H. (1975) Continuous breeding and associated behaviour in the Moorhen (*Gallinula chloropus*). *Ibis*, **117,** 102–109.
13.6.6

Sinclair, A. R. E. (1975) The resource limitation of trophic levels in tropical grassland ecosystems. *J. Anim. Ecol.*, **44,** 497–520.
6.14

Singh, N. B., Campbell, A. & Sinha, R. N. (1976) An energy budget of *Sitophilus oryzae* (Coleoptera: Curculionidae) *Ann. Ent. Soc. Amer.*, **69,** 503–512.
12.3

Sinha, R. C. & Kanungo, M. S. (1967) Effect of starvation on the scorpion *Palamnaeus bengalensis*. *Physiol. Zool.*, **40,** 386–390.
8.3

Skerlj, B. (1959) Age changes in fat distribution in the female body. *Acta Anat.*, **38,** 56–63.
8.7

Skerlj, B., Brozek, J. & Hunt, E. Jnr (1953) Subcutaneous fat and age changes in body build and body form in women. *Am. J. phys. Anthropol.*, **11,** 577–600.
8.7

Skutch, A. F. (1949) Do tropical birds rear as many birds as they can nourish? *Ibis*, **91,** 430–454.
10.4

Slatyer, R. O. (1967) *Plant–Water Relationships*. Academic Press, London.
12.3

Slobodkin, L. B. (1961) Calories/gm in species of animals. *Nature* (*London*), **191,** 299.
8.8

Slobodkin, L. B. (1962) Energy in animal ecology. In *Advances in Ecological Research*. Vol. 1. (Ed. Cragg J. B.) pp. 69–101. Academic Press, London.
8.8

Slobodkin, L. B. (1968) Towards a predictive theory of evolution. In *Population Biology and Evolution*. (Ed. Lewontin R. C.). Syracuse University Press, Syracuse, New York.
1.4.2

Smalley, R. L. & Dryer, R. L. (1967) Brown fat in hibernation. In *Mammalian Hibernation*. Vol. 3. (Eds. Fisher, K. C., Dawe, A. R., Lyman, C. P., Schönbaum, E. & South, F. E.) pp. 324–345. Oliver & Boyd, Edinburgh.
8.2

Smith, C. C. (1968) The adaptive nature of social organization in the genus of tree squirrels *Tamiasciurus Ecol. Monogr.*, **38,** 31–63.
10.1.2

Smith, C. C. & Fretwell, S. D. (1974) The optimal balance between size and number of offspring. *Amer. Natur.*, **108,** 499–506.
10.3, 12.1.2, 13.6.1

Smith-Sonneborn, J. (1979) DNA repair and longevity assurance in *Paramecium tetraurelia. Science*, **203,** 1115–1117.
7.6

Smythe, N. (1970) On the existence of pursuit invitation signals in mammals. *Amer. Natur.*, **104,** 491–494.
13.4.3

Snell, T. W. & King, C. E. (1977) Lifespan and fecundity patterns in rotifers: the cost of reproduction. *Evolution*, **31,** 882–890.
12.2.2

Snow, D. W. (1961) The natural history of the oilbird, *Steatornis caripensis* in Trinidad, W. I. Part 1. General behaviour and breeding habits. *Zoologica*, **46,** 27–47.
8.5

Snyder, N. F. R. (1967) *An alarm reaction of aquatic gastropods to intraspecific extract.* 122 pp. Cornell Univ. Agri. Exp. Sta. Memoir 403. Ithaca, New York.
6.18

Solbrig, O. T. (1971) The population biology of dandelions. *Amer. Sci.*, **59,** 686–694.
12.4

Solbrig, O. T. (in press) Studies on the population biology of the genus *Viola*. 5. Life history dynamics of two New England species growing in a variable environment. *Proc. II Cong. of Evol. and Syst.*, Vancouver.
12.4

Solbrig, O. T., Newell, S. J. & Kincaid, D. T. (1980) Studies on the population biology of the genus *Viola*. I. The demography of *V. sororia. J. Ecol.*, **68,** 521–546.
12.4

Solbrig, O. T. & Orians, G. H. (1978) The adaptive characteristics of desert plants. *Am. Sci.*, **65,** 412–421.
12.3

Solbrig, O. T. & Simpson, B. B. (1974) Components of regulation of a population of dandelions in Michigan. *J. Ecol.*, **62,** 473–486.
12.4

Solbrig, O. T. & Simpson, B. B. (1977) A garden experiment on competition between biotypes of the common dandelion (*Taraxacum officinale*). *J. Ecol.*, **65,** 427–430.
12.4

Solomon A. K. (1960) Compartmental methods of kinetic analysis. In *Mineral Metabolism*. Part A, Vol. 1. (Eds. Comar, C. L. & Bronner, F.) Academic Press, New York.
5.5.4

Sonneborn, T. M. (1930) Genetic studies on *Stenostomum incaudatum* n. sp. The nature and origin of the differences in individuals formed during vegetative reproduction. *J. exp. Zool.*, **57,** 57–108.
7.6

Sonneborn, T. M. (1954) The relation of autogamy to senescence and rejuvenescence in *Paramecium aurelia. J. Protozool.*, **1,** 38–53.
7.6, 7.10

Southern, H. N. (1970) Ecology at the crossroads. *J. Anim. Ecol.*, **39,** 1–11.
13.7.1

Southwood, T. R. E. (1976) Bionomic strategies and population parameters. In *Theoretical Ecology: Principles and Applications* (Ed. May R. M.) pp. 26–48. Blackwell Scientific Publications, Oxford.
2.8

Southwood, T. R. E. (1977) The relevance of population dynamic theory to pest status. In *Origins of Pest, Parasite, Disease and Weed Problems.* (Eds. Cherrett, J. M. & Sagar, G. R.) Blackwell Scientific Publications, Oxford.
9.4.1

Spinage, C. (1969) Naturalistic observations on the reproductive and maternal behaviour of the Uganda defassa waterbuck, *Kobus deffassa ugandae* Newmann. *Z. Tierpsychol.*, **26,** 39–47.
13.7.2

Spitzer, G. (1972) Jahreszeitliche Aspekte der Biologie der Bartmeise (*Panurus biarmicus*). *J. Orn. Lpz.*, **113,** 241–275.
5.3.1

Stanley, S. M. (1973) An explanation for Cope's rule. *Evolution*, **27,** 1–26.
2.2, 2.5, 2.6

Stanley, J. F., Pye, D. & MacGregor, A. (1975) Comparison of doubling numbers attained by cultured animal cells with lifespan of species. *Nature* (*London*), **255,** 158–159.
7.8

Stearns, S. C. (1976) Life-history tactics: a review of the ideas. *Quart. Rev. Biol.*, **51,** 3–47.
1.6, 7.5, 10.1, 10.2.2, 12.4

Stearns, S. C. (1977) The evolution of life-history traits: a critique of the theory and a

review of the data. *Ann. Rev. Ecol. Syst.*, **8,** 145–171.
1.4.2, 7.5, 10.1, 12.4, 13:7.1

Stocker, O. (1968) Physiological and morphological changes in plants due to water deficiency. *Arid Zone Res.*, **15,** 63–104.
12.3

Stowe, L. G. & Teeri, J. A. (1978) The geographic distribution of C_4 species of the Dicotyledonae in relation to climate. *Amer. Natur.*, **112,** 609–623.
3.8.1

Strong, D. R. (1972) Life-history variation among populations of an amphipod (*Hyalella azteca*). *Ecology*, **53,** 1103–1111.
9.5

Stenseth, N. C., Framstad, E., Migula, P., Trojan, P. & Wojciechowska-Trojan, B. (1980) Energy models for the common vole *Microtus arvalis*: energy as a limiting resource for reproductive output. *Oikos*, **34,** 1–22.
10.4

Sugiyama, T. & Boku, K. (1976) Differing sensitivity of pyruvate, orthophosphate dikinase to low temperature in maize cultivars. *Plant Cell Physiol.*, **17,** 851–854.
3.8.3

Sutton, B. G. (1975a) The path of carbon in CAM plants at night. *Aust. J. Plant Physiol.*, **2,** 377–387.
3.7.1

Sutton, B. G. (1975b) Glycolysis in CAM plants. *Aust. J. Plant Physiol.*, **2,** 389–402.
3.7.1

Sutton, B. G. (1975c) Kinetic properties of phosphorylase and 6-phosphofructokinase of *Kalanchoë daigremontianum* and *Atriplex spongiosa*. *Aust. J. Plant Physiol.*, **2,** 403–411.
3.7.1

Svardson, G. (1948) Natural selection and egg number in fish. *Rept. Inst. Freshwater Res., Drottningholm.*, **29,** 115–122.
1.6

Szarek, S. R., Johnson, H. B. & Ting, I. P. (1973) Drought adaptation in *Opuntia brasilaris*. Significance of recycling carbon through crassulacean acid metabolism. *Plant Physiol.*, **52,** 539–541.
3.7.3

Szilard, L. (1959) On the nature of the ageing process. *Proc. Nat. Acad. Sci. USA*, **45,** 30–45.
7.3

Taghon, G. L., Self, R. F. L. & Jumars, P. A. (1978) Predicting particle selection by deposit feeders: a model and its implications. *Limnol. Oceanogr.*, **23,** 752–759.
4.3.2, 5.6.3

Tanner, J. M. (1963) Regulation of growth in size in mammals. *Nature* (*London*), **199,** 845–850.
9.5

Taylor H. M., Gourley R. S., Lawrence C. E. & Kaplan R. S. (1974) Natural selection of life history attributes: an analytical approach. *Theor. Popul. Biol.*, **5,** 104–122.
7.5

Taylor, S. E. (1975) Optimal leaf form. In *Perspectives in Biophysical Ecology* (Eds. Gates, D. M. & Schmerl, R. B.). Springer-Verlag, New York.
12.3

Taylor, S. E. & Sexton, O. J. (1972) Some implications for leaf tearing in Musaceae. *Ecology*, **53,** 143–149.
12.3

Teeri, J. A. & Stowe, L. G. (1976) Climatic patterns and the distribution of C_4 grasses in North America. *Oecologia*, **23,** 1–12.
3.8.1

Teissier, G. (1931) Croissance des insectes. *Travaux de la Station biologique de Roscoff*, **9,** 29–238.
9.3.2

Telford, S. R. (1970) Seasonal fluctuations in liver and fat body weights of the Japanese lacertid *Takydromus tachydromoides* Schlegel. *Copeia*, 1970, 681–688.
8.5

Tenhunen, J., Yocum, C. S. & Gates, D. M. (1976) Development of a photosynthesis model with emphasis on ecological applications. *Oecologia*, **26,** 89–100.
12.3

Terri, J. A. (1979) The climatology of the C_4 photosynthetic pathway. In *Topics in Plant Population Biology*. (Eds. Solbrig, O. T. *et al.*) Columbia University Press, New York.
12.3

Thomson, A. J. & Holling, C. S. (1974) Experimental component analysis of blowfly feeding behaviour. *J. Insect Physiol.*, **20,** 1553–1563.
5.5.2 (Fig. 5.9 legend)

Thomson, R. J. (1979) Fecundity and reproductive effort in the Blue Mussel (*Mytilus edulis*), the Sea Urchin (*Stronglylocentrotus droebachiensis*) and the Snow Crab (*Chinoecetes opilio*) from populations in

Nova Scotia and Newfoundland. *J. Fish. Res. Bd. Can.*, **36,** 955–964.
10.2.2

Thorson, G. (1950) Reproductive and larval ecology of marine bottom organisms. *Biol. Rev.*, **25,** 1–45.
10.3

Tieszan, L. L., Seryinska, M. M., Imbamba, S. K. & Troughton, J. H. (1979) The distribution of C_3 and C_4 grasses and carbon isotope discrimination along an altitudinal and moisture gradient in Kenya. *Oecologia*, **37,** 337—350.
3.8.3

Tietz, A. (1965) Metabolic pathways in the insect fat body. In *Adipose Tissue*. Handbook of Physiology Section 5 (Eds. Renold, A. E. & Cahill, G. F.) pp. 45–54. American Physiological Society, Washington D. C.
8.3

Tikhomirov, E. A. (1971) Body growth and development of reproductive organs of the North Pacific phocids. In *Pinnipeds of the North*. (Eds. Arsen'ev, V. A. & Panin, K. I.) Israel Program for Scientific Translation, Jerusalem.
9.4.1

Tinbergen, N., Impekoven, M. & Franck, D. (1967) An experiment on spacing out as a defence against predation. *Behaviour*, **28,** 307.
13.4.2.

Ting, I. P. & Hanscom, Z. (1977) Induction of acid metabolism in *Portulacaria afra*. *Plant Physiol.*, **59,** 511–514.
3.7.3

Tinkle, D. W. (1969) The concept of reproductive effort and its relation to the evolution of life histories in lizards. *Amer. Natur.*, **103,** 501–516.
10.1, 10.2.2

Tinkle, D. W. & Hadley, N. F. (1975) Lizard reproductive effort: caloric estimates and comments on its evolution. *Ecology*, **56,** 427–434.
12.3, 13.7.1

Tonogawa, S., Maxam, A. M., Tizard, R., Bernard, O. & Gilbert, W. (1978) Sequence of a mouse germ-line gene for a variable region of an immunoglobin light chain. *Proc. Nat. Acad. Sci. USA*, **75,** 1485–1489.
12.1

Toth, R. S. & Chew, R. M. (1972) Development and energetics of *Notonecta undulata* during predation on *Culex tarsalis*. *Ann. Ent. Soc. America*, **65,** 1270–1279.
9.3.2

Townsend, C. R. & Hildrew, A. G. (1979) Resource partitioning by two freshwater invertebrate predators with contrasting foraging strategies. *J. Anim. Ecol.*, **48,** 909–920.
4.2

Townsend, C. R. & Hildrew, A. G. (1980) Foraging in a patchy environment by a predatory net-spinning caddis larva: a test of optimal foraging theory. *Oecologia*, **47,** 219–221.

Trebst, A. (1974) Energy conservation in photosynthetic electron transport of chloroplasts. *Ann. Rev. Plant Physiol.*, **25,** 423–458.
3.2

Treisman, M. (1975) Predation and the evolution of gregariousness. I. Models for concealment and evasion. *Anim. Behav.*, **23,** 779–800.
13.4.2

Trivers, R. L. (1972) Parental investment and sexual selection. In *Sexual Selection and the Descent of Man* (Ed. Campbell B.) pp. 136–179. Aldine-Atherton, Chicago.
13.6.1

Trivers, R. L. & Willard, D. E. (1973) Natural selection of parental ability to vary the sex ratio of offspring. *Science*, **179,** 90–91.
13.7.1

Tucker, V. A. (1975) Flight energetics. *Symp. Zool. Soc. Lond.*, **35,** 49–63.
8.4, 8.5

Turesson, G. (1922) The genotypic response of the plant species to the habitat. *Hereditas*, **3,** 211–350.
3.1

Turesson, G. (1925) The plant species in relation to habitat and climate. *Hereditas*, **9,** 81–101.
3.1

Turesson, G. (1931) The selective effect of climate upon plant species. *Hereditas*, **15,** 99–152.
3.1

Turner, E. R. A. (1964) Social feeding in birds. *Behaviour*, **24,** 1–46.
13.5.3

Utter, J. M. (1971) Daily energy expenditure of free-living purple martins and mockingbirds. Unpublished Ph.D. thesis, Rutgers

University, New Jersey.
10.4

Vadas, R. L. (1977) Preferential feeding: an optimization strategy in sea urchins. *Ecol. Monogr.*, **47,** 337–371.
9.3.2

Vague, J. (1953) *La différentiation Sexuelle Humaine: Ses incidences en Pathologie.* 386 pp. Masson et Cie, Paris.
8.5, 8.7

Vague, J., Boyer, J., Jubelin, J., Nicolino, C. & Pinto, C. (1969) Adipomuscular ration in human subjects. In *Physiopathology of Adipose Tissue.* pp. 360–386. Excerpta medica Foundation, Amsterdam.
8.3

van Emden H. F. (Ed.) (1973) *Insect/Plant Relationships.* 215 pp. Symposia of the Royal Entomological Society of London, No. 6. Blackwell Scientific Publications, Oxford.
6.1

van Lawick-Goodall, J. (1968) The behaviour of free-living chimpanzees in the Gombe Stream Reserve. *Anim. Behav. Monogr.*, **1,** 161–311.
13.6.1, 13.6.6

Verschuren, J. (1958) *Ecologie et biologie des grand mammiferes.* pp. 1–225. Exploration du Parc National de la Garamba, Institut des Parcs Nationaux du Congo Belge, Bruxelles.
13.6.1

Vine, I. (1973) Detection of prey flocks by predators. *J. theor. Biol.*, **40,** 207–210.
13.4.2

Vitt, L. J. & Congdon, J. D. (1978) Body shape, reproductive effort and relative clutch mass in lizards: resolution of a paradox. *Amer. Natur.*, **112,** 595–608.
10.2.1

Vitt, L. J., Congdon, J. D. & Dickson, N. A. (1977) Adaptive strategies and energetics of tail autotomy. *Ecology*, **58,** 326–337.
8.4

Vogt, F. (1947) *Das Rotwild.* Jagderei und Fischerei Verlag, Wien.
13.6.1

Waage, J. K. (1979) Foraging for patchily-distributed hosts by the parasitoid, *Nemeritis canescens. J. Anim. Ecol.*, **48,** 353–371.
4.2.1

Waddington, C. H. (1968) The paradigm for the evolutionary process. In *Population Biology and Evolution* (Ed. Lewontin R. C.) pp. 37–45. Syracuse University Press, Syracuse, New York.
1.4.2

Walker, D. A. (1976) Plastids and intracellular transport. In *Encyclopedia of Plant Physiology* (New Series) (Eds. Stocking, C. R. & Heber, U.) Vol. 3, pp. 85–136. Springer, Berlin.
3.7.1

Walker, E. P. (1964) *Mammals of the World.* 1500 pp. Johns Hopkins Press, Baltimore, Md.
6.16

Wallace, J. W. & Mansell, R. L. (Eds.) (1976) *Biochemical interaction between plants and insects.* 425 pp. *Recent advances in phytochemistry, Vol. 10.* Plenum Press, New York.
6.1

Walther, F. R. (1969) Flight behaviour and avoidance of predators in Thomson's gazelle (*Gazella thomsonii* Guenther 1884). *Behaviour*, **34,** 184–221.
13.4.3

Wang, A. H. J., Quigley, G. J., Kolpak, S. J., Crawford, T. L., Van Boom, J. H., Vander Morel, G. & Ritch, A. (1979) Molecular structure of a left-handed double helical DNA fragment at atomic resolution. *Nature (London)*, **282,** 680–686.
12.1

Warburg, O. & Negelein, E. (1920) Uber die Reduktion der salpetersäuren in grünen Zellen. *Biochem. Zeit.*, **110,** 66–115.
3.2

Wareing, P. F. & Patrick, J. (1975) Source-sink relations and the partition of assimilates in the plant. In *Photosynthesis and Productivity in Different Environments.* (Ed. Cooper, J. P.) Cambridge University Press, Cambridge.
12.1

Warner, R. G. & Flatt, W. P. (1965) Anatomical development of the ruminant stomach. In *Physiology of Digestion in the Ruminant.* (Ed. Dougherty, R. W.) Butterworths, London.
5.6.3

Watson, A. (1977) Population limitation and the adaptive value of territorial behaviour in the Scottish Red grouse. *Lagopus l. scoticus.* In *Evolutionary Ecology* (Eds. Stonehouse, B. & Perrins, C. M.) pp. 19–26. Macmillan

Press, London.
13.6.2

Watson, R. M. (1967) The population ecology of the wildebeest (*Connochaetes taurinus albojubatus* Thomas) in the Serengeti. Ph.D. thesis, Cambridge University.
13.4.1

Watts, C. R. & Stokes, A. W. (1971) The social order of turkeys. *Sci. Am.*, **224,** 112–118.
13.6.1

Weatherley, A. H. (1972) *Growth and Ecology of Fish Populations.* Academic Press, London.
10.2.1

Weiner J., Loud, A. V., Kimberg, D. V. & Spiro, D. J. (1968) A quantitative description of cortisone-induced alterations in the ultrastructure of rat liver parenchymal cells. *J. Cell. Biol.*, **37,** 47–61.
2.5

Weis-Fogh, T. (1954) Fat combustion and the metabolic rate of flying locusts (*Schistocerca gregaria* Fenshal). *Phil. Trans. Roy. Soc.*, **237,** 1–36.
9.2.3

Weis-Fogh, T. (1967) Metabolism and weight economy in migrating animals, particularly birds and insects. In *Insects and Physiology* (Eds. Beament, J. W. & Treherne, J. E.) pp. 143–159. Oliver & Boyd, Edinburgh.
8.2

Weisman, H. (1891) The duration of life. In *Essays on Heredity.* Oxford University Press, Oxford.
1.4.1, 7.8

Wells, M. J. (1978) *Octopus.* Chapman and Hall, London.
9.3.2

Wertheimer, H. E. (1965) Introduction—a perspective. In *Adipose tissue.* Handbook of Physiology, section 5 (Eds. Renold, A. E. & Cahill, G. F.) pp. 5–11. American Physiologial Society, Washington, D. C.
8.1, 8.3

West, C. G., Burns, J. J. & Modafferi, M. (1979) Fatty acid composition of pacific walrus skin and blubber fat. *Can. J. Zool.*, **57,** 1249–1255.
8.2

Western, D. (1979) Size, life history and ecology in mammals. *Afr. J. Ecol.*, **17,** 185–204.
2.8

Westerterp, K. (1977) How rats economize—energy loss in starvation. *Physiol. Zool.*, **50,** 331–362.
9.5

Westoby, M. (1974) An analysis of diet selection by large generalist herbivores. *Amer. Natul.*, **108,** 290–304.
4.3.1.1

White, J. & Harper, J. L. (1970) Correlated changes in plant size and number in plant populations. *J. Ecol.*, **58,** 467–485.
12.4

White, T. C. R. (1978) The importance of relative food shortage in animal ecology. *Oecologia*, **33,** 71–86.
1.5, 4.3

Widdowson, E. M., Dickerson, J. W. T. & McCance, R. A. (1960) Severe undernutrition in growing and adult animals. *Br. J. Nutr.*, **14,** 457–471
9.7.2

Wigglesworth, V. B. (1966) "Catalysomes" or enzyme caps on lipid droplets: an intracellular organelle. *Nature* (*London*), **210,** 759.
8.3

Wilbur, H. M. (1976) Life history evolution of seven milkweeds of the genus *Asclepias. J. Ecol.*, **64,** 225–240.
12.4

Wilbur, H. M., Rubenstein, D. I. & Fairchild, L. (1978) Sexual selection in toads: the roles of female choice and male body size. *Evolution*, **32,** 264–270.
13.6, 13.6.1

Wilbur, H. M., Tinkle, D. W. & Collins, J. P. (1974) Environmental certainty, trophic level, and resource availability in life history evolution. *Amer. Natur.*, **108,** 805–817.
1.6, 12.4

Williams, G. C. (1957) Pleiotropy, natural selection and the evolution of senescence. *Evolution*, **11,** 398–411.
7.8

Williams, G. C. (1966a) *Adaptation and Natural Selection.* Princeton University Press, Princeton.
1.6, 9.4.1

Williams, G. C. (1966b) Natural selection, the cost of reproduction and a refinement of Lack's principle. *Amer. Natur.*, **100,** 687–690.
10.4

Williams, G. C. (1979) The question of adaptive sex ratio in outcrossed vertebrates. *Proc. R. Soc. Lond.*, **205,** 497–580.
13.7.1

Williams R. T. (1959) *Detoxication Mechanisms*, 2nd edn. Chapman & Hall, London.
7.3

Willoughby, L. G. & Sutcliffe, D. W. (1976) Experiments on feeding and growth of the amphipod *Gammarus pulex* (L.) related to its distribution. *Fresh wat. Biol.* **6,** 577–586.
9.3.2

Wilson, E. O. (1975) *Sociobiology: the new synthesis.* Harvard University Press. Cambridge, Mass.
13.2, 13.6

Wilson, P. N. (1954) Growth analysis of the domestic fowl. II. Effect of plane of nutrition on carcass composition. *J. agric. Sci.*, **44,** 67–85.
8.5

Windell, J. T. (1967) Rates of digestion in fishes. In *The Biological Basis of Freshwater Fish Production.* (Ed. Gerking S. D.). Blackwell Scientific Publications, Oxford
5.5.2 (Fig. 5.9 legend)

Winter, K. (1974) Evidence for the significance of crassulacean acid metabolism as an adaptive mechanism to water stress. *Plant Sci. Lett.*, **3,** 279–281.
3.7.1

Winter, K. (1980) Day/night changes in the sensitivity of phosphoenolpyruvate carboxylase to malate during crassulacean acid metabolism. *Plant Physiol.*, **65,** 792–796.
3.7.1

Winter, K. & Troughton, J. H. (1978) Carbon assimilation pathways in *Mesembryanthemum nodiflorum* L. under natural conditions. *Z. Pflanzenphysiol.*, **88,** 153–162.
3.7.1

Wittenberger, J. F. (1979) A model for delayed reproduction in iteroparous animals. *Amer. Natur.*, **114,** 439–446.
10.2.2

Wodinsky, J. (1977) Hormonal inhibition of feeding and death in *Octopus*: control by optic gland secretion. *Science*, **198,** 948–951.
7.7

Wolf, L. L. & Hainsworth, F. R. (1971) Time and energy budgets of territorial hummingbirds. *Ecology*, **52,** 980–988.
12.2.5

Wolf, L. L., Hainsworth, F. R. & Stiles, F. G., (1972) Energetics of foraging: rate and efficiency of nectar extraction by hummingbirds. *Science*, **176,** 1351–1352.
12.2.5

Woolhouse H. W. (1967) The nature of senescence in plants. In *Aspects of the Biology of Ageing. Symposia of the Society for Experimental Biology No. XXI* (Ed. Woolhouse, H. W.) pp. 179–213. Cambridge University Press, Cambridge.
7.8

Woolhouse, H. W. (1969) DNA Polymerase, genetic variation and determination of the lifespan. In *Proc. 8th Int. Congr. Geront.* Vol. 1, pp. 162–166. Federation of American Societies for Experimental Biology, Washington, D. C.
7.3

Woolhouse, H. W. (1974) Longevity and senescence in plants. *Sci. Prog.*, **61,** 123–147.
7.7

Woolhouse, H. W. (1980) Possibilities for modification of the pattern of photosynthetic assimilation of CO_2 in relation to the growth and yield of crops. *Proc. 15th Int. Potash Inst. Colloquium, Wageningen*, 1.
3.8.4

Woolpy, J. H. (1968) The social organization of wolves. *Nat. Hist.*, **77,** 46–55.
13.6.1

Wootton, R. J. (1977) Effect of food limitation during the breeding season on the size, body components and egg production of female sticklebacks (*Gasterosteus aculeatus*). *J. Anim. Ecol.*, **46,** 823–834.
10.1.2

Wright, S. (1931) Evolution in Mendelian populations. *Genetics*, **16,** 97–159.
1.2, 1.3

Wright, S. (1968) *Evolution and the Genetics of Populations.* Vol. 1. University of Chicago Press, Chicago.
1.3

Wright, S. (1968–78) *Evolution and Genetics of Populations.* 4 Volumes. University of Chicago Press, Chicago.
12.1

Wynne-Edwards, V. C. (1962) *Animal Dispersion in Relation to Social Behaviour.* Oliver and Boyd, Edinburgh.
7.8

Young, R. A. (1976) Fat, energy and mammalian survival. *Am. Zool.*, **16,** 699–710.
8.3, 8.6

Zach, E. & Falls, J. B. (1978) Prey selection by Captive ovenbirds (Aves: Parulidae). *J.* Anim. Ecol., **47,** 929–943.
4.3.2.

Zahavi, A. (1975) Mate selection—a selection for a handicap. *J. theor. Biol.*, **53,** 205–214.
13.6.5

Zahavi, A. (1977) Reliability in communi-

cation systems and the evolution of altruism. In *Evolutionary Ecology* (Eds. Stonehouse, B. & Perrins C. M.) pp. 253–259. Macmillan, London.
13.6.5

Zamenhof, S. & Eichorn, H. (1967) Study of microbial evolution through loss of biosynthetic functions: establishment of "defective" mutants. *Nature (London)*, **216,** 456–458
9.7.3

Zelitch, I. (1975) Improving the efficiency of photosynthesis. *Science*, **188,** 626–633.
3.4

Zeuthen, E. (1947) Body size and metabolic rate in the animal kingdom with special regard to the marine micro-fauna. *C. r. Trav. Lab. Carlsberg, Ser. Chim.*, **26,** 17–161.
2.4

Zeuthen, E. (1953) Oxygen uptake as related to body size in organisms. *Q. Rev. Biol.*, **28,** 1–12.
2.4

Zeuthen, E. (1970) Rate of living as related to body size in organisms. *Pol. Arch. Hydrobiol.*, **17,** 21–30.
2.4, 2.5

Zlotkin, E. (1973) Chemistry of animal venoms. *Experientia*, **29,** 1453–1466.
6.1

Zs.-Nagy, I. (1978) A membrane hypothesis of ageing. *J. theor. Biol.*, **75,** 189–195.
7.8

Organism Index

Subject Index